Based on the Latest syllabus prescribed b
New Delhi and other State B

Learning Composite Mathematics

BOOK – 5
(For Class V)

AS PER NEW SYLLABUS

S.K. GUPTA
Principal (Retd.)
Birla Vidya Mandir
Nainital

ANUBHUTI GANGAL
M.A. (Gold Medalist), M.Ed.
Former, Senior Faculty Member
The Daly College, Indore

S. CHAND & COMPANY LTD.
(AN ISO 9001 : 2000 COMPANY)
RAM NAGAR, NEW DELHI - 110 055

S. CHAND & COMPANY LTD.

(An ISO 9001 : 2000 Company)

Head Office: 7361, RAM NAGAR, NEW DELHI - 110 055
Phone: 23672080-81-82, 9899107446, 9911310888
Fax: 91-11-23677446
Shop at: **schandgroup.com;** e-mail: **info@schandgroup.com**

Branches :

AHMEDABAD	: 1st Floor, Heritage, Near Gujarat Vidhyapeeth, Ashram Road, **Ahmedabad** - 380 014, Ph: 27541965, 27542369, ahmedabad@schandgroup.com
BANGALORE	: No. 6, Ahuja Chambers, 1st Cross, Kumara Krupa Road, **Bangalore** - 560 001, Ph: 22268048, 22354008, bangalore@schandgroup.com
BHOPAL	: 238-A, M.P. Nagar, Zone 1, **Bhopal** - 462 011, Ph: 4274723. bhopal@schandgroup.com
CHANDIGARH	: S.C.O. 2419-20, First Floor, Sector - 22-C (Near Aroma Hotel), **Chandigarh** -160 022, Ph: 2725443, 2725446, chandigarh@schandgroup.com
CHENNAI	: 152, Anna Salai, **Chennai** - 600 002, Ph: 28460026, 28460027, chennai@schandgroup.com
COIMBATORE	: Plot No. 5, Rajalakshmi Nagar, Peelamedu, **Coimbatore** -641 004, (M) 09444228242, coimbatore@schandgroup.com
CUTTACK	: 1st Floor, Bhartia Tower, Badambadi, **Cuttack** - 753 009, Ph: 2332580; 2332581, cuttack@schandgroup.com
DEHRADUN	: 1st Floor, 20, New Road, Near Dwarka Store, **Dehradun** - 248 001, Ph: 2711101, 2710861, dehradun@schandgroup.com
GUWAHATI	: Pan Bazar, **Guwahati** - 781 001, Ph: 2738811, 2735640 guwahati@schandgroup.com
HYDERABAD	: Padma Plaza, H.No. 3-4-630, Opp. Ratna College, Narayanaguda, **Hyderabad** - 500 029, Ph: 24651135, 24744815, hyderabad@schandgroup.com
JAIPUR	: A-14, Janta Store Shopping Complex, University Marg, Bapu Nagar, **Jaipur** - 302 015, Ph: 2719126, jaipur@schandgroup.com
JALANDHAR	: Mai Hiran Gate, **Jalandhar** - 144 008, Ph: 2401630, 5000630, jalandhar@schandgroup.com
JAMMU	: 67/B, B-Block, Gandhi Nagar, **Jammu** - 180 004, (M) 09878651464
KOCHI	: Kachapilly Square, Mullassery Canal Road, Ernakulam, **Kochi** - 682 011, Ph: 2378207, cochin@schandgroup.com
KOLKATA	: 285/J, Bipin Bihari Ganguli Street, **Kolkata** - 700 012, Ph: 22367459, 22373914, kolkata@schandgroup.com
LUCKNOW	: Mahabeer Market, 25 Gwynne Road, Aminabad, **Lucknow** - 226 018, Ph: 2626801, 2284815, lucknow@schandgroup.com
MUMBAI	: Blackie House, 103/5, Walchand Hirachand Marg, Opp. G.P.O., **Mumbai** - 400 001, Ph: 22690881, 22610885, mumbai@schandgroup.com
NAGPUR	: Karnal Bag, Model Mill Chowk, Umrer Road, **Nagpur** - 440 032, Ph: 2723901, 2777666 nagpur@schandgroup.com
PATNA	: 104, Citicentre Ashok, Govind Mitra Road, **Patna** - 800 004, Ph: 2300489, 2302100, patna@schandgroup.com
PUNE	: 291/1, Ganesh Gayatri Complex, 1st Floor, Somwarpeth, Near Jain Mandir, **Pune** - 411 011, Ph: 64017298, pune@schandgroup.com
RAIPUR	: Kailash Residency, Plot No. 4B, Bottle House Road, Shankar Nagar, **Raipur** - 492 007, Ph: 09981200834, raipur@schandgroup.com
RANCHI	: Flat No. 104, Sri Draupadi Smriti Apartments, East of Jaipal Singh Stadium, Neel Ratan Street, Upper Bazar, **Ranchi** - 834 001, Ph: 2208761, ranchi@schandgroup.com
VISAKHAPATNAM	: Plot No. 7, 1st Floor, Allipuram Extension, Opp. Radhakrishna Towers, Seethammadhara North Extn., **Visakhapatnam** - 530 013, (M) 09347580841, visakhapatnam@schandgroup.com

By the Same Authors – Composite Mathematics VI, VII, VIII

First Multicolour Edition 2008
Reprint With corrections 2008
Revised Edition 2009
Reprint 2010

ISBN : 81-219-2990-3 **Code :** 14A 561

PRINTED IN INDIA
By Rajendra Ravindra Printers Pvt. Ltd., 7361, Ram Nagar, New Delhi -110 055 and published by S. Chand & Company Ltd., 7361, Ram Nagar, New Delhi -110 055.

PREFACE

The Learning Composite Mathematics series (Books 0 to 5) is an exclusive course for classes Pre-primary to V meant for CBSE schools and all schools of other Boards following the latest **NCERT syllabus** based on the National Curriculum and the guidelines provided therein.

The content is student centered and activity based laying the utmost emphasis on developing problem solving skills, encouraging the child to think creatively and work independently. The methodology facilitates the teacher, student and the parent to ensure full involvement of child in classroom and at home.

All the mathematical concepts are presented in a very simple and encouraging form introducing the concepts by using motivating scripts and stories based on everyday life and daily experiences. The examples and Problems also make use of modern tools, gadgets and technology commonly used.

The approach of the authors has been designing the instruction in such a form that may create passion for mathematics in children rather than fear. It encourages them to enquire, explore and discover rather than learn by force.

Abundant use of visual tools such as diagrams, illustrations, cartoons, tables and charts would make learning fun and would help greater retention.

A **unique feature** of the series is the framing of separate sets of questions for class-work and home work to facilitate sustained learning. The problems given in **Skill Practice Exercises** given to reinforce the understanding and master the new concepts are to be done in the space provided in the book itself under the supervision of the teacher and problems in exercises meant for home-work are to be done in separate copybooks.

'Mental Maths Exercises' have been given to help the children acquire speed and gain confidence.

Puzzles, Quizzes, Did you know ? Fun with Numbers and Enrichment material have been included to widen the mental horizon of the children and give them a feel of the beauty of Mathematics:

'Maths Alert' given at various places point out the common mistakes and warn them against committing them themselves.

We are grateful to the CMD, Mrs. Nirmala Gupta, the Joint Managing Director, Mr. Himanshu Gupta and other directors, the CEO and Executive Director Mr. Amit Gupta, the Vice President publishing, Mr. Navin Joshi for their motivation to take up this project and continuous support. Thanks are also due to Mr. Amit Goel, the Sr. Manager (S & CS).

Our thanks are due to Mr. Shishir Bhatnagar, Manager (Pre-press) and Subject Editor, Dr. (Mrs.) Seema Chowdhary, Mr. Ajendra Negi, Mr. Dhan Singh and Mr. Pradeep Joshi for taking immense pains and seeing the books through various stages of preparation and printing. We express our appreciation and thanks to Mrs. Aradhana Gupta for preparing the beautiful cover design.

For more than 40 years, we have been setting trends in bringing unmatched content and pioneering Maths books for Indian Schools. It would not have been possible without support from teachers and students.

Suggestions and feedback for the improvement of the books would be most welcome from all the Principals, teachers, students and parents for continuous improvement of this newly introduced series.

January, 2008

e.mail : anubhutigangal @ hotmail. com.

S.K. GUPTA

NOTE ON THE 2009 EDITION

The text and answers have been thoroughly checked. The few misprints that we came across have been corrected.

LIST OF LAB ACTIVITIES AS SPECIFIED BY CBSE

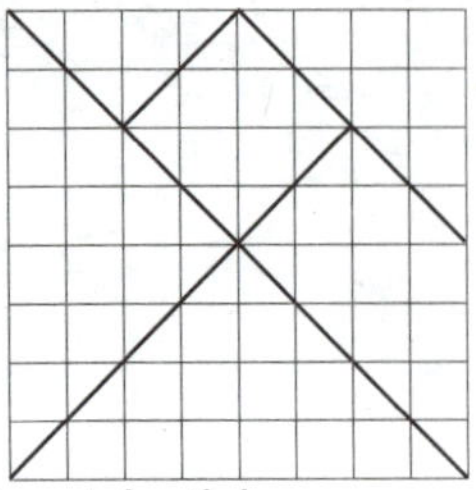

Activity 1 : To make a set of tangrams by paper folding using 8×8 grid.
Using the tangram pieces do the following :
(a) Try to make a triangle, a square and a rectangle using 2 pieces of the tangram set.
(b) Try do make a triangle, a square and a rectangle using 3 pieces of the tangram set.
(c) Calculate the area of each part by counting complete squares and half squares.

Activity 2 : (a) To make closed geometrical shapes using
(i) 3 match sticks (ii) 4 match sticks (iii) 5 match sticks (iv) 6 match sticks
(b) To observe the change in angles between any two adjacent match sticks.

Activity 3 : (a) To identify a right angle, angles less than a right angle and angle more than a right angle using body parts and to draw them using stick drawings.
(b) Write the word mathematics using straight lines and observe the number of right angles, number of angles more than a right angle and number of angles less than a right angle.
(c) To make a degree clock by paper folding.

Activity 4 : To observe hands of the clock at different time in a day and record four observations each of the following.
(a) right angle between hands of the clock
(b) angle less than a right angle between hands of the clock.
(c) angle more than a right angle between hands of the clock.

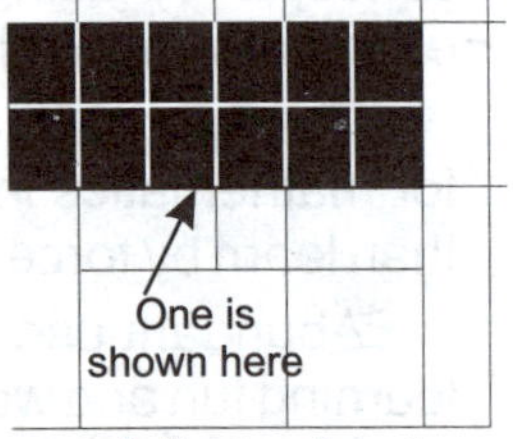

Activity 5 : (a) To make rectangles of different dimensions on squared paper using 12 adjacent squares.
(b) To calculate the perimeter and area of each of the rectangles so formed.
(c) To observe the fact that shapes having same area may have different perimeter.
(d) To observe the relation between dimensions of rectangles and the factors of 12.

Activity 6 : To calculate the perimeter of different shapes formed by shading six adjacent squares of dimension 1 cm each on a squared paper.
Some shapes are shown below

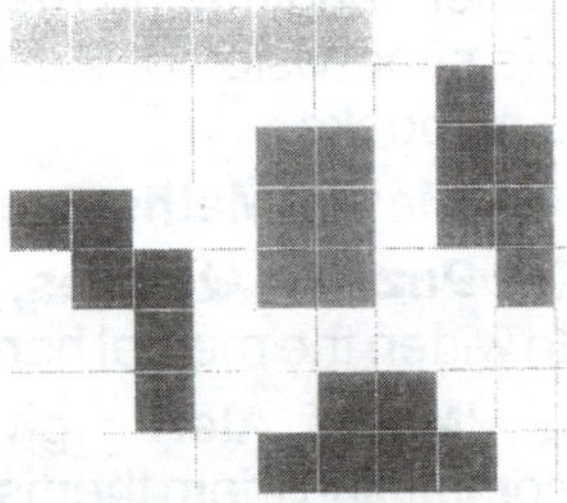

Activity 7 : (a) To represent the fractions $\frac{1}{2}$, $\frac{1}{3}$, $\frac{1}{4}$, 1/5, papers by paper folding.
(b) To represent the fractions $\frac{1}{2}$, $\frac{1}{4}$, $\frac{3}{8}$, $\frac{5}{8}$, $\frac{3}{4}$ etc. of a circular region by paper folding.

Activity 8 : To find the lines of symmetry in the following shapes by paper folding
(i) a square (ii) a rectangle

Activity 9 : To make a 3×3 magic square, using numbers from 1 to 9.
Hint :

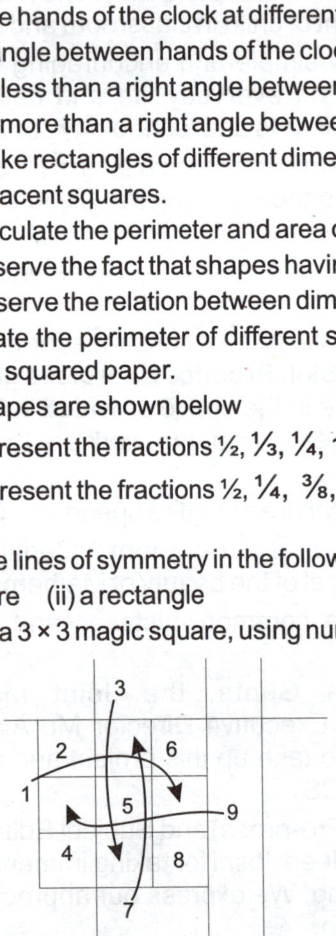

Activity 10 : (a) To make cube and cuboids of various dimensions using unit cubes
(i) $2 \times 3 \times 2$ cubic units (ii) $3 \times 3 \times 2$ cubic units (iii) $3 \times 3 \times 3$ cubic units
(b) To find the number of missing unit cubes in a given shape to obtain a given shape of desired dimensions.

How many units cubes are to be added to make a cuboid of dimensions $4 \times 4 \times 2$?

How many unit cubes are added to get a cuboid of dimension $5 \times 5 \times 3$?

Note : Teacher may ask the students to use unit cubes and explore.

CONTENTS

1.	Numeration	1 – 7
2.	Large Numbers	8 – 22
3.	Operations on Large Numbers	23 – 34
4.	Rounding Numbers	35 – 38
5.	Factors and Multiples	39 – 47
6.	Fractions	48 – 68
7.	Multiplication and Division of Fractions	69 – 81
8.	Decimals	82 – 93
9.	Multiplication and Division of Decimals	94 – 112
10.	Basics of Geometry	113 – 130
11.	Geometrical Shapes	131 – 149
12.	Measurement	150 – 171
13.	Time	172 – 182
14.	Temperature	183 – 188
15.	Percentages	189 – 195
16.	Money	196 – 204
17.	Perimeter and Area	205 – 214
18.	Volume	215 – 224
19.	Patterns	225 – 230
20.	Data Handling	231 – 236
21.	Answers	237 – 266

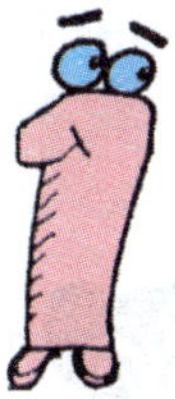

NUMERATION

NUMERATION AND NOTATION

1. The following numbers are the **counting numbers.**
   1, 2, 3, 4, 5, 6, 7, 8, 9, 10 ...
   When you add 0, to the list, you obtain the **whole numbers.**
   0, 1, 2, 3, 4, 5, 6, 7, 8, 9, 10, ...
2. Names or symbols for numbers are called **numerals.**
3. The diagram below shows the first few whole numbers on a **number line.**

4. The following are three ways of showing the number 3459.

Abacus

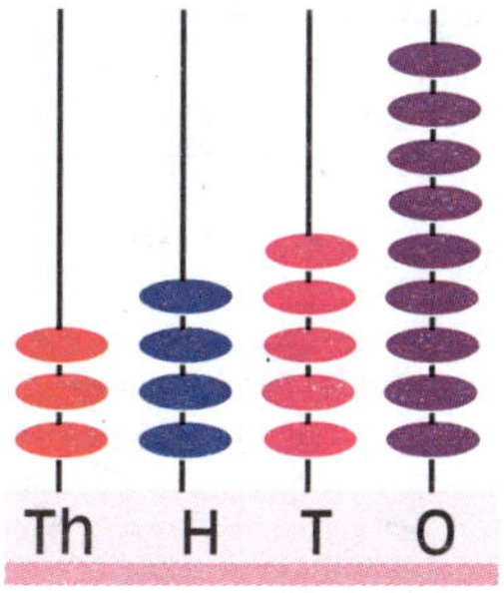

Place value chart

Th	H	T	O
3	4	5	9

Expanded numeral

3000 + 400 + 50 + 9

6. **Indian Place value chart**

Lakhs		Thousands		Ones		
Ten lakhs 10 00 000	lakhs 1 00 000	Ten Thousands 10 000	Thousands 1 000	Hundreds 100	Tens 10	Ones 1
7 digits	6 digits	5 digits	4 digits	3 digits	2 digits	1 digit

The first three digits from the right make **'ones'** or the **'units'** period, the next two make **'thousands'**, period and the next, two make **'lakhs'** period.

We write the numerals by leaving some space or by putting a comma between each period.

While reading a number, the digits in the same period are read together and the name of the period (except ones) is read along with them.

Numeral	Written as	Read as
2653	2 653 or 2,653	Two thousand six hundred fifty three
42576	42 576 or 42,576	Forty two thousand five hundred seventy six
237409	2 37 409 or 2,37,409	Two lakh thirty seven thousand four hundred nine

Caution : The word 'and' is not used in reading and writing whole numbers.

EXERCISE 1

1. What is the smallest counting number?
2. What is the smallest whole number?
3. **Complete the following table :**

Nature of counting numbers	Numbers from - to -	Smallest number	Greatest number
One digit numbers	1 – 9	1	9
Two digit numbers	10 – 99	10	99
Three digit numbers			
Four digit numbers			
Five digit numbers			
Six digit numbers			

4. **Write each numeral in figures.**

Number word	Numeral
(a) Thirteen thousand five hundred eighty six	13,586
(b) Fifty four thousand nine hundred seven	
(c) Eighty thousand seven hundred	
(d) Forty-nine thousand two hundred	
(e) Sixty thousand five	
(f) Five lakh forty three thousand nine hundred fifty six	
(g) Eight lakh ninety thousand three hundred	

5. **Write each of the following numbers in words :**

(a) 59,380 (b) 83,006 (c) 4,25,607

(d) 3,05,784 (e) 1,00,032 (f) 9,00,008 (g) 9,99,999

6. **Counting by fives write the numbers from :**

(i) 8723 to 8743 (ii) 15910 to 15930 (iii) 71837 to 71857 (iv) 80989 to 81014

7. **What is the place value of :**

(a) 7 in 57930 ? (b) 8 in 2801 ? (c) 6 in 63102 ? (d) 5 in 24053 ?

8. **Rewrite in ascending order :**

(i) 1835, 3815, 8135, 5813, 8153, 1853, 3581, 1538

(ii) 80808, 80088, 88008, 80008, 88808

9. **Arrange in descending order :**

(i) 76845, 87465, 67854, 86754 (ii) 420000, 427245, 423675, 421032

10. **What is the difference between the values of the**

(i) two sevens in the 70674 ? [**Hint :** 70 000 – 70]

(ii) two fours in 12494 ? (iii) two eights in 878513 ?

11. **Write the numeral for the largest possible number made from each of the following groups of digits.**

(a) 4, 3, 7, 0, 9 [**Hint :** 97,430] (b) 8, 0, 6, 2, 3, 7,

(c) 5, 1, 2, 9, 8, 3 (d) 0, 7, 4, 1, 8, 2

12. **Write the numeral for the smallest number you can make from each of the following groups of digits.**

(a) 3, 5, 1, 0, 4 (b) 7, 6, 9, 1, 2, 5 (c) 2, 3, 0, 7, 4, 5

ADDITION AND SUBTRACTION

EXERCISE 2

1. **Find the sum of :**

(a) 80396, 6194, 876, 75928 (b) 66578, 9728, 16736, 6500

(c)
```
  73257
   4372
  89499
   5876
+ 62903
```

(d)
```
  99142
   5364
  27943
  12397
+ 58430
```

(e)
```
   37784
   75576
    3987
     956
+ 536029
```

(f)
```
  398732
   59708
   62905
  298786
+ 100094
```

2. **Fill in the blanks :**

(a) 54637 + 0 = _____ (b) 21739 + 4103 = 4103 + _____

(c) 85619 + _____ = 85619 (d) _____ + 65980 = 65980 + 16307

3. In an election for a particular seat there were 4 candidates. They received votes as follows :

First : 42059, **Second** : 31847, **Third** : 9835, **Fourth** : 2801.

If 538 votes were found invalid, how many votes, in all were polled ?

4. For a primary health centre, Rs 28680 were given in the first year. In the second year, Rs 5850 more than the first year were given. In the third year, Rs 4700 more than the amount given in the second year were given. What was the total amount given to the centre in three years?

5. **Find the difference between :**

(i) 98403 and 47916 (ii) 72153 and 67438

(iii) 392881 and 308909 (iv) 326668 and 180803

(v) 98376 − 89667 (vi) 384691 − 279583 (vii) 600758 − 87961 (viii) 321097 − 188687

6. What must be added to 16003 to make it equal to 25000 ?

7. **Fill in the blanks :**

(a) 58437 − .... = 58437 (b) 30029 − 0 = ... (c) .... − 0 = 23568

8. The population of a town is 89762. How much less is it than one lakh ?

9. Geeta deposited Rs 30000 in her bank account. Later, she withdrew Rs 18237 from it. How much money is there in her account now ?

10. A sweet shop received the following amounts of money on Monday, Tuesday, Wednesday and Thursday : Rs 1507·36, Rs 2003·58, Rs 1923·91, Rs 2019·87. What was the total sale of the shop for those four days ?

MULTIPLICATION

EXERCISE 3

1. **Write the simplest multiplication numeral for the following :**

(Example : 4 + 4 + 4 + 4 + 4 = 5 × 4)

(a) 2 + 2 + 2 (b) 5 + 5 + 5 + 5 + 5 + 5

(c) 0 + 0 + 0 + 0 (d) 137 + 137 + 137 + 137

(e) 194 + 194 + 194 + 194 + 194.

2. **Multiply :**

(a) 53 by 10, 100, 1000, 10,000 (b) 510 by 80 (c) 3900 by 50

3. **Find the product :**

(a) 18 × 1 (b) 30 × 0 (c) 825 × 1 (d) 407 × 0

4. **Fill in the blanks :**

(a) 18 × 25 = 25 × ☐ (b) 48 × ☐ = 0

(c) 879 × 1 = ☐ (d) ☐ × 1 = 0

(e) 58 × 729 = ☐ × 58 (f) 59246 × ☐ = 0

5. **Multiply :**

(a) 709 × 526 (b) 654 × 376

(c) 847 × 9570 (d) 5608 × 298

6. A machine, on an average, manufactures 3786 screws in a day. How many screws does it manufacture in one year ?

7. A factory produces 470 toy-cars a day. How many cars did it produce in the year 1988, if there are 86 holidays ? [**Hint :** 1988 is a leap year]

8. An aeroplane, on an average, covers 7000 km in 12 hours. How much distance will it cover in 15 days ? [**Hint :** 24 hours = 1 day]

9. A factory packs 144 ball-pens in a carton. They have 395 cartons to be filled. They have already made 57000 ball-pens. Do they need to make more ball-pens in order to fill all the cartons ?

10. During an year, an engineering firm manufactured wrist watches at an average rate of 550 per week. Determine the total number of watches produced by the firm for the whole year. [**Hint :** 1 year = about 52 weeks]

DIVISION

Look at this

$$4\overline{)29}$$

Quotient: 7
Dividend: 29
−28
Remainder: 1
Divisor: 4

QUOTIENT ← 7
DIVIDEND ← 29
DIVISOR → 4
REMAINDER ← 1

Check : 4 x 7 + 1 = 29

Divisor x Quotient + Remainder = Dividend

EXERCISE 4

Divide :

1. (a) $67\overline{)5907}$ (b) $34\overline{)76521}$ (c) $48\overline{)65000}$ (d) $41\overline{)27854}$

2. (a) $42\overline{)360973}$ (b) $50\overline{)196997}$ (c) $63\overline{)274516}$ (d) $49\overline{)258277}$

3. **Divide and check :**

(a) 835 ÷ 39 (b) 3958 ÷ 57 (c) 30059 ÷ 48

[**Hint :** use divisor × quotient + remainder = dividend]

4. About 13200 babies are born in the world every hour. About how many are born in a minute ?

5. **Tell which of the following are meaningless operations :**

(a) $0 \div 5$ **(b)** $\frac{0}{17}$ **(c)** $21 \div (7 - 7)$ **(d)** $8 \div 0$ **(e)** $(9 - 9) \div 32$

6. If 638 marbles are shared among 53 boys so that each boy receives the same number, how many marbles will each boy get and how many will be left over ?
7. A girl knits 718 stitches. If there are 25 stiches in each row of knitting, how many complete rows have been knitted ?

8. A rocket travelled at a speed of 29760 kilometres an hour. How many kilometres would the rocket travel in one minute at this speed ?
9. The product of two numbers is 71424. If one number is 93, what is the other number ?
10. In a garden, there are 4891 trees in 67 rows. If each row has the same number of trees, how many trees are there in one row ?

MENTAL MATHS – 1

1. A boy is 10 years old now. How old will he be 15 years hence ? ____
2. The sum of two numbers is 120; the greater is 90. What is the lesser number ? ________
3. A child has a Rs 20 currency note and buys ice-cream for Rs 12. How much change will he receive back ? ________
4. What is the place value of 7 is 5,879 ? ________

5. Write down the value of 6 + 4 – 5 – 2. ________
6. By how much is one metre more than 33 cm ? ________
7. The price of a television costing Rs 8,950 is reduced by Rs 1,750. What is the new price?
8. Write the compact numeral for 9000 + 600 + 70 + 3. ________
9. Which is more ? Rs 800 or (Rs 880 – Rs 81) ________

10. 2,63,985 in words is ________.
11. 160 + ☐ – 60 = 150
12. Arrange the digits of 590371 to form the smallest number ________.
13. If it is the morning of the 11th of July, how many complete days are left in July ?____
14. I am 24 years old now ; in how many years will I be 65 ?____
15. What number is twenty times as big as 30 ?____

16. What is 25 less than six fifties ?____
17. Add 27 and 33 and find 20 times this sum.____
18. How many minutes are there in three hours?____
19. How many sixes are there in 54 ?____

20. Write down 10,000 paise in rupees.____

21. How many shares of Rs 10 can be bought out of a total of Rs 5,000 ?____

22. **Write > or < in box :**

550 ÷ 11 ☐ six nines.

23. In $16\overline{)920}$, the quotient is ___ and the remainder is ___.

24. How many twenties are there, in two hundred ___?

25. What is two hundred divided by five ? ____

KNOW YOUR NUMBERS

THE NUMBER ZERO

A **duck** is a score of zero in cricket. Love is a score of zero in tennis and badminton.

Zero is the number of children 7 metres tall.

When you subtract any number from itself, the result is zero.

You cannot divide any number by zero.

The number zero was invented in India about 2200 years back. It was one of the biggest inventions in the history of numbers. By allowing the use of place-value, it was made possible to write the same symbol, say 6 to represent different numbers, six, sixty, six hundred, six thousand etc. according to its place value.

LARGE NUMBERS

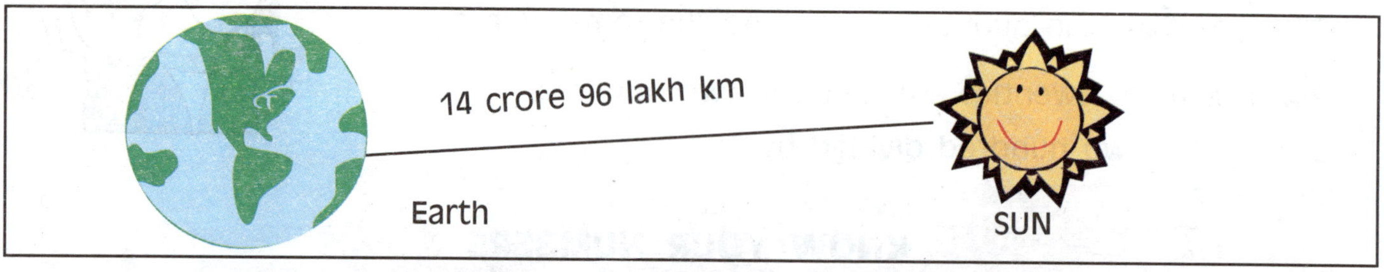

Sudhir read that the Population of Delhi is about 1 crore ninety lakh. The sun is about 14 crore 96 lakh kilometres from the earth. The population of India is more than 1 billion. He asked his teacher, "How much is a crore ?" The teacher told him that a crore is a big number. If you could spend Rs 100 a day for 250 years, you would not have spent a crore rupees.

You have read that the number 1 00 000 is one hundred thousand or 1 lakh. It is a **six digit** numeral.

The largest 6 digit number is	9,99,999	Nine lakh ninety nine thousand nine hundred ninety nine.
one more than this is	+1	
a seven digit number ⟶	**10,00,000**	**Ten lakh**
The largest 7 digit number is	99,99,999	Ninety nine lakh ninety nine thousand nine hundred ninety nine.
One more than that is an	+1	
eight digit number ⟶	**1,00,00,000**	**100 lakh or 1 crore.**

The number 10000000 (100 lakh) is the smallest number of eight digits. It is read as **1 crore**.

Similarly : 30000000 is read as 'three crore'

70000000 is read as 'seven crore'

90000000 is read as 'nine crore'

99999999 is the greatest number of 8 digits and is read as 'nine crore ninety nine lakh ninety nine thousand nine hundred ninety nine. If we add 1 to this number, we get the **smallest number of 9 digits**.

```
 99999999
+       1
100000000
```

Ten crore

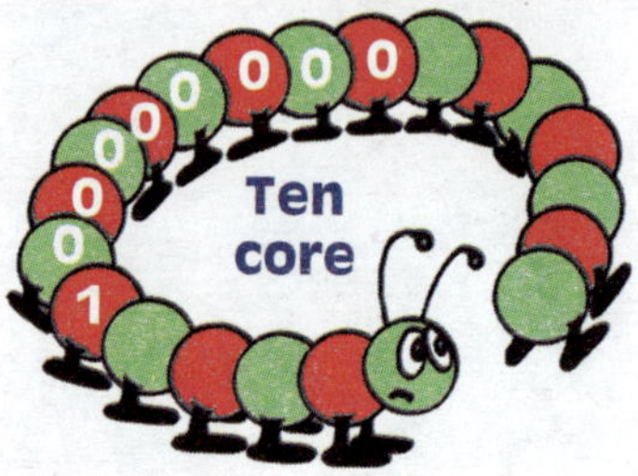

It is read as '**ten crore**'

Similarly : 300000000 is read as thirty crore.

500000000 is read as fifty crore.

900000000 is read as ninety crore.

999999999 is the **greatest number of 9 digits** and is read as 'ninety nine crore ninety nine lakh ninety nine thousand nine hundred ninety nine.

If you add one to this number you will get **the smallest number of 10 digits.**

$$\begin{array}{r} 999999999 \\ +\ 1 \\ \hline 1000000000 \end{array}$$

Hundred crore

You may read this number as **one hundred** crore. This is also read as '**one arab**' but the modern practice is to read it as one hundred crore. Similarly, we read,

3000000000 as 3 hundred crore.

7000000000 as 7 hundred crore.

You also come across such numbers as 1000 crore (one thousand crore) or 30,000 crore or 50,000 crore or fifty thousand crore in the newspapers these days.

1	=	Unit or one	(1 digit)
10	=	ten	(2 digits)
100	=	hundred	(3 digits)
1000	=	thousand	(4 digits)
10000	=	ten thousand	(5 digits)
100000	=	1 lakh (= one hundred thousand)	(6 digits)
1000000	=	ten lakh	(7 digits)
10000000	=	1 crore (= 100 lakh)	(8 digits)
100000000	=	ten crore	(9 digits)

Here is a place value chart for the first nine digits in our (Indian) system of numeration.

Crores		Lakhs		Thousands		Ones		
Ten Crore	one Crore	Ten Lakh	One Lakh	Ten Thousands	One Thousand	Hundred (100)	Ten (10)	One (1)
10,00,00,000	1,00,00,000	10,00,000	1,00,000	10,000	1,000			
T.C	**C**	**T.L.**	**L**	**T.Th**	**Th**	**H**	**T**	**O**

To make it easy to read a large number, we first divide it into periods starting from the right. For making periods, we put a comma after **three** digits starting from the right and then we put comma after every **two** digits. The first period consisting of the 3 digits is called the **units or the ones period**, the next period consisting of two digits is called the **thousands period**. Next period also consisting of two digits is called the **lakh period** and the next period which also consists of two digits is called the **crores period**.

The modern convention is to leave space to separate different periods instead of putting a comma between them. Thus the number 15, 83,72,609 would be written as 15 83 72 609

15,83,72,609

While reading a numeral, all the digits in the same period are read together and the name of the period (except the ones) is read together.

Thus the above number is read as **fifteen crore, eighty three lakh, seventy two thousand, six hundred nine.**

Note : that we should not write the word 'and' here as the common practice is.

We do not use the words indicating the periods in plural form. Thus we write fifteen crore and not fifteen crores.

The population of Japan in the year 2007 reached 12,86,46,345.

- A place value chart can help to explain what the number means.

Crores		Lakhs		Thousands		Ones		
Ten crore	Crore	Ten lakh	Lakh	Ten Thousand	Thousand	Hundred	Tens	Ones
1	2	8	6	4	6	3	4	5

- The different ways in which 12, 86, 46, 345 can be written are:

Standard Form	12, 86, 46, 345 **or** 12 86 46 345
Expanded Form	1 × 100000000 + 2 × 10000000 + 8 × 1000000 + 6 × 100000 + 4 × 10000 + 6 × 1000 + 3 × 100 + 4 × 10 + 5
Word Form	Twelve crore eighty six lakh forty six thousand three hundred forty five.

Example 1. Write the numeral (i) 593274081 (ii) 57301690 by separating the periods.

Solution. **1.** Start from the right.

2. Make first group of 3 digits, then each group of 2 digits.

3. Put commas between different groups.

Thus we write

	By Putting Commas	Leaving Spaces
(i) 593274081	as 59,32,74,081	59 32 74 081
(ii) 57301690	as 5,73,01,690	5 73 01 690

Example 2. Write the number names for each of the following numerals :

(i) 4 73 28 506 **(ii) 8 00 03 690** **(iii) 52 10 00 004**

(iv) 60,05,00,200 **(v) 90,40,50,700** **(vi) 80,00,00,008**

Solution.

	T.C	C	T.L	L	T.Th	Th	H	T	O	
(i)		4	7	3	2	8	5	0	6	Four crore seventy three lakh twenty eight thousand five hundred six.
(ii)		8	0	0	0	3	6	9	0	Eight crore three thousand six hundred ninety.
(iii)	5	2	1	0	0	0	0	0	4	Fifty two crore ten lakh four.
(iv)	6	0	0	5	0	0	2	0	0	Sixty crore five lakh two hundred
(v)	9	0	4	0	5	0	7	0	0	Ninety crore forty lakh fifty thousand seven hundred.
(vi)	8	0	0	0	0	0	0	0	8	Eighty crore eight.

Example 3. Write 357890216 in expanded form.

Solution. The given number is 35 crore 78 lakh 90 thousand 2 hundred sixteen. The given numeral is

T-C	C	T-L	L	T-Th	Th	H	T	O
3	5	7	8	9	0	2	1	6

= 3 ten crore + 5 crore + 7 ten lakh + 8 lakh + 9 ten thousand + 0 thousand + 2 hundred + 1 ten + 6 ones

= 3 × 100000000 + 5 × 10000000 + 7 × 1000000 + 8 × 100000 + 9 × 10000 + 0 × 1000 + 2 × 100 + 1 × 10 + 6

↑ Total 9 digits 1 and 8 zeros (100000000)
↑ Total 8 digits 1 and 7 zeros (10000000)
↑ Total 7 digits 1 and 6 zeros (1000000)
↑ Total 6 digits 1 and 5 zeros (100000)
↑ Total 5 digits 1 and 4 zeros (10000)

= 300000000 + 50000000 + 7000000 + 800000 + 90000 + 0 + 200 + 10 + 6

Example 4. In which period and place is the box digit in each of the following :

(a) 7 [8] 60 542 **(b) [2] 83 00 671** **(c) 81 [7] 4, 20, 509**

(d) 59 27 4 [3] 036 **(e) 34, 17, 29, [6] 40** **(f) 1, 49, 60 05 [8]**

Solution. Let us first write all these numbers in a place value chart and encircle the given digit.

Periods	Crore		Lakhs		Thousands		Ones		
	TC	C	TL	L	T-Th	Th	H	T	O
(a)			7	8	6	0	5	4	2
(b)		2	8	3	0	0	6	7	1
(c)	8	1	7	4	2	0	5	0	9
(d)	5	9	2	7	4	3	0	3	6
(e)	3	4	1	7	2	9	6	4	0
(f)		1	4	9	6	0	0	5	8

Now reading the period and place value of each of the given digits we have

Solution.

	Digit	Period	Place Value		
(a)	8	Lakhs	8 × 1 lakh	=	8 00 000
(b)	2	Crores	2 × 1 crore	=	2 00 00 000
(c)	7	Ten Lakhs	7 × 10 lakh	=	70 00 000
(d)	3	Thousands	3 × 1 thousand	=	3000
(e)	6	Hundreds	6 × 1 hundred	=	600
(f)	8	Ones	8 × 1	=	8

1. Write the numeral 723446960 in the standard form using commas and leaving spaces.

Using commas | **Leaving spaces**

2. Write the number names for the following numerals.

(a) 9, 25, 86, 190 ____________________

____

(b) 16, 48, 00, 009 ____________________

____

3. Write the numeral for each of the following:

(a) Twelve crore seven lakh two thousand ninety four.

(b) Fifty Crore five hundred five

4. **Write the place value of each encircled digit.**

(a) 45,(6)2, 93, 200

(b) 1(6), 00, 00, 090

5. **Write in expanded form:** 

(a) 675 86 412 ______________________

(b) 2348 00 007 ______________________

EXERCISE 5

1. **Write a 7 digit numeral in the 'Indian system'.**

(a)

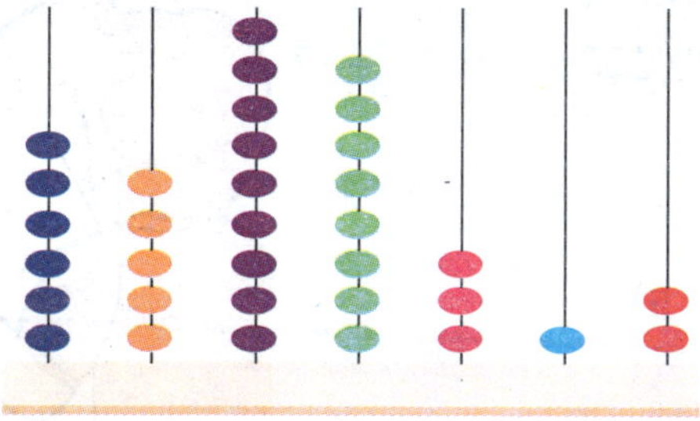

(b)

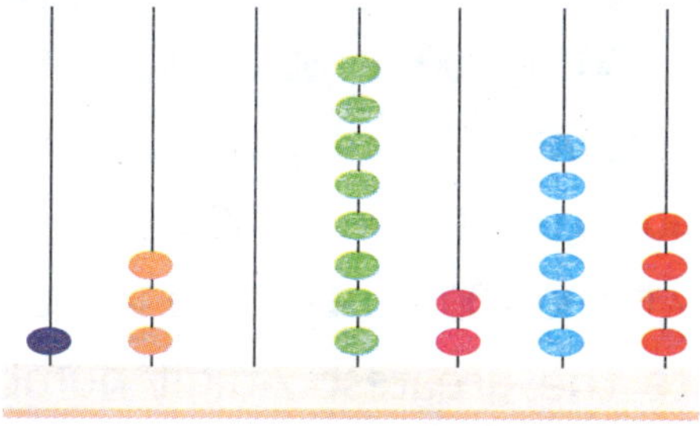

2. **Write number names of the following numbers in words:**

(i) 4,57,29,318 (ii) 8,00,00,060 (iii) 50,00,26,003

(iv) 72,80,66,550 (v) 83,00,22,049 (vi) 70,06,00,080

(vii) 48,99,85,361 (viii) 90,02,00,590 (ix) 90,00,00,009

3. **Write numeral for each of the following numbers:**

(i) Fifty eight lakh seven thousand nine

(ii) Seven crore two lakh seventy five thousand nineteen

(iii) Nineteen crore eighty four lakh seven thousand three hundred

(iv) Five crore eight lakh four

(v) Eighty nine crore sixty five lakh fifty nine thousand six hundred eighteen

(vi) Five crore five (vii) Ninety nine crore

(viii) Seventy seven crore seven hundred seventeen.

4. **The sun is 14,88,00,000 km from the earth. Write this number in words in the Indian system.**

5. **Write the period of the digits encircled in each of the following :**

(i) (5)39 72 406 (ii) 75 1(3)59 286 (iii) 6 88 2(7)105

(iv) 3 49(6)2 579 (v) (6)0 53 29 446 (vi) 8(4)00 45 967

6. **What is the place value of the encircled digit in each of the following numbers :**

(a) (4)83 20 196 (b) 59(2)3 71 606 (c) 8 29(6)0 001

(d) 22 98 69(3)07 (e) (8)7 00 19 512 (f) 10 33(5)4 219

7. Write the period and place value of the encircled digit in each of the following numbers :

(a) 5370 268 (b) 24 90 358 (c) 174 28 350

(d) 62 00 00 384 (e) 66 20 70 839 (f) 5 03 62008

8. Write in the expanded form :

(i) 7 52 418 (ii) 36 12 920 (iii) 6 98 45 137

(iv) 26 18 44 509 (v) 73 59 65 614 (vi) 6 98 05 714

(vii) 9 00 87 023 (viii) 24 00 03 106

9. Write the numbers whose expanded forms are

(a) 50,00,00,000 + 8,00,00,000 + 20,00,000 + 4,00,000 + 90,000 + 7,000 + 600 + 30 + 1

(b) 70,00,00,000 + 2,00,00,000 + 10,00,000 + 5,00,000 + 80,000 + 3,000 + 800 + 40 + 6

(c) 90,00,00,000 + 50,00,000 + 4,00,000 + 90,000 + 600 + 7

(d) 6,00,00,000 + 40,00,000 + 8,000 + 90 + 0

(e) 40,00,00,000 + 70,000 + 8

(f) 7,00,00,000 + 9,00,000 + 60,000 + 80 + 5

10. Write the greatest 7-digit number.

11. Write the smallest 8 digit number.

12. Write the smallest 8-digit number having three different digits.

13. Write the greatest 8-digit number having three different digits.

14. Write the successor of each of the following numbers:

Method : Successor of a given number = Given number + 1

Ex. The successor of 41,87,56,789 is 41,87,56,790

(i) 59,60,73,875 (ii) 6,79,83,969

(iii) 5,80,76,275 (iv) 72,99,99,999

(v) 6,87,99,999 (vi) 23,48,32,899

15. Write the predecessor of the following numbers:

Method : Predecessor of a given number = Given number – 1

Ex. The predecessor of 43,00,00,000 is 42,99,99,999

(i) 59,00,00,000 (ii) 7,56,87,600

(iii) 20,71,57,830 (iv) 70,00,36,000

(v) 9,90,00,000 (vi) 45,26,99,990

16. Counting by twos, write the numbers from 5009998 to 5010002.

17. Counting by hundreds, write the numbers from 7009909 to 7010209.

18. Counting by thousands, write the numbers from 8659317 to 8663317.

19. Counting by threes, write the numbers in reverse order from 10000004 to 999995.

20. Find the difference of the place values of two 7's in 47085790.

21. Find the product of the place values of two 4's in 5694741.

COMPARING AND ORDERING LARGE NUMBERS

Rule:

1. Line up the numbers by the place values.
2. In numbers with unequal number of digits, the number having more digits is greater than a number having fewer digits.
3. In numbers with equal number of digits start from the left and compare the digits until they differ.

Example 1. Which is greater 6,73,10,009 or 12,89,73,640 ?

Solution. Lining up the numbers by place values, we have

TC	C	TL	L	T-Th	Th	H	T	O
	6	7	3	1	0	0	0	9
1	2	8	9	7	3	6	4	0

This shows,12, 89, 73, 640 > 6, 73, 10, 009

(9-digits) (8-digits)

Example 2. Compare 73846205 and 73857619.

Solution. Line up the numbers by place values and compare the digits from the left most place until they differ.

C	T-L	L	T-Th	Th	H	T	O
7	3	8	4	6	2	0	5
7	3	8	5	7	6	1	9

same

5 > 4

The digits differ at the T-TH place.

5 > 4. ∴ 7,38,57619 > 7,38,46,205.

Example 3. Arrange the following numbers in descending order: 7,36,01,295, 80,34,51,276, 3,04,63,589, 81,72,345, 91,00,32,947.

Solution. Step 1. Line up the numbers by place values.

Step 2. There are two 9-digit numbers, two 8-digit numbers and one 7-digit number.

Comparing 9-digit numbers: ∴ 9 Ten Crore > 8 Ten Crore, ∵ 910032947 > 803451276

Comparing 8-digit numbers, ∵ 7 Crore > 3 Crore ∴ 73601295 > 30463589

7-digit number is the smallest.

TC	C	TL	L	T-Th	Th	H	T	O
	7	3	6	0	1	2	9	5
8	0	3	4	5	1	2	7	6
	3	0	4	6	3	5	8	9
		8	1	7	2	3	4	5
9	1	0	0	3	2	9	4	7

Hence the numbers written in descending order are:
910032947, 803451276, 73601295, 30463589, 8172345.

Example 4. Arrange the following numbers in ascending order.
2345678, 607810542, 694317, 84120079, 498900351, 902347016.

Solution. Step 1. Line up the given numbers by place values.

T-C	C	T-L	L	T-Th	Th	H	T	O
		2	3	4	5	6	7	8
6	0	7	8	1	0	5	4	2
			6	9	4	3	1	7
	8	4	1	2	0	0	7	9
4	9	8	9	0	0	3	5	1
9	0	2	3	4	7	0	1	6

Step 2. Comparing the three 9-digit numbers,

$\therefore$ 4TC < 6TC < 9TC

$\therefore$ 498900351 < 607810542 < 902347016

Also $\because$ 6 digit number < 7 digit number < 8 digit number

$\because$ 694317 < 2345678 < 84120079

Therefore, written in ascending order, the numbers are
694317, 2345678, 84120079, 498900351, 607810542, 902347016.

1. Compare write >, < or = for each

(a) 72341725 ◯ 2358931 **(b)** 64704789 ◯ 64735891

2. Arrange the following numbers in ascending order.

4763875, 3289762, 4893970, 12587689 ______________________

3. Order the following set of numbers from greatest to least.

64812661, 82811121, 82935871, 581234890 ______________________

EXERCISE 6

1. Compare the following numbers and put > or < in the blanks:

(i) 3841920 ___ 3850098 **(ii)** 49103689 ___ 48996778
(iii) 50176003 ___ 60892144 **(iv)** 70134625 ___ 71987836
(v) 15387260 ___ 9987634 **(vi)** 66326019 ___ 65210986
(vii) 80088008 ___ 80088009 **(viii)** 77172765 ___ 77162765

2. Arrange the following numbers in ascending order :

(a) 2828335, 3537286, 1995764, 2989632 **(b)** 54619717, 39400316, 8798422, 90100510
(c) 49182766, 9956130, 891300, 76180542

3. Arrange the following numbers in descending order :

(a) 2003568, 19872003, 279868, 20016930 **(b)** 39876542, 4000196, 22398761, 5018960
(c) 8765014, 3610303, 55109872, 5688764

4. **Write the smallest and the greatest numbers using each of the following digits only once :**

(a) 4, 8, 0 , 1, 7, 6, 5 **(b)** 2, 0, 8, 9, 6, 5, 4, 1

(c) 0, 6, 5, 9, 1, 2, 3, 4, 8 **(d)** 0, 2, 4, 6, 9, 7, 3, 8

5. **Choose the smallest number and the largest number in each group.**

(a) 20561937, 418260329, 5463978, 14053610, 325809647

(b) 213345207, 96354357, 19501872, 220100043, 95468600

(c) 8048784, 13427561, 30303080, 95173694, 10082987

(d) 8000008, 80000007, 8853648, 96880808, 77777777

6. The given table shows the projected population of 5 metropolitan cities in the year 2015.

Metropolitan City	Projected population for the year 2015
Mumbai, India	2, 61, 38, 000
Mexico city, Mexico	1, 91, 80, 000
Sao Paulo, Brazil	2, 03, 97, 000
Tokyo, Japan	2, 64, 44, 000
New York City U.S.	1, 74, 32, 000

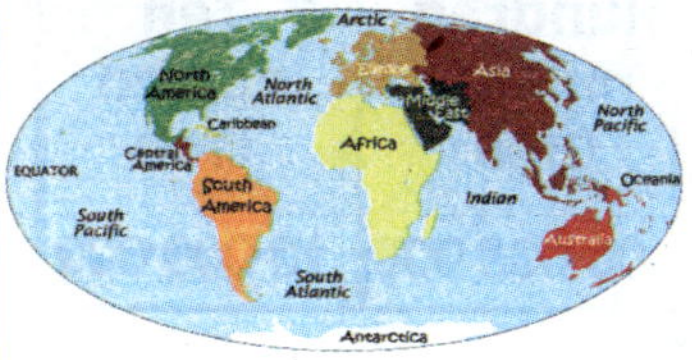

Use the above table to answer the following questions.

(a) Which city is projected to have the greatest population in 2015? Which city will have the least population?

(b) Arrange the cities in order from least population to greatest population according to the projected populations in 2015.

(c) Which cities will have population $>$ 2 crore in 2015?

(d) In 2015 which cities will have the closest population to 2 crore?

7. **Fill in a number in the box to make the given statement true.**

(a) 64, □ 93 > 64, 486 **(b)** 1, 24, 56, 789 < □, 00, 23, 960

(c) 76, 45, 89, 000 > 7 □ , 45, 89, 000 **(d)** 14, 64, 920 < 1 □, 26, 870

INTERNATIONAL SYSTEM OF WRITING NUMBERS

You have read that the number 100000 is one hundred thousand or 1 lakh. It is a six digit numeral. In the English system (or International system), the number 10 lakh or ten hundred thousand, that is, a thousand thousand is called a **million.** It is written as under :

1 000 000 1 million

This space separates the million from the hundred thousand

This space separates the thousand from the hundred

Instead of lakhs, crores etc as periods, we use millions, billions etc. We have

The smallest 8 digit number is 1,00,00,000 called one crore.

10 lakh	**=**	**1 million**
1 crore	**=**	**100 lakh = 10 million**
10 crore	=	100 million
100 crore	**=**	**1,000 million = 1 billion**
1000 crore	=	10,000 million = 10 billion

The smallest 8 digit numeral is 10,000 000 and is called 10 million

The international place value chart is given below.

Millions			Thousands			Ones		
Hundred million (HM)	**Ten million (TM)**	**One million (M)**	**Hundred thousand (HTh)**	**Ten thousand (TTh)**	**One thousand Th**	**Hundred H**	**Ten T**	**One O**
100 000 000	10 000 000	1 000 000	100 000	10 000	1 000	100	10	1

The chart can be extended to its left to include more than 9 places. The period just to the left of the **millions period** is called the **billions period**. Then we have the **trillions period**. The places left to the hundred billions places are called one trillion place, ten trillions place and hundred trillions place.

1000 million = 1 billion (= 100 crore)

1000 billion = 1 trillion (= 1 lakh crore)

Note that in the Indian system starting from the right after the first period of 3 digits all other periods are of 2 digits but in the International system **each period consists of three digits.** We place comma after every 3 digits. In the modern convention instead of putting a comma, we leave space after every 3 digits.

The abacus shows the number five million, eight hundred thirty nine thousand, two hundred forty seven. To make the task of reading and writing such large numbers easy, **start counting digits from the right by threes** and leave some space after every three digits as shown below.

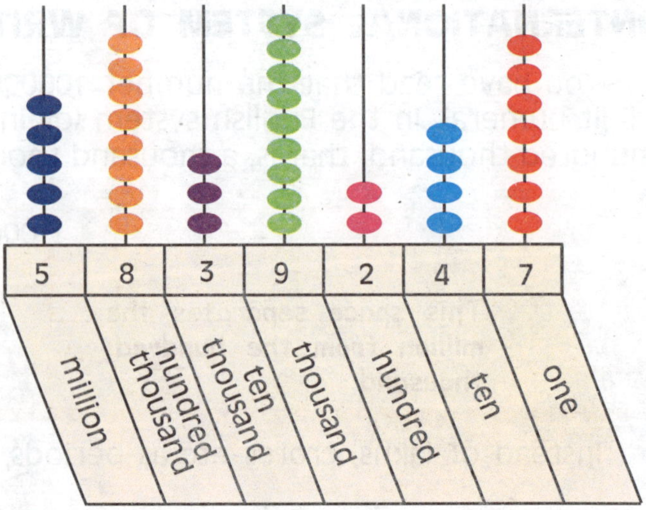

How can you write the above number in Indian System?

58, 39, 247

Example 1. Write in numerals:

(a) 3 million **(b) 10 million**

Solution. **(a)** 3 million 3 000 000 or 3,000,000

(b) 10 million 10 000 000 or 10,000,000

Example 2. Write the following numbers in the International system and write their number names. (a) 37426589 (b) 549118766

Solution. Starting from the right hand side and counting by threes, the numbers in the International system can be written as:

(a) 37 426 589, i.e., Thirty seven million, four hundred twenty six thousand, five hundred eighty nine.

(b) 549 118 766 i.e., Five hundred forty nine million, one hundred eighteen thousand, seven hundred sixty six.

Example 3. Using International place-value chart write the numeral for each of the following.

(i) Ninety six million, three hundred forty thousand, two hundred fifty six.

(ii) Sixty eight million, seven hundred thousand, three hundred twenty nine.

(iii) Two hundred nine million, one hundred seventy six thousand, five hundred thirty four.

Solution. **Writing in the place value chart,**

Periods	Millions			Thousands			Ones		
Place values	HM	TM	M	HTh	TTh	TH	H	T	O
(i)		9	6	3	4	0	2	5	6
(ii)		6	8	7	0	0	3	2	9
(iii)	2	0	9	1	7	6	5	3	4

The numerals are **(i)** 96 340 256 **(ii)** 68 700 329 **(iii)** 209 176 534.

Example 4. Write 590341627 first in the Indian system and then in the International system. Read them both ways.

Solution.

Indian System

59, 03, 41, 627

Fifty nine crore three lakh forty one thousand six hundred twenty seven.

International System

590, 341, 627

Five hundred ninety million three hundred forty one thousand six hundred twenty seven.

Class Work

Skill Practice

1. Rewrite the following numerals using the periods and places in the international place value chart. Also write the number names.

	Numeral	Number name
(a) 23487900	____________	________________________________
(b) 370008003	____________	________________________________

2. Write in numeral form in the International place value system.

(a) Sixty five million, two hundred five thousand, seven hundred forty six. ____________

(b) Two hundred million, seventy thousand thirty two. ____________

3. Write 814008250 both in (i) Indian system and (ii) International system. Also write the number names:

	Indian System	International System
Numeral	____________	____________
Number Name	____________	____________

Home Work

EXERCISE 7

1. Rewrite each of the following numerals using periods and places in the International place value chart. Also write the number name of each.

(a) 183497026 **(b)** 463721985 **(c)** 5086432

(d) 46300529 **(e)** 700083460 **(f)** 810000008

2. Write the following in numeral form in the International system.

(a) Eighty five million, four hundred five thousand, nine hundred.

(b) Two hundred million, eight hundred seventeen thousand five hundred ten.

(c) Five hundred thirty million, six hundred eighty five thousand, nine hundred thirty nine.

(d) Eight million, four hundred fifty five thousand, one hundred two.

(e) Five hundred seven million, six thousand, two.

3. **Write the following numbers both in the Indian way and in the International way. Also write the number names for both the ways.**

(a) 429012795 **(b)** 501043768 **(c)** 321543006

(d) 349100860 **(e)** 92600512 **(f)** 99000017

4. **The estimated population in the year 2007 of some Asian countries is given below.**

Afghanistan	27089593
Bangladesh	137493990
Indonesia	224481720
Japan	128646345
Malaysia	28294120
Nepal	25874519
Pakistan	167806831
Sri Lanka	19796874
Singapore	3654103
Thailand	67249456

Use the table to answer the following questions.

(a) Write each number both in the Indian as well as the International system. Also write the number name for each number in both the systems.

(b) Write the names of countries in descending order of their population.

(c) Which of the above given countries is most populous ?

(d) Which of these countries is least populated?

5. **Use the digits 2, 4, 8, 9, 3, 7, 4, 1 and 0 to write in both the systems,**

(a) the largest possible number **(b)** the smallest possible number. Also, write their number names.

6. **Write the counting numbers in natural order beginning with 999 995 and ending with 1 000 010.**

DO YOU KNOW?

The population of India in 2007 is 1129667528.

Written and read in the Indian system it is **1, 12, 96, 67, 528** One Arab, twelve crore ninety six lakh, sixty seven thousand, five hundred twenty eight.

Written and read in the International system it is **1, 129, 667, 528** One billion, one hundred twenty nine million six hundred sixty seven thousand five hundred twenty eight.

MENTAL MATHS – 2

1. Read out the number, 589 039, 274.
2. The standard form of the number 80, 00, 000 + 40, 000 + 200 + 10 + 3 is _____.
3. What is the place value of the underlined digit 6, <u>8</u>9, 10, 434 ?
4. Divide the place value of 5 in 62571 by the successor of 999.
5. Write the predecessor of 718, 264, 500.
6. How many five thousands are there in one lakh?
7. Multiply two million by fifty. Read the product aloud. What number is it in the Indian system ?
8. 7000 lakh = ____________ crore.
9. Divide the place value of 2 by the place value of 1 in 76215.
10. One-fifth of a million = ____________ lakh

KNOW YOUR NUMBERS

(THE NUMBER ONE)

One, unit, unity, single, solo.......

There is only one President of India, there is only one prime minister, there is only one river 'Ganga'. There is only one you. All of these are unique words beginning with **uni**-meaning, 'There is one of something.' **Mono** can also mean that there is one of something. A monorail is a railway, where the track consists of a single rail. Whatever number you multiply by 1, or divide by 1, remains unchanged.

37 × 1 = 37
592 ÷ 1 = 592
No change in the number

One important feature of number 1 is its ability to create *palindromes*.

A palindrome is a number that **reads the same backwards or forwards.**
For example :

$$1 \times 1 = 1$$
$$11 \times 11 = 121$$
$$111 \times 111 = 12321$$
$$1111 \times 1111 = 1234321$$

3 OPERATIONS ON LARGE NUMBERS

You have learnt the addition and subtraction of 5-digit and 6-digit numbers in class 4. Here, we will extend our study to the addition of 7-digit, 8-digit and 9-digit numbers.

ADDITION

Two or more numbers called the **addends** are added to give an answer called the **sum.**

Example 1. The projected population in 2007 of Mumbai is 1, 30, 73, 926; Kolkata is 46, 48, 011; New Delhi is 1, 15, 05, 196 and that of Chennai is 43, 76, 400. What will be the total projected population of the four metros ?

Solution.

	C	T-L	L	T - Th	Th	H	T	O
	1	1	2	2	1	1	1	
	1	3	0	7	3	9	2	6
		4	6	4	8	0	1	1
	1	1	5	0	5	1	9	6
+		4	3	7	6	4	0	0
	3	3	6	0	3	5	3	3

1. Write in vertical form.
2. Add column wise-ones, tens, hundreds, and so on.
3. Regroup and carry, wherever, necessary.

The total projected population in 2007 of the four metros is **3, 36, 03,533.**

SUBTRACTION

When a number is subtracted **(subtrahend)** from another number **(minuend)**, the result is their **difference.**

Example 2. According to 1991 census, the number of people speaking Bengali is 6, 95, 95, 738 and that of speaking Gujarati is 4, 06, 73, 814. How many more people speak Bengali than Gujarati?

Solution.

	C	T-L	L	T-Th	Th	H	T	O
		8	15		4	17		
	6	~~9~~	~~5~~	9	~~5~~	~~7~~	3	8
−	4	0	6	7	3	8	1	4
	2	8	9	2	1	9	2	4

1. Write in vertical form – Subtrahend below the minuend.
2. Here minuend = 6, 95, 95, 738 and subtrahend = 4, 06, 73, 814
3. Subtract columnwise, regrouping (borrowing), if necessary.

You can check subtraction as :

		1			1			
	4	0	6	7	3	8	1	4
+	2	8	9	2	1	9	2	4
	6	9	5	9	5	7	3	8

Subtrahend + Difference
= Minuend.

Skill Practice

Class Work

Add:

1.
```
   2, 85, 93, 614
 + 8, 93, 41, 123
```

2.
```
   14, 93, 87, 914
    2, 00, 35, 153
 + 25, 87, 13, 234
```

Subtract:

3.
```
   2, 43, 15, 902
 − 2, 05, 12, 973
```

4.
```
   56, 43, 00, 975
 − 25, 65, 12, 856
```

Home Work

EXERCISE 8

Copy and add:

1.
```
   85,70,392
 + 74,95,287
```

2.
```
   79,18,895
 + 51,74,526
```

3.
```
   6,58,90,107
 + 2,73,56,849
```

4.
```
   16, 59, 38, 903
 + 28, 16, 27, 862
```

5.
```
   76, 54, 28, 097
 +  8, 93, 56, 893
```

6.
```
    9, 54, 56, 378
 + 39, 72, 84, 650
```

7.
```
   97643238
   50967642
 +   859837
```

8.
```
   73564121
   92473787
 + 13894296
```

9.
```
   890341265
   218126459
 + 742958003
```

10. 341698853 + 37519162 + 599678437 + 172564

11. 34857696 + 27502394 + 61927880 + 8067617 + 9520

12. 92568214 + 83256329 + 4327568 + 643263 + 71952875

13. 86412765 + 32194320 + 7397918 + 658014 + 62783 + 8019

Copy and subtract. Check your answer with addition

14.
```
   82754380
 − 10817593
```

15.
```
   447147907
 −  36518698
```

16.
```
   558258014
 −  72102162
```

17. 8,75,00,190 − 99,64,385

18. 2,77,63,500 − 1,88,90,000

19. 819528756 − 97921242

20. 73124732 − 54792465

21. How much is 50 million more than nine million, nine hundred ninety nine thousand nine hundred ninety nine?

22. How much is 6 crore more than five lakh seventy eight thousand ?

23. What must be added to 79,62,360 to make it equal to 80,00,000 ?

Copy and simplify :

24. 4,35,06,789 + 59,67,892 + 25,76,19,517 − 19,76,28,635.

25. 2,89,46,837 + 7,96,85,498 − 9,99,86,999.

26. **The population of different states of India as per 2001 census is as under :**

1.	Andhra Pradesh	7,57,27,541	15.	Maharashtra	9,67,52,247
2.	Arunachal Pradesh	10,91,117	16.	Manipur	23,88,634
3.	Assam	2,66,38,407	17.	Meghalaya	23,06,069
4.	Bihar	8,28,78,796	18.	Mizoram	8,91,058
5.	Chattisgarh	2,07,96,956	19.	Nagaland	19,88,636
6.	Goa	13,43,998	20.	Orissa	3,67,06,920
7.	Gujarat	5,06,71,017	21.	Punjab	2,42,89,296
8.	Haryana	2,10,82,989	22.	Rajasthan	5,64,73,122
9.	Himachal Pradesh	60,77,248	23.	Sikkim	5,40,493
10.	Jammu and Kashmir	1,00,69,917	24.	Tamilnadu	6,21,10,839
11.	Jharkhand	2,69,09,428	25.	Tripura	31,91,168
12.	Karnataka	5,27,33,958	26.	Uttarakhand	84,79,562
13.	Kerala	3,18,38,619	27.	Uttar Pradesh	16,60,52,859
14.	Madhya Pradesh	6,03,85,118	28.	West Bengal	8,02,21,171

Use this chart to answer the following questions.

(i) Which state has more population, Madhya Pradesh or West Bengal and by how much?

(ii) What is the combined population of

(a) Uttar Pradesh, Punjab and Rajasthan ?

(b) Manipur, Meghalaya, Mizoram, Nagaland and Tripura?

(c) Kerala, Karnataka, Tamilnadu and Maharashtra ?

True or False ?

(iii) (a) The population of Uttar Pradesh is more than the combined population of West Bengal, Orissa and Punjab.

(b) The state of Sikkim has the least population.

(iv) Name the states whose population lies between 10 million and 50 million. What is their combined population ?

(v) Name the state which has the most population and write the figure in numeral and words in the international system.

27. Write the number which is

(a) 5000 more than 3 475 639. **Ans: 3 480 639**

(b) 2000 more than 4 396 137.

(c) 7000 more than 2 508 913.

(d) 800 more than 71,95,638.

(e) 12000 more than 5,76,83,104.

(f) 19000 more than 285 819 702.

(g) 40000 more than 66,135,862.

28. Fill in the missing digits

(a)

	4	6	□	2	□	8
+	□	□	6	□	7	□
	9	8	9	1	9	0

(b)

		7	□	9	□	2	□
+		□	7	□	6	□	8
	1	6	2	4	8	0	2

(c)

	5	□	□	9	□	2
−	□	5	8	□	2	□
	1	1	0	2	2	1

(d)

	9	□	□	□	□	□
−	8	□	□	8	9	9
		1	1	1	0	1

29. The number of people who travelled by Delhi Metro on Wednesday was 5, 15, 273; on Thursday was 4,23,894 and on Friday was 6,05,009. How many people in all travelled by Delhi Metro on all these three days ? Express your answer in words.

30. The cost of a car and a flat together is Rs 1,04,78,930. If the car costs Rs 15,23,485, what is the cost of the flat?

MULTIPLICATION

You know that multiplication is **repeated addition.** The number to be multiplied is called the **multiplicand** and the number by which it is multiplied is called the **multiplier.** The result of multiplication is called the **product.**

$$4 + 4 + 4 + 4 + 4 = 5 \times 4 = 20$$

4 added repeatedly five times. (5 → Multiplier; 4 → Multiplicand; 20 ← Product)

5 times 4 is 20
or
5 fours are 20

Note : The multiplier and multiplicand are also called the factors. Let us quickly revise the properties of multiplication you have studied in Class IV.

Properties of Multiplication	
★ Changing the order of factors does not change the product 13 × 7 = 7 × 13 = 91	★ The product of any number and the number 1 is the number itself 518 × 1 = 1 × 518 = 518.
★ Changing the grouping of factors does not change the product. 2 × (5 × 9) = (2 × 5) × 9 = 90. 2 × 45 = 90 10 × 9 = 90	★ The product of any number and 0 is 0. 12,314 × 0 = 0. 

MULTIPLICATION BY MULTIPLES OF 10

When one or both factors are multiples of 10, you can multiply mentally. Look at these patterns to revise how:

6 × 10 = 60	4 × 80 = 320
6 × 100 = 600	4 × 800 = 3200
6 × 1000 = 6000	4 × 8000 = 32000
60 × 100 = 6000	40 × 80 = 3200
600 × 100 = 60,000	400 × 800 = 320,000

6 × 1 tens = 6 tens
= 60
4 × 8 hundreds = 32 hundreds
= 3200

To multiply by multiples of 10, first find the product of the front non-zero digits. Then write as many zeros after the product as there are zeros in both the factors.

Following the same pattern, we can mentally multiply a number by 10,000 or 1,00,000 and write the product

8693 × 10,000 = 8693 × 1 ten-thousand = 8693 ten-thousands = 86930000

Insert 4 zeros at the right

7583 × 1,00,000 = 7583 × 1 lakh = 7583 lakh = 758300000

Insert 5 zeros at the right

Example 1. Find the products :

(a) 5378 × 400 **(b) 6985 × 7000** **(c) 23950 × 40,000**

Solution. (a) 5378 × 400 = 5378 × (4 × 100)
= (5378 × 4) × 100
= 21512 × 100 = **21,51,200.**

Multiply by 4 and insert 2 zeros at the right

(b) 6985 × 7000 = 6985 × (7 × 1000)
= (6985 × 7) × 1000
= 48895 × 1000 = **4,88,95,000.**

Multiply by 7 and insert 3 zeros at the right.

(c) 23950 × 40000 = 23950 × (4 × 10,000)
= (23950 × 4) × 10,000
= 95800 × 10000 = **95,80,00,000.**

Multiply by 4 and insert 4 zeros at the right.

Fill in the blanks

1. 42 × 10,000 = ____________

2. 165 × 1,00,000 = ____________

3. 12 × 2000 = ____________

4. 14 × 40,000 = ____________

5. 500 × 30,000 = ____________

6. 1500 × 4000 = ____________

7. ____________ × 90,000 = 2,70,00,000

8. 11,000 × ____________ = 55,00,00,000

MULTIPLYING LARGE NUMBERS

Example 2. **Multiply (a) 78,935 by 86 (b) 32,893 by 975.**

Solution.

(a)

78935	
× 86	
473610	← 78935 × 6
+ 6314800	← 78935 × 80
6788410	

Multiply by 6 ones,
Multiply by 8 tens
Add the partial products

(b)

32893	
× 975	
164465	← 32893 × 5 ones (5)
2302510	← 32893 × 7 tens (70)
+ 29603700	← 32893 × 9 hundreds (900)
32070675	

Example 3. **Multiply 5983 by 2769.**

Solution.

5983	
× 2769	
53847	← 5983 × 9
358980	← 5983 × 60
4188100	← 5983 × 700
11966000	← 5983 × 2000
16566927	**Add**

Example 4. **Multiply 5896 by 9760.**

Solution.

5896	
× 9760	
0000	5896 × 0
353760	5896 × 60
4127200	5896 × 700
53064000	5896 × 9000
57544960	

Short form

5896
× 9760
353760
4127200
53064000
57544960

Write '0' to indicate multiplication by 0 ones

Example 5. **Multiply 49568 by 7094.**

Solution.

49568	
× 7094	
198272	49568 × 4
4461120	49568 × 90
0000000	49568 × 0
346976000	49568 × 7000
351635392	

Short form

49568
× 7094
198272
4461120
346976000
351635392

Write '0' at the hundreds place to indicate that the product obtained on multiplication by 0 hundreds is 0

Example 6. **Fair Sky Travels charges Rs 80,563 for a 7-day trip to Europe. If 8790 passengers take that trip in a year, how much has the travel agency earned during that year ?**

Solution.

Cost for 1 passenger

= Rs 80,563

∴ Cost for 8790 passengers

= Rs 80,563 × 8790

```
     80563
  ×   8790
   7250670
  56394100   ← 80563 × 700
 644504000   ← 80563 × 8000
 708148770
```

Write '0' to indicate multiplication by '0' ones. Then multiply by 9

Note. As shown below, if care is taken in placing the numerals, the zeros may be dropped without changing the product.

```
     5896    |      49568    |     80563
    ×9760    |     ×7094     |    × 8790
   353760    |     198272    |   7250670
   41272 ←   |    446112     |  563941
  53064  ←   |   3469760     | 644504
 57544960    |   351635392   | 708148770
Drop zeros.  |               |
```

Multiply:

1.	2.	3.
4 7 8 9 2 3 × 2 8 9	6 5 8 4 × 1 0 3 4	5 0 9 8 × 7 0 4 0

EXERCISE 9

1. Write the product.

(i) 845 × 10 **(ii)** 9572 × 100 **(iii)** 605 × 1000

(iv) 987 × 10,000 **(v)** 549 × 1,00,000 **(vi)** 237 × 10,00,000

(vii) 695 × 30 **(viii)** 732 × 900 **(ix)** 923 × 12000.

Multiply

2. 99706 × 73 **3.** 89659 × 93 **4.** 69725 × 846 **5.** 723896 × 958

6. 596832 × 790 **7.** 99990 × 970 **8.** 7851 × 6925 **9.** 391525 × 861

10. 65839 × 6007 **11.** 65839 × 627 **12.** 249753 × 601 **13.** 90039 × 9001

14. 47030 × 5600 **15.** 65000 × 3800 **16.** 70095 × 950 **17.** 90990 × 9090

18. 59900 × 4700 **19.** 24703 × 6020

20. A residential school has 780 students on its rolls. If the annual fee per student is Rs 50,000, what is the total fee collected annually by the school ?

21. A motor cycle costs Rs 25,980. If 6958 motor cycles are sold by a dealer, what is the total money earned by him ?

22. A builder constructs 430 flats in a society. If each flat costs Rs 18,90,560, what is the combined cost of all the flats ?

23. 25,978 copies of the latest Harry Potter book were sold. If each books costs Rs 975, how much is earned by the sale of these books ?

24. The membership fee for a particular club is Rs 5225 per year. If 9789 people are its members, how much does the club earn in a year ?

DIVISION

Division is the opposite of multiplication. It means sharing equally or distributing equally. The number to be divided is called **dividend.** The number which divides is called the **divisor** and the result obtained is called the **quotient.** The number remaining in the end is called **remainder.**

Divide 320 by 15

```
                 2 1   ←—— Quotient
Divisor ——→ 15 ) 3 2 0 ←—— Dividend
               - 3 0↓
               -----
                 2 0
               - 1 5
               -----
                   5   ←—— Remainder
```

Do you remember the Divison Algorithm? We use it to check division.

Divisor × Quotient + Remainder = Dividend.

For the above example, **15 × 21 + 5** = 315 + 5 = **320.**

Important facts About Division	
★ Dividing a number by itself is equal to 1 4394 ÷ 4394 = 1 ★ Dividing a number by 1 is equal to the number itself. 65,895 ÷ 1 = 65895	★ Dividing zero by any number is equal to 0. 0 ÷ 12,156 = 0 ★ Dividing any number by zero is not defined. 2893 ÷ 0 (not permissible)

DIVISION BY MULTIPLES OF 10

You have studied about division by multiples of 10 in class *4* in detail. Let us revise the same here.

296 ÷ 10 ⇒ Quotient = 29, Remainder = 6

4594 ÷ 100 ⇒ Quotient = 45, Remainder = 94

6893 ÷ 1000 ⇒ Quotient = 6, Remainder = 893

Thus, we see that :

	Quotient	Remainder
Any number ÷ 10	All the digits except digit at ones place	Digit at ones place.
Any number ÷ 100	All the digits except digits at tens and one place	Digits at tens and ones place
Any number ÷ 1000	All the digits except digits at hundreds, tens and ones place.	Digits at hundreds, tens and ones place.

Following the same pattern.

85,963 ÷ 10,000 ⇒ Quotient = 8, Remainder = 5963
2,13,895 ÷ 10,000 ⇒ Quotient = 21, Remainder = 3895
4,05,920 ÷ 1,00,000 ⇒ Quotient = 4, Remainder = 5920
13,85,123 ÷ 1,00,000 ⇒ Quotient = 13, Remainder = 85,123.

Fill in the table.

		Quotient	Remainder
1.	85,930 ÷ 10,000	______	______
2.	4,15,158 ÷ 10,000	______	______
3.	93,895 ÷ 100	______	______
4.	1,23,612 ÷ 10	______	______
5.	32,45,325 ÷ 1000	______	______
6.	14,983 ÷ 1000	______	______
7.	8,93,425 ÷ 1,00,000	______	______

DIVISION BY A 2-DIGIT DIVISOR

Study the following example:

Example 1. Divide 6590893 by 95.

```
       69377
95 ) 6590893
    -570
     890      Bring down 0
    -855      Subtract 9 × 95
      358     Bring down 8
      285     Subtract 3 × 95
       739    Bring down 9
       665    Subtract 7 × 95
        743   Bring down 3
       -665   Subtract 7 × 95
         78
```

Take the first digit 9 of the divisor and divide 65 by 9. 9 sevens are 63 but 7 × 95 = 665 is greater than 659. So we take 6 in the quotient and subtract 6 × 95 = 570 from 659. Now bring down 0. Take the first two digits of 890, *i.e.*, 89. Divide by 9. 9 nines are 81 and 9 × 95 = 855. Subtract 855 from 890. Take down 8. Now proceed further in a similar manner till the remainder is less than 95.

Ans : 69377 *r* 78.

Dividing by a 3-digit divisor.

Example 2. : Divide 371654 by 968.

```
       383
968)371654
   - 2904
     8125
    -7744
     3814
    -2904
      910
```

Explanation. The divisor contains 3 digits, so we should consider 3 digits of the dividend that is 371. But since 371 is smaller than 968, we have to consider 4 digits, that is 3716. Take the first digit 9 of the divisor and divide 37 by 9. The trial quotient is 4. But 4 × 968 gives 3872 which is greater than 3716, so take 3 as the quotient, 3 × 968 = 2904. Subtract 2904 from 3716 and bring down the next digit 5. Now complete the division as shown.

Ans. 383 *r* 910.

Example 2. Divide 212589 by 697.

```
       305
697)212589
   -2091
      348
     -000
     3489
    -3485
        4
```

Short form

```
       305
697)212589
   - 2091
     3489
    -3485
        4
```

Ans. 305 *r* 4.

> We bring down 8 but the divisor 697 cannot divide 348 even once, so we put 0 after 3 in the quotient and bring down 9. Complete as shown above.

Note : Follow the short form.

Example 4. Divide 1886526 by 627.

```
        3008
627)1886526
   - 1881
       5526
     - 5016
        510
```

The first digit in the quotient is 3. On subtracting 3 × 627 = 1881 from 1886 you get 5. Bring down 5, you have 55. The divisor 627 cannot divide it. Put 0 after 3 in the quotient and bring down 2. You now have 552. The divisor 627 cannot divide this also. So again put a 0 after the previous 0 in the quotient and bring down 6. Now complete the division as shown here.

Ans. 3008 *r* 510.

Skill Practice — **Class Work**

Divide

1. 23)51958 **2.** 312)1567892 **3.** 937)4693475

EXERCISE 10

Divide:

1. 14,375 ÷ 16
2. 2,79,845 ÷ 65
3. 11,70,019 ÷ 79
4. 79,03,019 ÷ 99
5. 5919 ÷ 132
6. 6847 ÷ 198
7. 8166 ÷ 368
8. 14235 ÷ 347
9. 17,868 ÷ 470
10. 36,940 ÷ 648
11. 4,09,183 ÷ 623
12. 1,83,800 ÷ 769
13. 5,32,641 ÷ 691
14. 70,17,071 ÷ 259
15. 53,82,695 ÷ 870

WORD PROBLEMS

Example 5. 239 bags of cement can be loaded in a wagon. How many wagons are required to load 8365 bags ?

Solution. Number of wagons to load 239 bags = 1

∴ Number of wagons required to load 8365 bags
= 8365 ÷ 239
= 35.

Ans: 35 wagons are required to load 8365 bags of cement.

$$\begin{array}{r} 35 \\ 239\,\overline{)8365} \\ -717 \\ \hline 1195 \\ 1195 \\ \hline 0 \end{array}$$

EXERCISE 11

1. The auditorium of a school can seat 417 people at a time. The principal wants all the 3,336 students of all the local schools to see a special film on wild life. How many times must the film be shown, so that every student sees the film once ?
2. Find the number of pages in a book which has on an average 207 words on a page, and contains 2,01,411 words all together.
3. 7,65,564 books are to arranged equally in shelves. If 487 books are arranged in each shelf, how many shelves will be needed.
4. The total cost of air tickets of 52 people from New Delhi to London is Eight lakh twenty nine thousand six hundred eight. What will be the cost for 1 person ?

5. An engine pumps out 875 litres of water per minute. How much time will it take to pump out 1312500 litres of water ?
6. Divide the greatest number of 8-digits by the greatest number of 3-digits.

7. What must 789 be multiplied by to give 7897890 ?
8. What is 532th part of 1004416 ?
9. How many heaps of bricks, each containing 139 bricks, could be made out of half a million bricks ?

10. In a division sum, the divisor is 63, the quotient is 857, and remainder is 27. What is the dividend ?

MENTAL MATHS – 3

1. Add together 5, 15 and 25 and divide the sum by 9.
2. Multiply three thousand two hundred by twenty.
3. Divide the successor of 6499 by 13.
4. Add twenty sixes to thirty eights.
5. Multiply eighty thousand by nine hundred. Write and read it in both Indian and International system.
6. The remainder in 367 ÷ 6 is ____________.
7. The quotient in 60 ÷ 7 is ____________.
8. If 2800 kg of potatoes are put into sacks and each sack holds 70 kg, how many sacks are required ?
9. What number when multiplied by the successor of 69 gives 21000 ?
10. What is the sum of one lakh, ninety nine thousand, nine hundred two and eight ? Express in words.

KNOW YOUR NUMBERS

THE NUMBER TWO

A deuce, a couple, a duo or a pair, ... are all 2 in number. People have two hands. Bi- means two. A bicycle has two wheels. A biceps has two tongs. In binary code, numbers are written to the base two. It uses just two symbols, 0 and 1. In binary code, the numbers 1, 2, 3, 4, 5, 6..... become 1, 10, 11, 100, 101, 110 ......

2 is a prime number and is the only even prime number.

Two has a very special property because 2 + 2 = 4 and also 2 × 2 = 4.

An Interesting sum associated with 2

```
   1 2 3 4 5 6 7 8 9
 + 1 2 3 4 5 6 7 8 9
 + 9 8 7 6 5 4 3 2 1
 + 9 8 7 6 5 4 3 2 1
 +                 2
 -------------------
 2 2 2 2 2 2 2 2 2 2
```

4 ROUNDING NUMBERS

LOOKING BACK

- ★ To round a number to the nearest 10, we use the nearest multiple of 10.
- ★ To round a number to the nearest 100, we use the nearest multiple of 100.
- ★ To round a number to the nearest 1000, we use the nearest multiple of 1000.

Example 1. Round 2,126 to the nearest 10, 100 and 1000.

Solution. The multiples of 10 nearest to 2,126 are 2,120 and 2,130.

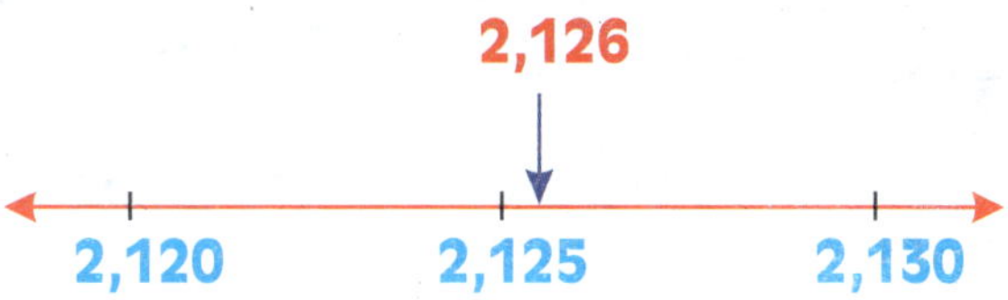

(nearest 10 before 2,126) (nearest 10 after 2,126)

2,126 as can be seen is closer to 2,130, so it is rounded up to 2,130.

The short cut is as under :

Rule 1 : To round to the nearest 10, check the digit at the one's place.

Greater than 5

2,126 — **Rounded to** ⟹ **2130**

Replace by zero

Increase to 3

> **Rounded up:** Since 6 > 5 add 1 to digit at tens place, replace the ones place by 0.

Rule 2 : To round to the nearest 100, check the digit at the ten's place.

Less than 5

2 , 1 2 6 — **Rounded to** ⟹ **2100**

Replace by zeros

No change

> **Rounded down:** 2 < 5 : No change in the digit at the hundreds place, replace digits at tens and ones place by zeros.

Rule 3 : To round to the nearest 1000, check the digit at the hundred's place.

Less than 5

2 , 1 2 6 — **Rounded to** ⟹ **2000**

Replace by zeros

No change

Example 2. Round 38,500 to the nearest 1000.

Solution. 38,500 lies midway between 38,000 and 39,000.

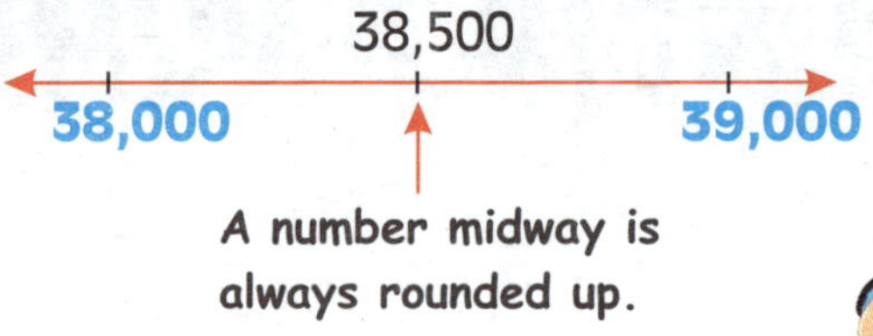

38,500 $\xrightarrow[\text{to}]{\text{Rounded}}$ 39,000

You can also check by the short cut method.

38,500 (5: Equal to 5; 8: Increase to 9; 5, 0, 0: Replace by zeros) $\xrightarrow[\text{to}]{\text{Rounded}}$ 39,000

ROUNDING OFF LARGE NUMBERS

The same method as discussed above is applied to rounding off large numbers. The following examples will help you to understand it more fully.

Example 1. Round off each of the following numbers to the nearest ten thousand:

(i) 682437 (ii) 4287532.

Solution. To round off to the nearest ten thousand, examine the digit at the thousands place.

(i) In the number 682437 the digit at the thousands place is 2, which is less than 5, therefore, we do not make any change in the digit 8 at the ten-thousands place, and replace the digits 2, 4, 3 and 7 by zeros. Hence, rounded off to ten-thousand, 682437 = 680000.

6 8 2 4 3 7 (2: Less than 5; 8: No change; 2 4 3 7: Replace by zeros) $\xrightarrow[\text{nearest 10,000}]{\text{Rounded to}}$ 680000

(ii) In the number 4287532, the digit at the thousands place is 7, which is greater than 5. Therefore, we increase the digit 8 at the ten-thousands place to 9, and replace the digits 7, 5, 3, 2 by zeros.

Hence, rounded off to ten-thousands place 4287532 = 4290000

4 2 8 7 5 3 2 (7: Greater than 5; 8: Increase to 9; 7 5 3 2: Replace by zeros) $\xrightarrow[\text{nearest 10,000}]{\text{Rounded to}}$ 4290000

Example 2. Round off each of the following numbers to the nearest lakh:

(a) 25,37,981 **(b) 1,43,81,507**

Solution. To round off to the nearest lakh, examine the digit at the ten-thousands place.

(a) In the number 25,37,981, the digit at the ten-thousands place is 3, which is less than 5. Therefore, we do not make any change in the digit 5 and replace the digits 3, 7, 9, 8, 1 by zeros.

Less than 5

2 5 3 7 9 8 1 —Rounded to nearest lakh⟹ 25,00,000

Replace by zeros

No change

(b) In the number 1,43,81,507, the digit at the ten-thousands place is 8 which is greater than 5. Therefore, we increase the digit 3 at the lakhs place to 4 and replace the digits 8, 1, 5, 0, 7 by zeros.

Greater than 5

1 4 3 8 1 5 0 7 —Rounded to nearest lakh⟹ 1,44,00,000

Replace by zeros

Increase to 4

Example 3. Round off 52,91,37,846 to the nearest (a) ten-lakh (b) crore.

Solution. (a) To round off to the nearest ten-lakh **examine the digit at the lakhs place.** It is 1 and is less than 5. Therefore, we do not make any change in the digit 9 at the ten-lakhs place and replace the digits 1, 3, 7, 8, 4, 6 by zeros.

Less than 5

5 2 9 1 3 7 8 4 6 —Rounded to nearest ten lakh⟹ 52,90,00,000

Replace by zeros

No change

(b) To round off to the nearest crore, **examine the digit at the ten lakhs place.** It is 9 and is greater than 5. Therefore, we increase the digit 2 at the crores place to 3, and replace the digits 9, 1, 3, 7, 8, 4, 6 by zeros.

Greater than 5

5 2 9 1 3 7 8 4 6 —Rounded to nearest crore⟹ 53,00,00,000

Replace by zeros

Increase to 3

Class Work

Round off to the nearest ten-thousand.

1. 63,809 ____________ **2.** 3,25,523 ____________

Round off to the nearest lakh.

3. 4,61,223 ____________ **4.** 16,25,106 ____________

Round off to the nearest ten-lakh.

5. 52,63,145 ____________ **6.** 67, 89, 289 ____________

Round off to the nearest crore.

7. 6,57,23,215 ____________ **8.** 12,03,45,012 ____________

Home Work

EXERCISE 12

1. Round off each of the following numbers to the nearest ten-thousand.

(a) 46398 **(b)** 83716 **(c)** 15392 **(d)** 277162

(e) 539641 **(f)** 1648302 **(g)** 2506427 **(h)** 1474895

2. Round off each of the following numbers to the nearest lakh.

(a) 375012 **(b)** 625431 **(c)** 3486107 **(d)** 8735482

(e) 2479863 **(f)** 3806792 **(g)** 7513085 **(h)** 2358964

3. Round off each of the following numbers to the nearest ten-lakh.

(a) 1632450 **(b)** 3487162 **(c)** 1051237 **(d)** 20867430

(e) 75283410 **(f)** 61535621 **(g)** 31570369 **(h)** 92065428

4. Round off each of the following numbers to the nearest crore.

(a) 74965231 **(b)** 26407865 **(c)** 15034912 **(d)** 26719543

(e) 107146352 **(f)** 51386792 **(g)** 315443201 **(h)** 297784249

5. Round off 593702164 to the nearest

(a) ten **(b)** hundred **(c)** thousand **(d)** ten-thousand

(e) lakh **(f)** ten-lakh **(g)** crore

6. The population of some of the Indian states (as per 2001 census) are given below. Round these to the nearest crore.

(a) Uttar Pradesh 16,60,52,859 **(b)** West Bengal 8,02,21,171

(c) Andhra Pradesh 7,57,27,541 **(d)** Bihar 8,28,78,796

(e) Orissa 3,67,06,920 **(f)** Rajasthan 5,64,73,122

7. The population of India as per 2001 census was 1,02,70,15,247. Round this to the nearest crore.

5 FACTORS AND MULTIPLES

Let us revise the basics of factors and multiples which you have already studied in Class IV.

★ **A factor** is a whole number that is multiplied by another counting number to name a product. In the mathematical sentence $6 \times 8 = 48$, 6 and 8 are **factors** and 48 is the product. You can also see that $48 \div 6 = 8$ and $48 \div 8 = 6$ and there is no remainder in each case. Therefore, we can also say that if a number divides another number exactly (without any remainder), the divisor is a factor of the dividend.

Example 1. List all the factors of (a) 48 (b) 11.

Solution.

(a) $48 = 1 \times 48$
$= 2 \times 24$
$= 3 \times 16$
$= 4 \times 12$
$= 6 \times 8$

∴ The factors of 48 are
1, 2, 3, 4, 6, 8, 12, 16, 24 and 48

(b) $11 = 1 \times 11$

∴ The factors of 11 are 1 and 11

From the above example it is clear that,

1. **Every number has at least two factors 1 and the number itself.**
2. **All the factors of a number are less than or equal to the number, not greater than it.**

★ A prime number is any natural number which has exactly two factors, 1 and itself:

$11 = 1 \times 11$. There are no other factors of 11, so 11 is a prime number.

★ Every natural number except 1 that has more than two different factors is called a composite number.

> The number 1 is neither prime nor composite.
> It's not prime because it does not have exactly two factors and it's not composite because it does not have more than two factors.

Example 2. Tell whether each number is composite or prime : 56, 89, 98, 67

Solution.

$56 = 1 \times 56 = 2 \times 28 = 4 \times 14 = 8 \times 7$ } 8 factors, therfore, composite

$89 = 1 \times 89$ } exactly 2 factors therefore, Prime

$67 = 1 \times 67$ } exactly 2 factors, therefore Prime

$98 = 1 \times 98 = 2 \times 49 = 7 \times 14$ } 6 factors, therefore composite

★ **Twin Primes** are prime numbers that differ by 2 like 3 and 5, 5 and 7, 11 and 13 etc.

★ **Co-primes :** Two natural numbers are said to be co-primes if they have only 1 as a common factor, e.g., 2 and 3, 5 and 7, 3 and 4, 4 and 9 etc.

1. List the factors of 54.

2. (a) Cross out all prime numbers from the given list:
4, 7, 9, 11, 23, 25, 31, 36, 37, 42, 43, 51

(b) From the above list make 5 pairs of co-primes.

PRIME FACTORIZATION

Any composite number can be written as a product of prime numbers. An expression written as a product of prime factors is called the **prime factorization** of the number.

Example 3. Find the prime factorization of 90.

Method 1 (Using a factor tree)

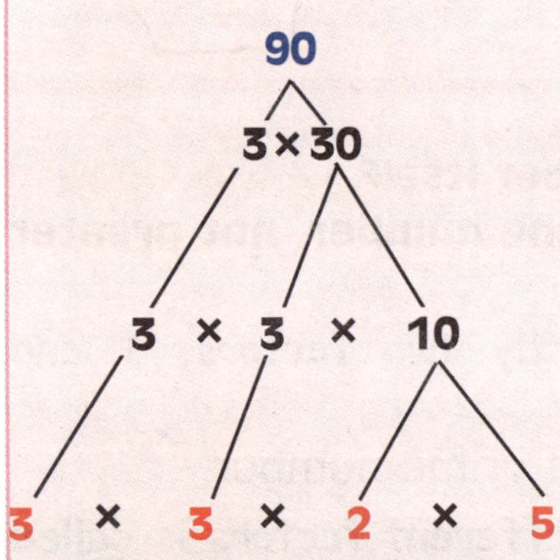

Write the number to be factorised at the top.

Choose any pair of factors as branches. If either of these factors is composite, factorize again.

Choose a pair of factors of each composite number : Continue the branches for prime factor(s).

Keep factoring till you have a row of prime factors.

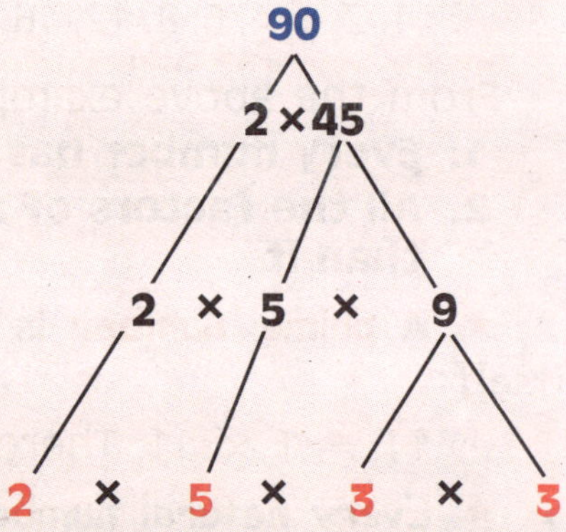

Conventionally the prime factors are written in ascending order.

$\therefore$ **90 = 2 × 3 × 3 × 5.**

You can have more factor trees for the number 90. Try! But at the end, the prime factors will be the same.

Method II (Continuous Divison)

2	90
3	45
3	15
5	5
	1

The prime factorization is 2 × 3 × 3 × 5.

Skill Practice — **Class Work**

Find the prime factorization of 120 by both factor tree method and continuous division method. Try to make 2 factor trees.

Factor Tree 1	Factor Tree 2	Continuous Division

HIGHEST COMMON FACTOR (HCF)

If a number is a factor of two or more whole numbers, it is called a **common factor** of these numbers. $3 \times 7 = 21$ and $3 \times 2 = 6$. So 3 is a common factor of 6 and 21. The **Highest Common Factor (HCF)** of two or more numbers is the common factor that is greater than any other common factor.

Find the HCF of 8 and 40

Different ways to find the HCF of two or more numbers.

Method I. By listing factors find the HCF of 8 and 40

Factors of 8 : (1), (2), (4), (8)

Factors of 40: (1), (2), (4), 5 , (8), 10, 20 and 40

The common factors are 1, 2, 4 and 8

Out of these 8 is the largest ∴ The HCF of 8 and 40 is **8.**

> **Step 1 :** List all the factors of each number.
>
> **Step 2 :** Identify common factors.
>
> **Step 3 :** Identify the largest number in the list of common factors.

Method II. By Prime Factorization

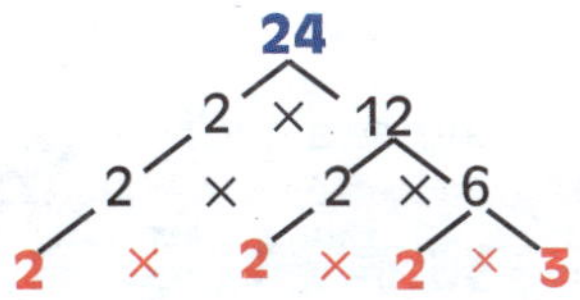

or

2	24
2	12
2	6
3	3
	1

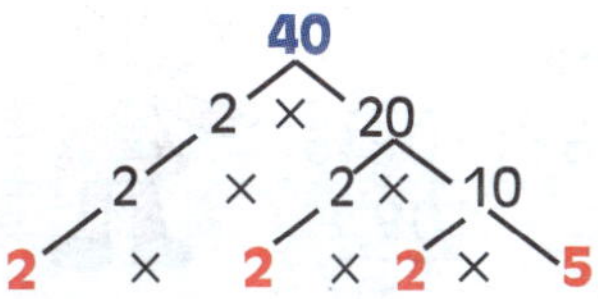

or

2	40
2	20
2	10
5	5
	1

$24 = (2) \times (2) \times (2) \times 3$

$40 = (2) \times (2) \times (2) \times 5$

$\text{H.C.F} = 2 \times 2 \times 2 = \mathbf{8}$

You can get the prime factorization by making factor trees or by continuous division.

Method III. Find the HCF of 12, 18 and 42.

Common Factors		
	2	12, 18, 42
	3	6, 9, 21
		2, 3, 7

∴ $\text{HCF} = 2 \times 3 = \mathbf{6}.$

2, 3 and 7 do not have any common factor except 1, so stop division.

> **Step 1:** Divide the numbers by common factors only.
>
> **Step 2:** Stop dividing when there are no more common factors except 1.
>
> **Step 3:** Find the product of common factors.

You can use any of the three methods listed above to find the HCF of two or more numbers.

Skill Practice — **Class Work**

Find the HCF of :

1. 20, 28 **(By listing factors)**	**2.** 12 and 30 **(By prime factorisation – factor tree method)**
3. 27 and 45 **(By Prime factorization)**	**4.** 18, 42 and 48 **(By continuous division method)**

H.C.F OF LARGE NUMBERS

It is difficult to find the HCF of large numbers by the above methods, so we use the division method. You will be able to understand better with the help of an example.

Example 1. Find the HCF of 144 and 252.

Solution.

```
144)252(1
   -144
    108)144(1
       -108
         36)108(3
           -108
              0
```

Step 1: Divide the greater number by the smaller number.

Step 2: Make the remainder the new divisor and the original divisor the new dividend. Repeat the process until you get 0 as the remainder.

Step 3: The last divisor is the required HCF.

Ans. HCF of 144 and 252 = **36.**

Example 2. Find the HCF of 275, 525 and 830.

Solution. In case of three numbers, first find the HCF of any two numbers. Then take that HCF and the remainder number and find their HCF.

First we take 275 and 525. Now we find the HCF of 25 and 830

```
275)525(1
   -275
    250)275(1
       -250
         25)250(10
           -250
              0
```

The HCF of 275 and 525 is 25.

```
25)830( 33
   75
   80
   75
    5)25( 5
      25
       0
```

∴ The HCF of 275, 525 and 830 is **5.**

Find the HCF of :

1. 217 and 686	2. 320 and 736	3. 450, 675, 825

MULTIPLES

A multiple of a whole number is the product of the number and any counting number. We can state some multiples of 3 and 12 as :

$1 \times 3 = 3$
$2 \times 3 = 6$
$3 \times 3 = 9$
$4 \times 3 = 12$
$5 \times 3 = 15$

$1 \times 12 = 12$
$2 \times 12 = 24$
$3 \times 12 = 36$
$4 \times 12 = 48$
$5 \times 12 = 60$

If a number is a multiple of two or more numbers, it is called a **common multiple** of the numbers.

Note:
1. Zero is a multiple of every number.
2. The smallest multiple of any number (other than zero) is the number itself.
3. The number of multiples of a given number is infinite.

LEAST COMMON MULTIPLE (LCM)

If a number is a multiple of two or more numbers, it is called a **common multiple** of the numbers.

$2 \times 9 = 18.$ $\quad \therefore$ 18 is a common multiple of 2 and 9.

The smallest number (other then zero) that is a multiple of two or more whole numbers is the **least common multiple** (LCM) of the numbers.

Different ways to find the LCM of two or more numbers.

Example : Find the LCM of 6, 8, and 12.

Method 1. By listing the multiples

The multiples of 6 are 6, 12, 18, **24**, 30, 36, 42, **48**, ...

The multiples of 8 are 8, 16, **24**, 32, 40, **48**, ...

The multiples of 12 are 12, **24**, 36, **48**, ...

Step 1: List some multiples of each number.

Step 2: Identify common multiple.

Step 3: Pick up the least number from the list of common multiples.

Some common multiples are 24 and 48, 72, ...

out of these 24 is the least, so the least common multiple of 6, 8 and 12 is **24.**

Method 2. By Prime Factorization

Example : Find the LCM of 18 and 24.

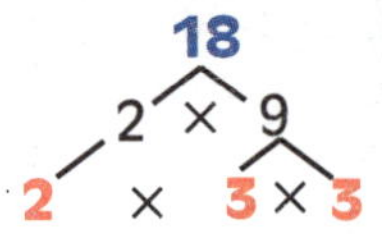

or

2	18
3	9
	3

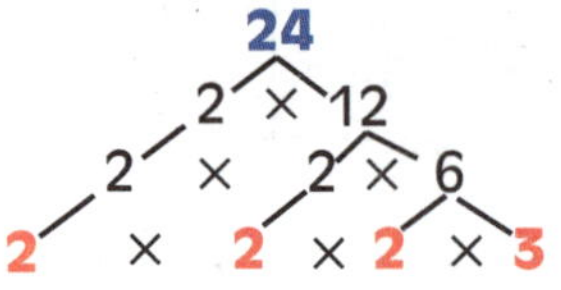

or

2	24
2	12
2	6
	3

18 = 2 × 3 × 3

24 = 2 × 2 × 2 × 3

LCM = 2 × 3 × 3 × 2 × 2 = **72.**

Common factors (2, 3)   Extra factors (3, 2, 2)

Step 1: Find the prime factorization of each number.

Step 2: Identify the common factors.

Step 3: Multiply the common factors and the extra factors.

Method 3. By division method

Example : Find the LCM of 16, 28 and 40.

2	16, 28, 40
2	8, 14, 20
2	4, 7, 10
	2, 7, 5

Step 1: Divide the numbers by a prime factor common to at least 2 of the given numbers. Copy the number as such if it is not completely divisible by the prime factor.

Step 2: Stop dividing when there is no further common factor except 1.

Step 3: Find the products of the numbers in the left column and the last remainders.

∴ LCM = 2 × 2 × 2 × 2 × 5 × 7 = **560.**

Find the L.C.M of:

1. 9 and 12 **(By listing multiples)**

2. 15 and 48 **(By prime factorization)**

3. 18, 42 and 48 **(By division method)**

4. 15, 25 and 30 **(By division method)**

EXERCISE 13

1. **Write all the factors of each of the following numbers and then tell whether it is a prime number or composite number.**

   (a) 11 **(b)** 32 **(c)** 40 **(d)** 19 **(e)** 18 **(f)** 41

2. Given the three prime numbers 13, 37 and 43, find the sum of these prime numbers. Is the sum a prime number ?

3. **Complete the factor tree. Then write the prime factorization.**

   (a) 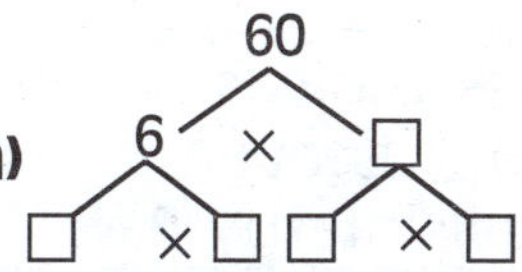

   (b) 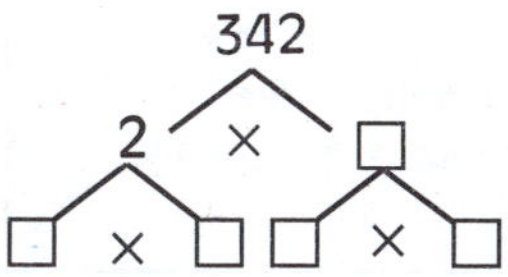

   (c) 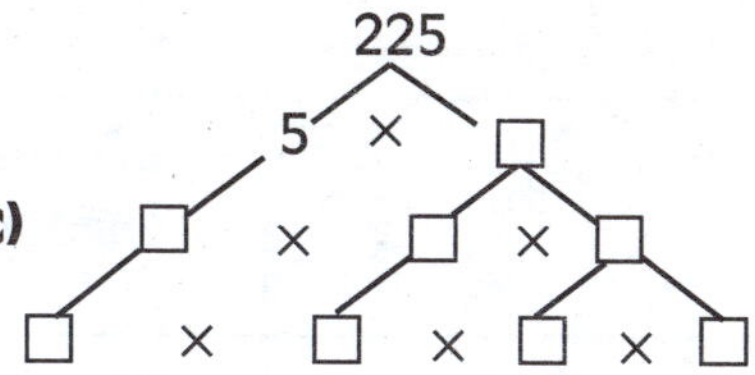

4. **Find the prime factorization of each of the following numbers.**

   (a) 216 **(b)** 150 **(c)** 84 **(d)** 1080

5. **List the common factors of each of the following sets of numbers. Hence find their HCF.**

   (a) 15, 35 **(b)** 16, 27 **(c)** 30, 42 **(d)** 18, 32, 48

6. **Find the HCF of the following by Prime factorization method.**

   (a) 36, 60 **(b)** 81, 64 **(c)** 18, 24, 42 **(d)** 15, 35, 45

   (e) 56, 72, 140 **(f)** 36, 45, 90 **(g)** 28, 42, 56 **(h)** 38, 76, 114

7. **Find the HCF of the following by long division method.**

   (a) 128 and 192 **(b)** 145 and 325 **(c)** 168 and 345

   (d) 495 and 945 **(e)** 84, 192 and 216 **(f)** 102, 408, 850

   (g) 130, 520, 845 **(h)** 125, 180, 500 **(i)** 74, 320, 384

8. **List the first eight common multiples for each pair of numbers. Hence find their LCM.**

   (a) 5 and 7 **(b)** 6 and 15 **(c)** 12 and 16

   (d) 10 and 25 **(e)** 14 and 21 **(f)** 18 and 30

   (g) 8 and 14 **(h)** 9 and 15

9. **Find the LCM of the following by prime factorization method.**

   (a) 8 and 12 **(b)** 6 and 10 **(c)** 9 and 12

   (d) 24 and 32 **(e)** 11 and 20 **(f)** 26 and 20

   (g) 48 and 32 **(h)** 25 and 40

10. Find the LCM of the following by division method.

(a) 4, 12 and 20 **(b)** 10, 20, 45 **(c)** 16, 28, 44 **(d)** 39, 45, 54

(e) 25, 40, 80, 120 **(f)** 130, 150, 210 **(g)** 60, 90, 105 **(h)** 24, 36, 56, 81.

DIVISIBILITY RULES

One natural number is divisible by another natural number, if on dividing, you get a remainder equal to zero. There are some rules or patterns that make it easier to tell whether a given number is divisible by another given number or not. You have been introduced to divisibility rules in class IV. We will revise them here,

A number is divisible by:	Rule	Test with 360
2	If the number is even, *i.e.*, the ones digit is 0, 2, 4, 6 or 8	360: the ones digit is 0, the number is even. So, 360 is divisible by 2.
3	If the sum of the digits is divisible by 3.	360: 3 + 6 + 0 = 9. 9 is divisible by 3. So, 360 is divisible by 3
4	If the number formed by the last two digits is divisible by 4.	360: the last two digits are 6 and 0. 60 is divisible by 4. So, 360 is divisible by 4.
5	If the ones digit is 0 or 5.	360: The ones digit is 0 So, 360 is divisible by 5.
6	If the number is divisible by 2 and 3.	360: is divisible by 2 and 3. So, 360 is divisible by 6.
9	If the sum of the digits is divisible by 9.	360: 3 + 6 + 0 = 9 9 is divisible by 9. So, 360 is divisble by 9.
10	If the ones digit is 0	360: the ones digit is 0. So, 360 is divisible by 10.
15	If the number is divisible by 3 and 5	360: is divisible by 3 and 5. So, 360 is divisible by 15.

From the above table we can conclude that :

Divisibility can help us to determine the factors of a number. From the example given above, we know that 2, 3, 4, 5, 6, 9 and 10 are some of the factors of 360.

Also, you can see

A number is divisible by 6 if it is divisible by 2 and 3,

A number is divisible by 10 if it is divisible by 2 and 5,

A number is divisible by 15 if it is divisible by 3 and 5.

2 and 3 are the co-prime factors of 6
2 and 5 are the co-prime factors of 10
3 and 5 are the co-prime factors of 15

Hence,

A number is divisible by another number, if it is divisible by its co-prime factors.

∴ A number divisible by 12 will also be divisible by 3 and 4. Check for the number **300.**

Example : Check the divisibility of 11718 by 18.

Solution. Think of the co-prime factors of 18. The co-prime factors of 18 are 2 and 9. Therefore, to check the divisibility of 11718 by 18, we check whether the given number is divisible by 2 and by 9.

11718 : digit at ones place is 8, so 11718 divisible by 2

: 1 + 1 + 7 + 1 + 8 = 18, 18 is divisible by 9, so 11718 is divisible by 9.

Hence 11718 is divisible by 2 × 9 = 18.

Fill in the blanks:

1. 1920 is divisible by 20. Therefore some of the factors of 1920 are ______, ______, ______ and ______ .
2. 2080 is divisible by 32. Therefore, it is also divisible by its ______ .
3. A number is divisible by 33. It is also divisible by its co-prime factors ______ and ______ .
4. There are 142 people attending a party. Can the host fill tables of 5 ? ______ (Yes/No).

EXERCISE 14

1. Check the divisibility of each of given first number by the second number, applying the rules of divisibility.

(a) 5695 by 5 **(b)** 32900 by 10 **(c)** 3978 by 5 **(d)** 4236 by 6
(e) 12345 by 3 **(f)** 68709 by 9 **(g)** 13416 by 4 **(h)** 100008 by 9
(i) 36720 by 12 **(j)** 17360 by 6 **(k)** 3540 by 15 **(l)** 1975 by 15.

2. Check if the first number is divisible by the second. (Think of the co-prime factors of the divisor:)

(a) 17500 by 20 **(b)** 6759 by 12 **(c)** 18864 by 36
(d) 3000 by 40 **(e)** 4410 by 45 **(f)** 3150 by 90

MENTAL MATHS–4

1. Round Rs 15987 to the nearest 1 rupee.
2. Round Rs 38792 to the nearest ten rupees.
3. Round 5937850 to the nearest thousand.
4. Name the prime numbers between 30 and 40.
5. Name the least prime number which is also an even number.
6. 3rd multiple of 8 +4th multiple of 4 = ______ multiple of 10.
7. Name the multiples of 7 between 20 and 40.
8. State the prime factors of 70.
9. What is the H.C.F. of 12 and 18 ?
10. What is the LCM of 15 and 25 ?

FRACTIONS

WHAT IS A FRACTION ?

A **fraction** is a part of a whole. If a whole thing is divided into equal parts, each part is a fraction.

It is made up of two parts.

Fraction of a cake

The **denominator** tells us how many equal parts are there in the whole. → $\frac{1}{4}$ ← The **numerator** tells how many of the equal parts you have taken.

FRACTIONS CAN BE USED TO :

★ To name a part of a single thing.

1 part out of 4 equal parts.

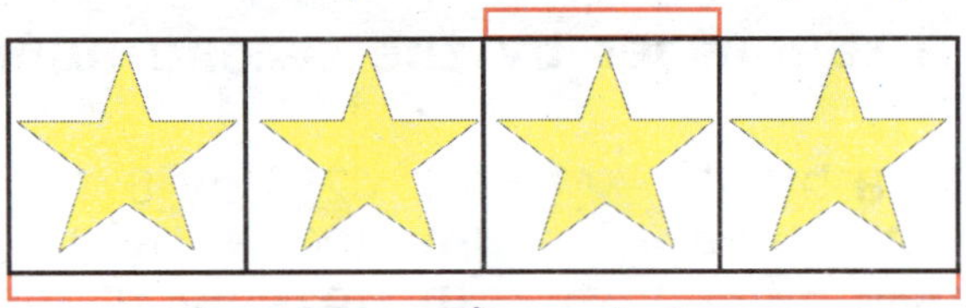

4 equal parts

$\frac{1}{4}$ of the pizza has mushroom topping.

★ To name a part of a group of things.

In this mini zoo, you have 8 animals out of which 3 are zebras.

So, we write $\frac{3}{8}$ are zebras.

REPRESENTATION OF A FRACTION ON A NUMBER LINE

The fraction $\frac{2}{7}$ is shown on the number line as :

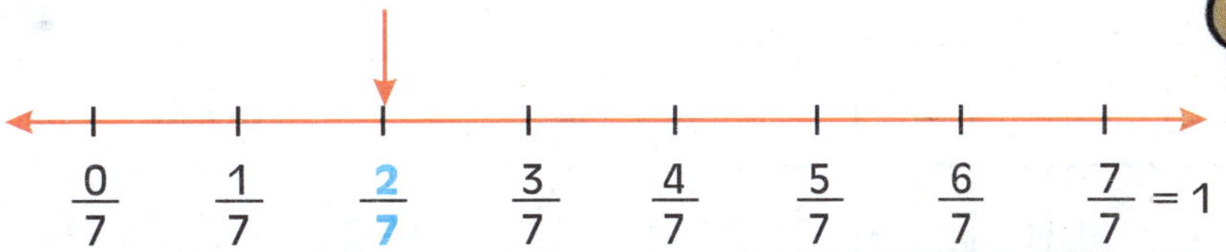

$\frac{2}{7}<1$, so we draw the number line from 0 to 7 and divide it into 7 equal parts. ($\because$ denominator = 7). The arrow shows the fraction $\frac{2}{7}$.

TYPES OF FRACTIONS

Proper Fractions : The fractions in which the numerator is less than the denominator are called proper fractions. Thus, $\frac{0}{2}, \frac{1}{4}, \frac{2}{5}, \frac{3}{7}, \frac{81}{100}$ etc., all name proper fractions. **All proper fractions are less than 1.**

Improper Fractions : The fractions in which the numerator is greater than or equal to the denominator are called improper fractions, e.g., $\frac{6}{6}, \frac{7}{4}, \frac{21}{20}, \frac{121}{100}$ all name improper fractions. **All improper fractions are greater than or equal to 1.**

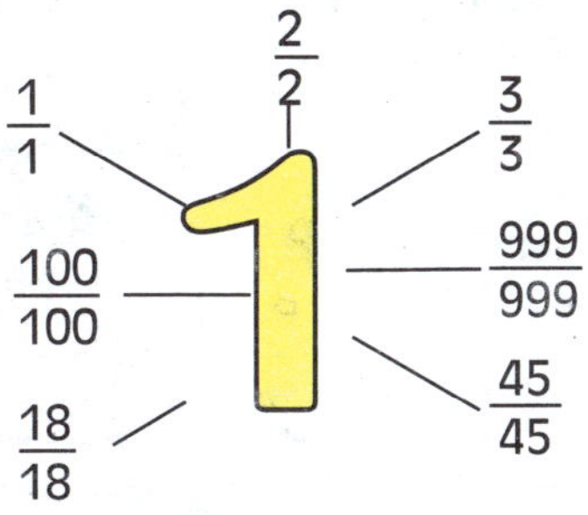

MIXED NUMBERS

A mixed number has a whole number (except 0) mixed with a fraction, e.g., $2\frac{1}{2}, 3\frac{1}{4}, 6\frac{1}{3}, 4\frac{1}{100}$ all name mixed numbers.

Whole number ← ② $\left(\frac{1}{2}\right)$ → Proper fraction

$$2\frac{1}{2}=2+\frac{1}{2}$$

$2\frac{1}{2}$ glass of orange juice

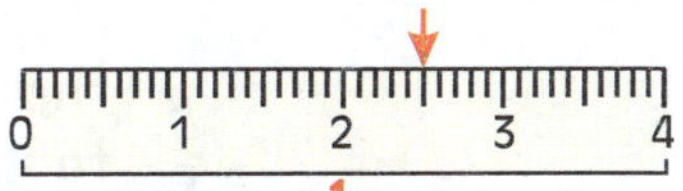

$2\frac{1}{2}$ cm.

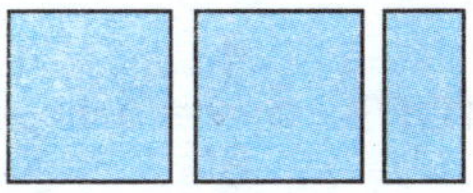

$2\frac{1}{2}$ Sheets of chart paper.

We can write a mixed number as an improper fraction and an improper fraction as a mixed number.

Example 1. **Write $\frac{8}{5}$ as a mixed number.**

Solution. Pictorially, it can be shown as :

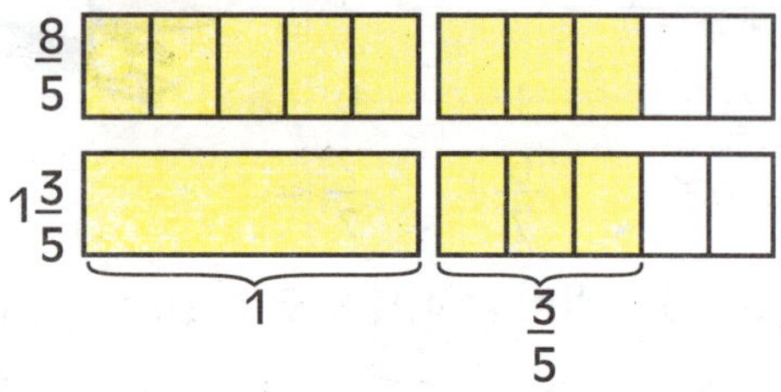

$\therefore \quad \frac{8}{5} = 1\frac{3}{5}$

Method II

$$5 \overline{)8} = 1 \text{ (quotient (Q))}, \quad -5, \quad 3 \text{ (remainder (R))}$$

(D) Divisor → 5) 8 → 1 ← quotient (Q)
− 5
3 ← remainder (R)

$\therefore \quad \frac{8}{5} = 1\frac{3}{5} = Q\frac{R}{D}$

Note that $1\frac{3}{5}$ means $1+\frac{3}{5}$.

It can also be shown on the number line as :

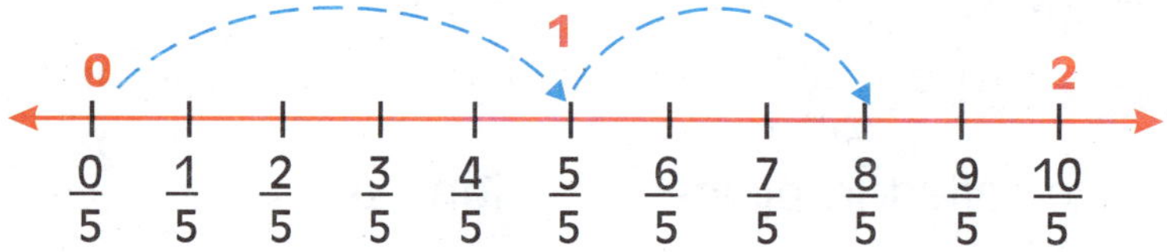

Jump 5-fifths to land on 1. Jump 3 more fifths to land on $\frac{8}{5}$. This is same as jumping by 1 and then by another $\frac{3}{5}$. $\quad \therefore \quad \frac{8}{5} = 1\frac{3}{5}$.

Example 2. **Write $3\frac{1}{6}$ as an improper fraction.**

Solution. Pictorially, it can be shown as :

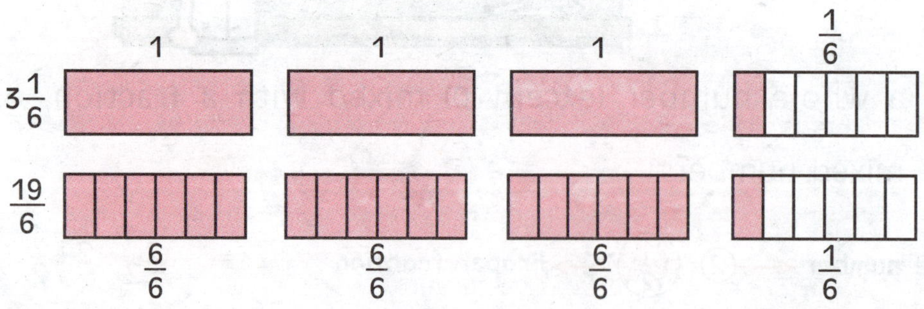

Therefore, $3\frac{1}{6} = 3 + \frac{1}{6} = \frac{19}{6}$

Multiply the whole number part by the denominator and add numerator to the product.

Alternatively ,

$3\frac{1}{6} = \frac{3 \times 6 + 1}{6} = \frac{18+1}{6} = \mathbf{\frac{19}{6}}$.

Write an improper fraction and mixed number for the shaded part

		Mixed Number	Improper Fraction
1.	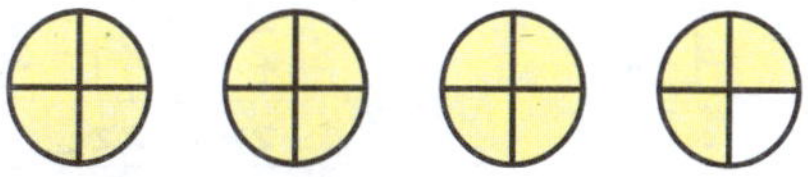	________	________
2.	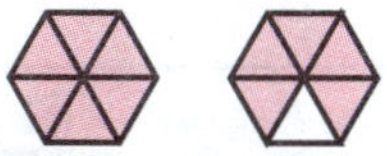	________	________
3.	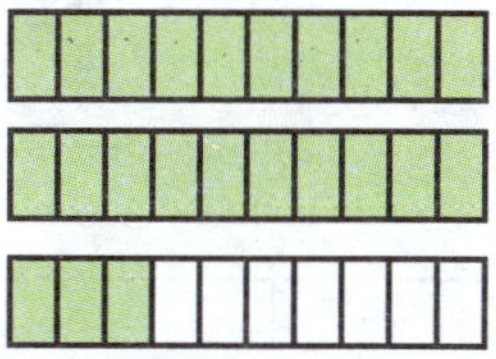	________	________

EQUIVALENT FRACTIONS

Consider the following fraction bar :

If you look carefully at these figures given alongside, you will notice that the same part of the rectangle is shaded, *i.e.*, $\frac{1}{3}$ of the big rectangle is shaded. Thus, the fractions $\frac{1}{3}, \frac{2}{6}$ and $\frac{3}{9}$ name the same amount.

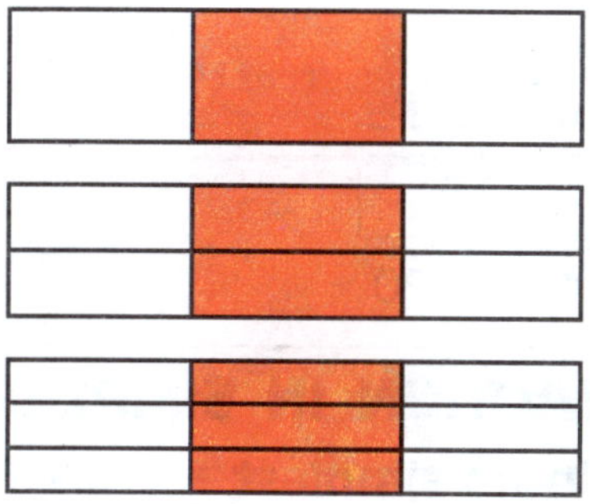

Shaded Part

$\frac{1}{3}$

$\frac{2}{6}$

$\frac{3}{9}$

Fractions that name the same part of the whole are called **equivalent fractions.**

Therefore, $\frac{1}{3}, \frac{2}{6}$ and $\frac{3}{9}$ are equivalent fractions and we write $\frac{1}{3} = \frac{2}{6} = \frac{3}{9}$.

Note that : $\frac{1}{3} = \frac{1}{3} \times \frac{2}{2} = \frac{2}{6}$ or $\frac{2}{6} = \frac{2 \div 2}{6 \div 2} = \frac{1}{3}$

$\frac{1}{3} = \frac{1}{3} \times \frac{3}{3} = \frac{3}{9}$ or $\frac{3}{9} = \frac{3 \div 3}{9 \div 3} = \frac{1}{3}$

$\frac{1}{3}$, $\frac{2}{6}$, $\frac{3}{9}$

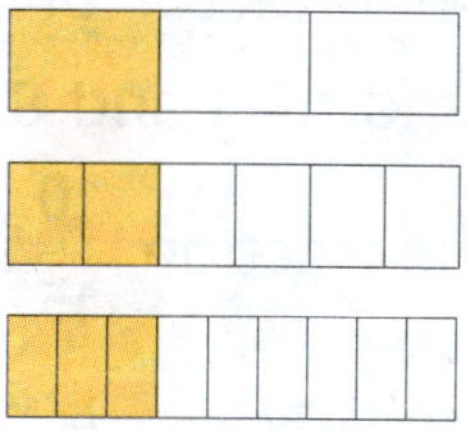

Thus, to find equivalent fractions, we either multiply or divide the numerator and denominator by the same number.

Example 1. Fill in the missing numbers.

(a) $\frac{3}{5} = \frac{15}{\square}$ (b) $\frac{2}{3} = \frac{\square}{18}$ (c) $\frac{20}{50} = \frac{2}{\square}$ (d) $\frac{21}{33} = \frac{\square}{11}$

Solution. (a) $\frac{3}{5} \overset{\times 5}{=} \frac{15}{\square}$ $\quad \therefore \frac{3}{5} = \frac{3 \times 5}{5 \times 5} = \frac{\mathbf{15}}{\mathbf{25}}$. $\qquad$ (b) $\frac{2}{3} \underset{\times 6}{=} \frac{\square}{18}$ $\quad \therefore \frac{2}{3} = \frac{2 \times 6}{3 \times 6} = \frac{\mathbf{12}}{\mathbf{18}}$.

(c) $\frac{20}{50} \overset{\div 10}{=} \frac{2}{\square}$ $\quad \therefore \frac{20}{50} = \frac{20 \div 10}{50 \div 10} = \frac{\mathbf{2}}{\mathbf{5}}$. $\qquad$ (d) $\frac{21}{33} \underset{\div 3}{=} \frac{\square}{11}$ $\quad \therefore \frac{21}{33} = \frac{21 \div 3}{33 \div 3} = \frac{\mathbf{7}}{\mathbf{11}}$.

1. Name the equivalent fractions shown

(a)

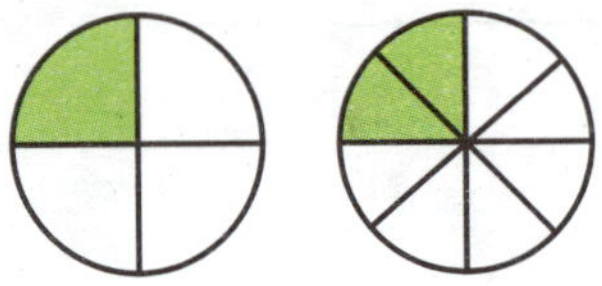

$\frac{\square}{4} = \frac{\square}{8}$

(b)

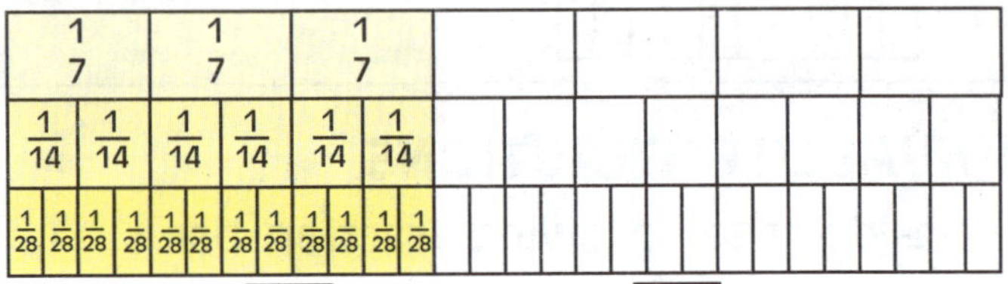

$\frac{3}{7} = \frac{\square}{\square} = \frac{\square}{\square}$

2. Fill in the blanks:

(a) $\frac{2}{5} = \frac{\square}{10} = \frac{\square}{15} = \frac{\square}{20}$ $\qquad$ (b) $\frac{5}{8} = \frac{\square}{16} = \frac{15}{\square} = \frac{\square}{32}$

WRITING A FRACTION IN SIMPLEST FORM

A fraction is in its simplest form when its numerator and denominator have no common factors other than 1.

Fractions such as $\frac{1}{2}, \frac{2}{5}, \frac{3}{8}$ are all fractions in their simplest form. In $\frac{1}{2}$, there is no common factor of 1 and 2 except 1. Likewise, the only common factor of 2 and 5 in $\frac{2}{5}$ is 1 and of 3 and 8 in $\frac{3}{8}$ is 1.

The fraction $\frac{10}{15}$ is not in its simplest form because 5 is a common factor of 10 and 15.

Similarly, $\frac{6}{21}$ is not a fraction in its lowest form because 3 is a common factor of both 6 and 21.

Example. Express the following fractions in their simplest form.

(a) $\frac{6}{8}$ $\qquad$ **(b)** $\frac{18}{30}$ $\qquad$ **(c)** $\frac{210}{330}$

Solution. (a) Divide both numerator and denominator by a common factor.

Factors of 6 are 1, **2**, 3, 6
Factors of 8 are 1, **2**, 4, 8 } 2 is a common factor of 6 and 8.

Dividing 6 and 8 by 2, we have,

$\frac{6}{8} = \frac{6 \div 2}{8 \div 2} = \frac{\mathbf{3}}{\mathbf{4}}$. Now, 3 and 4 have no common factor except 1.

(b) $\frac{18}{30} = \frac{18 \div 2}{30 \div 2} = \frac{\mathbf{9}}{\mathbf{15}}$. 2 is a common factor of 18 and 30

$\frac{9}{15} = \frac{9 \div 3}{15 \div 3} = \frac{\mathbf{3}}{\mathbf{5}}$. 3 is a common factor of 9 and 15

Alternative Method

$\frac{\cancel{18}^{\cancel{9}^{3}}}{\cancel{30}_{\cancel{15}_{5}}} = \frac{3}{5}$

Cancelling indicates dividing

Hence, $\frac{18}{30}$ in its simplest form $= \frac{3}{5}$ because 3 and 5 have no common factor except 1.

Method : To reduce a fraction to its simplest form, keep dividing by common factors until the only common factor is 1.

(c) $\frac{210}{330} = \frac{\cancel{210}^{\cancel{105}^{\cancel{35}^{7}}}}{\cancel{330}_{\cancel{165}_{\cancel{55}_{11}}}} = \frac{\mathbf{7}}{\mathbf{11}}$.

Think !

Start dividing by the least common factor. First divide by 2, then by 3 and lastly by 5.

Reduce the following fractions to their lowest terms.

1. $\frac{16}{18} =$	**2.** $\frac{60}{105} =$	**3.** $\frac{84}{98} =$	**4.** $\frac{55}{70} =$

COMPARING FRACTIONS

Comparing Fractions with like denominators

Example. **For a birthday party, $\frac{6}{15}$ of the children are girls and $\frac{9}{15}$ are boys. Whose number is greater in the party, of boys or of girls ?**

Solution. To answer this question we have to compare $\frac{6}{15}$ and $\frac{9}{15}$.

We can use two ways for comparing them and then state a general rule.

I. Using Models	II. Using number line
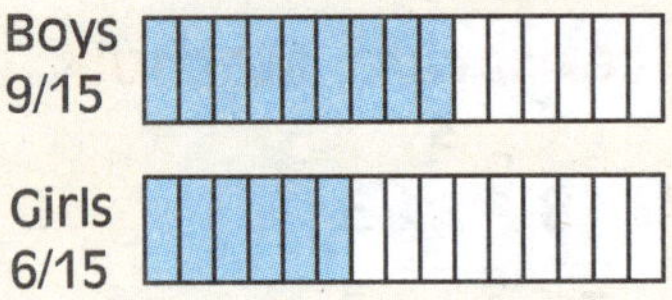 On comparing the shaded parts you can see that $\frac{9}{15} > \frac{6}{15}$.	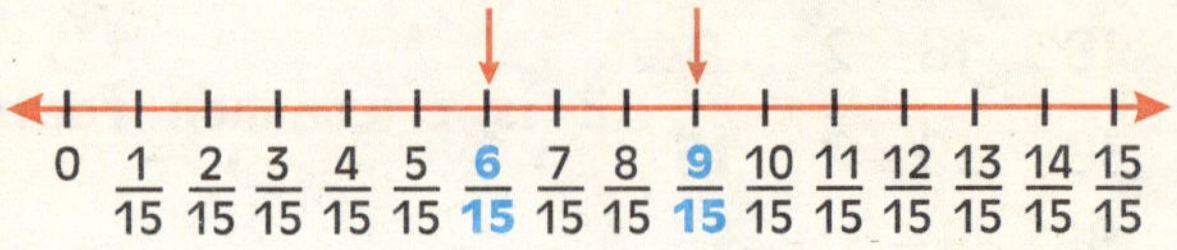 $\frac{9}{15}$ is to the right of $\frac{6}{15}$, therefore $\frac{9}{15} > \frac{6}{15}$.

Thus, If two or more fractions have the same denominator, the fraction with the greater numerator names the greater number. $\frac{6}{7} > \frac{5}{7}$ or $\frac{5}{7} < \frac{6}{7}$.

Comparing Fractions with Unlike denominators

Which is greater : $\frac{2}{3}$ or $\frac{4}{5}$?

Pictorially, $\frac{2}{3}$ and $\frac{4}{5}$ can be compared as:

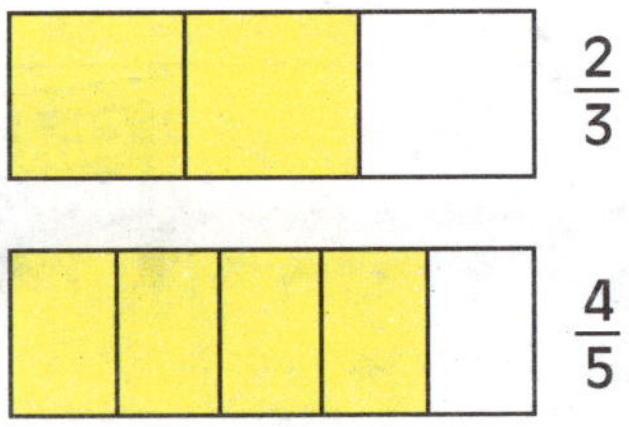

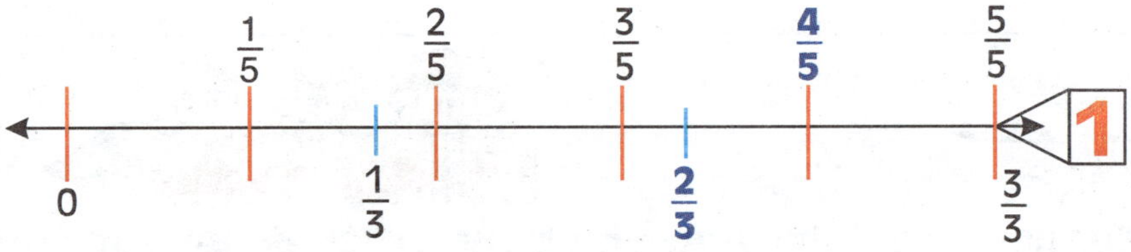

On comparing the shaded parts, you can see that $\frac{4}{5} > \frac{2}{3}$

$\frac{4}{5}$ is to the right of $\frac{2}{3}$ on the number line, therefore, $\frac{4}{5} > \frac{2}{3}$.

By finding equivalent fractions using LCM

To compare $\frac{2}{3}$ and $\frac{4}{5}$, we convert them to equivalent fractions both having the same denominator.

The common denominator in this case will be taken as the LCM of 3 and 5.

LCM (3, 5) = 3 × 5 = 15.

$\frac{2}{3} = \frac{2 \times 5}{3 \times 5} = \frac{10}{15}$; $\frac{4}{5} = \frac{4 \times 3}{5 \times 3} = \frac{12}{15}$. Since $\frac{12}{15} > \frac{10}{15}$, therefore, $\frac{4}{5} > \frac{2}{3}$.

Note : This is the most commonly used method used for comparing and ordering unlike fractions.

EXERCISE 15

Evaluate the following, giving you answers in the simplest form.

1. $\frac{6}{11}+\frac{2}{11}$ 2. $\frac{1}{7}+\frac{6}{7}$ 3. $\frac{7}{12}+\frac{1}{12}$ 4. $\frac{5}{16}+\frac{7}{16}$ 5. $\frac{10}{13}-\frac{7}{13}$

6. $\frac{8}{9}-\frac{5}{9}$ 7. $\frac{11}{18}-\frac{2}{18}$ 8. $\frac{16}{21}-\frac{11}{21}$ 9. $\frac{4}{17}+\frac{6}{17}+\frac{3}{17}$

10. $\frac{3}{14}+\frac{5}{14}+\frac{9}{14}$ 11. $\frac{15}{22}-\frac{5}{22}-\frac{1}{22}$ 12. $\frac{14}{25}-\frac{3}{25}-\frac{6}{25}$ 13. $\frac{17}{21}-\frac{4}{21}-\frac{1}{21}$ 14. $\frac{41}{45}-\frac{14}{45}-\frac{6}{45}$

TO ADD OR SUBTRACT FRACTIONS WITH DIFFERENT DENOMINATORS

Example 1. Hourly rainfall was recorded by Ravi on a certain day. The first hour $\frac{1}{4}$ cm of rain fell. The second hour $\frac{1}{2}$ cm of rain fell. How much rain fell during the two hours ?

Solution. Total rain $= \frac{1}{4}\text{cm} + \frac{1}{2}\text{cm}.$

To be able to add these, first change one or both of the fractions to fractions having **common denominator.**

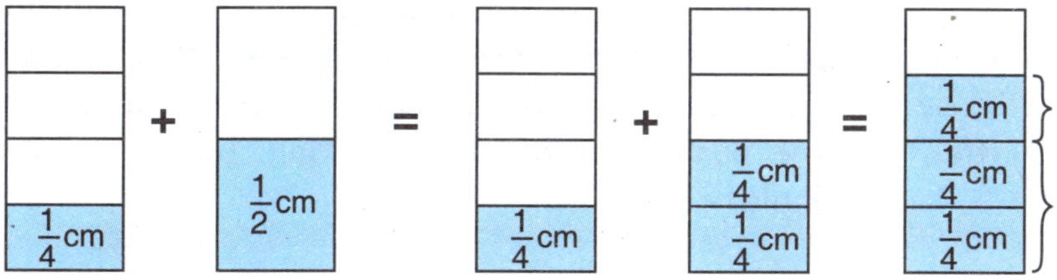

Therefore, $\frac{3}{4}$ cm of rain fell during the two hours.

$$\frac{1}{4} + \frac{1}{2} = \frac{1}{4} + \frac{2}{4} = \frac{3}{4}$$

1 fourth + 1 half = 1 fourth + 2 fourths = 3 fourths

When we add fractions with different denominators, we change the fractions to fractions having the same denominator. This common denominator is the least common denominator of the given fractions. What is meant by least common denominator ?

Study the two rows of fractions given below:

$$\frac{1}{6} = \frac{2}{12} = \frac{3}{18} = \frac{4}{24} = \frac{5}{30} = \frac{6}{36} = \frac{7}{42} = \frac{8}{48} = \frac{9}{54} = \frac{10}{60}$$

$$\frac{3}{4} = \frac{6}{8} = \frac{9}{12} = \frac{12}{16} = \frac{15}{20} = \frac{18}{24} = \frac{21}{28} = \frac{24}{32} = \frac{27}{36} = \frac{30}{40}$$

The numbers 12, 24 and 36 are some of the common denominators of $\frac{1}{6}$ and $\frac{3}{4}$. Of these, **12** is **the least common denominator.** Also we can see that 12 is the LCM of 6 and 4.

Thus, the least common denominator of two fractions is the same as the least common multiple of their denominators.

Therefore, to add or subtract unlike fractions, write the given fractions as equivalent fractions with the least common denominator (LCM of the denominators) and then add or subtract like fractions.

Example 2. Add: (a) $\frac{2}{3}+\frac{3}{4}$ **(b)** $\frac{5}{6}+\frac{1}{10}$.

Solution. (a) LCM of 3 and 4 = 3 × 4 = 12

$$\frac{2}{3}+\frac{3}{4}=\frac{2\times4}{3\times4}+\frac{3\times3}{4\times3}$$

Change $\frac{2}{3}$ and $\frac{3}{4}$ to equivalent fractions having the same denominantor 12.

$$=\frac{8}{12}+\frac{9}{12}=\frac{17}{12}=1\frac{5}{12}.$$

If the sum is an improper fraction, change to mixed number.

You may also set your work as under :

$$\frac{2}{3}+\frac{3}{4}=\frac{4\times2+3\times3}{12}$$

$$=\frac{8+9}{12}=\frac{17}{12}=1\frac{5}{12}.$$

Explanation

Find LCM of the denominators. Here it is 12. Divide 12 by the denominator of the first fraction *i.e.*, 3. The quotient is 4. Multiply this quotient by the numerator of this fraction 4 × 2 = 8. Similarly divide 12 by the denominator of the second fraction and multiply the quotient by the numerator of this fraction. 12 ÷ 4 = 3 and 3 × 3 = 9. Now complete as shown.

(b) Method I

LCM of 6 and 10 = 30

Changing $\frac{5}{6}$ and $\frac{1}{10}$ to equivalent fractions having the same denominator 30.

$$\frac{5}{6}+\frac{1}{10}=\frac{5\times5}{6\times5}+\frac{1\times3}{10\times3}$$

$$=\frac{25}{30}+\frac{3}{30}=\frac{\cancel{28}^{14}}{\cancel{30}_{15}}=\frac{14}{15}.$$

Reduce this fraction to simplest form.

Method II

$$\frac{5}{6}+\frac{1}{10}=\frac{5\times5+1\times3}{30}$$

$$=\frac{25+3}{30}=\frac{\cancel{28}^{14}}{\cancel{30}_{15}}$$

$$=\frac{14}{15}.$$

2 | 6, 10
  | 3, 5

LCM of 6 and 10
= 2 × 3 × 5
= 30

Think !
30 ÷ 6 = 5, 5 × 5 = 25
30 ÷ 10 = 3, 1 × 3 = 3

Example 2. **Subtract :** (a) $\frac{3}{4}-\frac{5}{8}$ (b) $\frac{4}{7}-\frac{5}{11}$

Solution.

(a) Method I

LCM of 8 and 4 = 8

$\frac{5}{8}=\frac{5}{8}$ and $\frac{3}{4}=\frac{3\times2}{4\times2}=\frac{6}{8}$

$\frac{3}{4}-\frac{5}{8}=\frac{6}{8}-\frac{5}{8}=\frac{6-5}{8}=\frac{1}{8}$.

Method II

$\frac{3}{4}-\frac{5}{8}=\frac{2\times3-1\times5}{8}$

$=\frac{6-5}{8}=\frac{1}{8}$.

(b) Method I

LCM of 7 and 11 is 7 × 11 = 77.

$\frac{4}{7}-\frac{5}{11}=\frac{4\times11}{7\times11}-\frac{5\times7}{11\times7}$

$=\frac{44}{77}-\frac{35}{77}=\frac{44-35}{77}=\frac{9}{77}$.

Method II

$\frac{4}{7}-\frac{5}{11}=\frac{11\times4-7\times5}{77}$

$=\frac{44-35}{77}=\frac{9}{77}$.

Evaluate, giving the answer in the simplest form.

1. $\frac{1}{2}+\frac{1}{6}$	**2.** $\frac{2}{5}+\frac{1}{6}$	**3.** $\frac{1}{5}-\frac{1}{6}$	**4.** $\frac{7}{12}-\frac{4}{9}$

EXERCISE 16

Add: (Give your answer in simplest form).

1. $\frac{2}{5}+\frac{1}{10}$ **2.** $\frac{2}{3}+\frac{1}{6}$ **3.** $\frac{5}{9}+\frac{1}{3}$ **4.** $\frac{5}{12}+\frac{1}{2}$

5. $\frac{1}{3}+\frac{2}{7}$ **6.** $\frac{3}{5}+\frac{1}{3}$ **7.** $\frac{3}{7}+\frac{2}{5}$ **8.** $\frac{8}{9}+\frac{4}{6}$

9. $\frac{7}{8}+\frac{5}{6}$

10. $\frac{9}{10}+\frac{7}{8}$

11. $\frac{6}{7}+\frac{3}{4}$

12. $\frac{13}{15}+\frac{15}{30}$

13. $\frac{5}{16}+\frac{7}{18}$

14. $\frac{1}{3}+\frac{1}{2}+\frac{1}{12}$

15. $\frac{3}{5}+\frac{7}{10}+\frac{1}{2}$

16. $\frac{2}{5}+\frac{4}{15}+\frac{3}{10}$

17. $\frac{2}{3}+\frac{3}{4}+\frac{9}{12}$

18. $\frac{1}{3}+\frac{4}{5}+\frac{11}{15}$

19. $\frac{5}{14}+\frac{19}{21}+\frac{47}{84}$

Subtract : (Give your answer in simplest form).

20. $\frac{1}{3}-\frac{1}{4}$

21. $\frac{5}{6}-\frac{5}{8}$

22. $\frac{7}{8}-\frac{5}{12}$

23. $\frac{9}{10}-\frac{4}{5}$

24. $\frac{6}{13}-\frac{5}{26}$

25. $\frac{9}{10}-\frac{11}{20}$

26. $\frac{4}{9}-\frac{10}{27}$

27. $\frac{11}{12}-\frac{5}{9}$

28. $\frac{5}{12}-\frac{3}{16}$

29. $\frac{3}{14}-\frac{4}{21}$

30. $\frac{5}{9}-\frac{17}{36}$

31. $\frac{17}{24}-\frac{19}{36}$

32. $\frac{9}{10}-\frac{1}{5}-\frac{1}{2}$

33. $\frac{5}{9}-\frac{1}{3}-\frac{1}{6}$

34. $\frac{11}{15}-\frac{2}{5}-\frac{1}{3}$

ADDITION AND SUBTRACTION OF MIXED NUMBERS

Type I. With same denominators

Example 1. When released, a gas balloon first rose by $1\frac{1}{4}$ m and then again by $2\frac{2}{4}$ m. What is the total height by which the gas balloon rose?

Solution. Pictorially, the problem can be represented as:

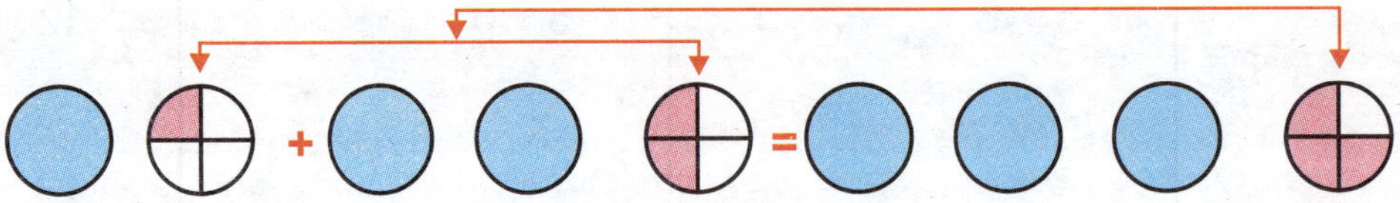

or $\quad 1+\frac{1}{4} \quad + \quad 2+\frac{2}{4} \quad = \quad 3+\frac{3}{4}$

i.e., $\quad 1\frac{1}{4} \quad + \quad 2\frac{2}{4} \quad = \quad 3\frac{3}{4}$

Thus, we can add mixed numbers by first adding whole numbers and then proper fractions.

Therefore, the above work can be set as:

$$1\frac{1}{4}+2\frac{2}{4}=(1+2)+\left(\frac{1}{4}+\frac{2}{4}\right)=3+\frac{3}{4}=\mathbf{3\frac{3}{4}}.$$

Example 2. **Mr. Varma bought $3\frac{5}{6}$ litres of petrol in the morning and $4\frac{3}{6}$ litres in the evening. How many litres of petrol did he buy that day ?**

Solution. Total quantity of petrol bought $= 3\frac{5}{6} + 4\frac{3}{6} = 3 + \frac{5}{6} + 4 + \frac{3}{6}$

$$= (3+4) + \left(\frac{5}{6} + \frac{3}{6}\right) = 7 + \frac{8}{6} = 7 + 1\frac{2}{6} = 7 + 1 + \frac{\cancel{2}^{1}}{\cancel{6}_{3}} = 8\frac{1}{3}.$$

Thus Mr. Varma bought $8\frac{1}{3}$ litres of petrol in all.

Note: You have used two more concepts here. Changing the improper fraction $\frac{8}{6}$ to a mixed number and then reducing $\frac{2}{6}$ to its simplest form.

Add.

1. $1\frac{1}{3} + 3\frac{1}{3}$

2. $4\frac{2}{5} + 6\frac{1}{5}$

3. $3\frac{3}{4} + 2\frac{3}{4}$

4. $1\frac{7}{8} + 2\frac{3}{8}$

Type II. With different denominators.

Example 1. **Add : $3\frac{1}{4} + 4\frac{1}{2}$.**

Solution. Method I

$3\frac{1}{4} + 4\frac{1}{2} = 3 + 4 + \frac{1}{4} + \frac{1}{2}$

LCM of 2 and 4 = 4

$= 7 + \frac{1}{4} + \frac{2}{4} = 7 + \frac{3}{4} = 7\frac{3}{4}.$

Method II

(By changing into improper fractions)

$3\frac{1}{4} + 4\frac{1}{2} = \frac{13}{4} + \frac{9}{2}$

$= \frac{13 \times 1 + 9 \times 2}{4} = \frac{13 + 18}{4} = \frac{31}{4} = 7\frac{3}{4}.$

Example 2. **Add: $\frac{3}{4} + 5\frac{1}{6} + 2\frac{1}{12}$.**

Solution. LCM of 4, 6, 12 = 2 × 2 × 3 = 12

2	4, 6, 12
2	2, 3, 6
3	1, 3, 3
	1, 1, 1

Method I

$\frac{3}{4}+5\frac{1}{6}+2\frac{1}{12}$

$=5+2+\frac{3}{4}+\frac{1}{6}+\frac{1}{12}$

$=7+\frac{9}{12}+\frac{2}{12}+\frac{1}{12}$

$=7+\frac{9+2+1}{12}=7+\frac{12}{12}$

$= 7 + 1 = \mathbf{8}.$

Method II

$\frac{3}{4}+5\frac{1}{6}+2\frac{1}{12}$

$=\frac{3}{4}+\frac{31}{6}+\frac{25}{12}$

$=\frac{3\times 3+31\times 2+25\times 1}{12}$

$=\frac{9+62+25}{12}=\frac{96}{12}=\mathbf{8}.$

Skill Practice

Class Work

Add:

1. $2\frac{1}{2}+1\frac{1}{3}$
2. $5\frac{3}{4}+3\frac{5}{6}$
3. $1\frac{1}{4}+2\frac{2}{3}$
4. $4\frac{3}{5}+2\frac{1}{2}$

Home Work

EXERCISE 17

Add : (Give your answer in simplest form)

1. $3\frac{1}{2}+4\frac{1}{2}$
2. $1\frac{1}{4}+5\frac{3}{4}$
3. $2\frac{1}{5}+3\frac{3}{5}$
4. $12\frac{1}{7}+11\frac{3}{7}$
5. $1\frac{2}{5}+7\frac{3}{5}$
6. $9\frac{3}{8}+10\frac{5}{8}$
7. $105\frac{7}{15}+220\frac{2}{15}$
8. $6\frac{1}{19}+7\frac{3}{19}+10\frac{9}{19}$
9. $21\frac{4}{21}+5\frac{8}{21}+3\frac{9}{21}$
10. $12+12\frac{7}{18}+12\frac{17}{18}$
11. $4\frac{1}{3}+2\frac{1}{6}$
12. $3\frac{3}{5}+2\frac{1}{2}$
13. $1\frac{1}{7}+6\frac{1}{14}$
14. $1\frac{7}{8}+4\frac{5}{6}$
15. $1\frac{3}{10}+2\frac{7}{100}$
16. $4\frac{4}{5}+5\frac{1}{8}$
17. $5\frac{3}{5}+2\frac{4}{15}$
18. $1\frac{1}{6}+2\frac{4}{9}$

SUBTRACTING MIXED NUMBERS

Example 1. **Subtract $6\frac{7}{8}-4\frac{3}{8}$.**

Solution. $6\frac{7}{8}-4\frac{3}{8}=(6-4)+\left(\frac{7}{8}-\frac{3}{8}\right)$

$$=2+\frac{4}{8}=2+\frac{1}{2}=\mathbf{2\frac{1}{2}}.$$

Subtract the whole numbers
Subtract proper fractions

Example 2. **Subtract $3\frac{1}{4}$ from $9\frac{5}{12}$.**

Method I

Solution. $9\frac{5}{12}-3\frac{1}{4}=(9-3)+\left(\frac{5}{12}-\frac{1}{4}\right)$

$$=6+\left(\frac{5}{12}-\frac{3}{12}\right) \quad \text{LCM of 4 and 12 = 12}$$

$$=6+\frac{\cancel{2}^{1}}{\cancel{12}_{6}}=6+\frac{1}{6}=\mathbf{6\frac{1}{6}}.$$

1. Subtract the whole numbers
2. Subtract fractions
3. Change fractions to equivalent fractions by finding LCM.

Method II. Convert to improper fractions

$$9\frac{5}{12}-3\frac{1}{4}=\frac{113}{12}-\frac{13}{4}=\frac{113\times1-13\times3}{12}=\frac{113-39}{12}=\frac{\cancel{74}^{37}}{\cancel{12}_{6}}=\frac{37}{6}=\mathbf{6\frac{1}{6}}.$$

Example 3. **Subtract $9\frac{7}{8}$ from $11\frac{1}{6}$.**

Solution.

Method I.

$$11\frac{1}{6}-9\frac{7}{8}=(11-9)+\left(\frac{1}{6}-\frac{7}{8}\right)=2+\frac{4}{24}-\frac{21}{24}=2+\frac{4-21}{24}$$

You cannot subtract 21 from 4 so work out as under :

2	6, 8
	3, 4

LCM = 2 × 3 × 4 = 24

$$11\frac{1}{6}-9\frac{7}{8}=(10+1)+\frac{1}{6}-9\frac{7}{8}=\left(10+\frac{6}{6}+\frac{1}{6}\right)-9\frac{7}{8}$$

$$=10+\frac{7}{6}-9\frac{7}{8}=(10-9)+\left(\frac{7}{6}-\frac{7}{8}\right)=1+\frac{28}{24}-\frac{21}{24}=1+\frac{28-21}{24}=\mathbf{1\frac{7}{24}}.$$

Method II.

$$11\frac{1}{6}-9\frac{7}{8}=\frac{67}{6}-\frac{79}{8}=\frac{67\times4-79\times3}{24}=\frac{268-237}{24}=\frac{31}{24}=\mathbf{1\frac{7}{24}}.$$

Subtract:

1. $4\frac{3}{7}-2\frac{1}{7}$	**2.** $12\frac{8}{13}-9\frac{5}{13}$	**3.** $2\frac{1}{2}-1\frac{1}{4}$	**4.** $5\frac{1}{3}-2\frac{1}{4}$

SUBTRACTING FROM WHOLE NUMBERS

Subtracting from 1

Example 1. **From a jug containing 1 litre of pineapple juice I emptied $\frac{5}{8}$ litre into glasses. How much do I still have in the jug ?**

Solution. Quantity remaining in jug $= 1-\frac{5}{8}$

$$= \frac{8}{8}-\frac{5}{8}=\frac{8-5}{8}=\frac{3}{8}$$

or

$$1-\frac{5}{8}=\frac{1}{1}-\frac{5}{8}=\frac{1\times 8-5\times 1}{8}=\frac{8-5}{8}=\frac{\mathbf{3}}{\mathbf{8}}.$$ LCM of 1 and 8 = 8

I still have $\frac{\mathbf{3}}{\mathbf{8}}$ litre of pineapple juice in the jug.

Example 2. **Ajit jumped 5 metres in a broad jump. Shyam jumped $\frac{6}{7}$ metre less than Ajit. How long was Shyam's jump ?**

Solution. **Method I**

Shyam's jump $= 5-\frac{6}{7}=4+1-\frac{6}{7}=4+\frac{7}{7}-\frac{6}{7}$

$$=4+\frac{7-6}{7}=4+\frac{1}{7}=\mathbf{4\frac{1}{7}}$$

Shyam's jump $=4\frac{1}{7}$ metres.

Method II

$$5-\frac{6}{7}=\frac{5}{1}-\frac{6}{7}=\frac{5\times 7-6\times 1}{7}$$

$$=\frac{35-6}{7}=\frac{29}{7}=\mathbf{4\frac{1}{7}}.$$

Example 3. **Subtract :** $8-3\frac{2}{5}$.

Solution. **Method I**

$$8-3\frac{2}{5}=7\frac{5}{5}-3\frac{2}{5}=(7-3)+\left(\frac{5}{5}-\frac{2}{5}\right)=4+\frac{3}{5}=4\frac{3}{5}.$$

Note : $8=7+1=7+\frac{5}{5}=7\frac{5}{5}$

Method II

$$8-3\frac{2}{5}=\frac{8}{1}-\frac{17}{5}=\frac{8\times5-17\times1}{5}$$

$$=\frac{40-17}{5}=\frac{23}{5}=4\frac{3}{5}.$$

It is Method II which is generally preferred.

Skill Practice — **Class Work**

Subtract

1. $1-\frac{2}{5}$
2. $1-\frac{8}{21}$
3. $3-\frac{4}{7}$
4. $9-2\frac{3}{8}$
5. $4-1\frac{5}{7}$

Home Work

EXERCISE 18

Subtract :

1. $1-\frac{3}{4}$
2. $1-\frac{7}{12}$
3. $1-\frac{9}{31}$
4. $2-\frac{3}{7}$
5. $3-\frac{5}{12}$
6. $5-\frac{7}{8}$
7. $6-\frac{1}{3}$
8. $3-1\frac{7}{10}$
9. $11-8\frac{4}{5}$
10. $7-3\frac{6}{13}$
11. $6-4\frac{2}{3}$
12. $9-3\frac{7}{9}$
13. $16-10\frac{12}{13}$
14. $17-16\frac{5}{8}$
15. $100-66\frac{2}{3}$.

PROPERTIES OF ADDITION OF FRACTIONS

★ **Changing the order of two addends does not change the sum.**

For example, $\frac{1}{5}+\frac{2}{5}=\frac{2}{5}+\frac{1}{5}=\frac{3}{5}$

Pictorially, it can be shown as :

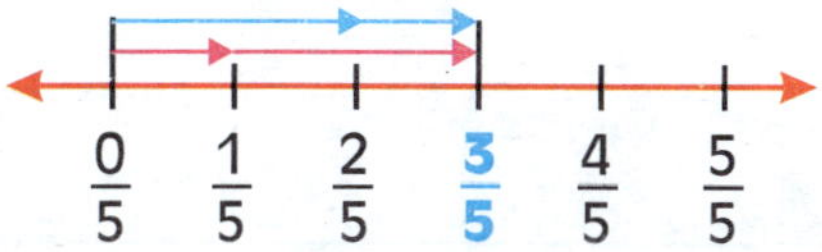

or

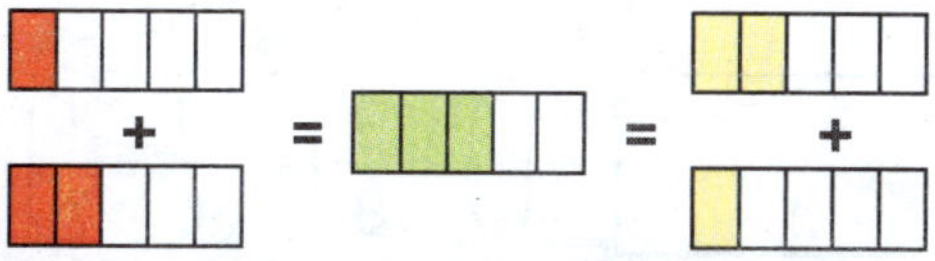

By actual addition you can yourself verify that :

$$\frac{1}{4}+\frac{1}{5}=\frac{1}{5}+\frac{1}{4} \qquad 3\frac{1}{8}+2\frac{1}{12}=2\frac{1}{12}+3\frac{1}{8}$$

★ **Changing the way in which we group the addends does not change the sum.**

For example, $\left(\frac{2}{10}+\frac{3}{10}\right)+\frac{4}{10}=\frac{2+3}{10}+\frac{4}{10}$

$=\frac{5}{10}+\frac{4}{10}=\frac{5+4}{10}$

$=\mathbf{\frac{9}{10}}.$

$\frac{2}{10}+\left(\frac{3}{10}+\frac{4}{10}\right)$

$=\frac{2}{10}+\frac{3+4}{10}$

$=\frac{2}{10}+\frac{7}{10}=\frac{2+7}{10}$

$=\mathbf{\frac{9}{10}}.$

This shows that $\left(\frac{2}{10}+\frac{3}{10}\right)+\frac{4}{10}=\frac{2}{10}+\left(\frac{3}{10}+\frac{4}{10}\right)$

This is called the grouping property of addition of fractions.

★ **Additive property of 0**

$\frac{5}{8}+0=\frac{5}{8}$ and $0+\frac{5}{8}=0.$

Thus, the sum of a fraction and 0 is the fraction itself.

Fill in the blanks

1. $\frac{2}{7}+\frac{5}{7}=\frac{5}{7}+$ ______

2. ______ $+\frac{5}{22}=\frac{5}{22}+\frac{3}{22}$

3. $3\frac{1}{8}+0=$ ______

4. $0+11\frac{1}{2}=$ ______

5. $3\frac{1}{5}+$ ______ $=3\frac{1}{5}$

6. $\left(\frac{1}{3}+\frac{1}{4}\right)+\frac{1}{5}=$ ______ $+\left(\frac{1}{4}+\frac{1}{5}\right)$

7. ______ $+\left(\frac{1}{7}+\frac{1}{8}\right)=\left(\frac{1}{6}+\frac{1}{7}\right)+\frac{1}{8}$

WORD PROBLEMS

Example 1. Anju adds $\frac{5}{9}$ l of apple juice to $\frac{1}{2}$ l of orange juice to make a fruit punch. How many litres of fruit punch does she have ?

Solution. Quantity of fruit punch (in litres) $= \frac{5}{9} + \frac{1}{2}$

$$= \frac{5 \times 2 + 1 \times 9}{18}$$

LCM of 9 and 2 = 9 × 2 = 18

$$= \frac{10+9}{18} = \frac{19}{18} = 1\frac{1}{18}$$

> Do not leave the answer as an improper fraction. Convert to mixed number.

$\therefore$ Anju has $1\frac{1}{18}$ l of fruit punch. **Ans.**

Example 2. Pinki had $3\frac{1}{4}$ m of red ribbon. She used $2\frac{1}{2}$ m of ribbon to wrap some gifts. How much ribbon has she left ?

Solution. Ribbon left (in metres) $= 3\frac{1}{4} - 2\frac{1}{2}$

$$= \frac{13}{4} - \frac{5}{2} = \frac{13 \times 1 - 5 \times 2}{4} = \frac{13-10}{4} = \frac{3}{4}$$

Hence, she has $\frac{3}{4}$ **m** of ribbon left. **Ans.**

Example 3 : A student spent $\frac{5}{8}$ of his summer vacations at home, $\frac{1}{10}$ at a relative's house and the rest at a hill station. What part of the summer vacations did he spend at the hill station ?

Solution : Total part of the summer vacations = 1

Part of summer vacations spent at home and at a relative's place together

$$= \frac{5}{8} + \frac{1}{10} = \frac{5 \times 5 + 1 \times 4}{40} = \frac{25+4}{40} = \frac{29}{40}$$

2	8, 10
	4, 5

$\therefore$ LCM of 8, 10 = 2 × 4 × 5 = 40

Therefore, part of summer vacations spent at the hill station

$$= 1 - \frac{29}{40} = \frac{40}{40} - \frac{29}{40} = \frac{40-29}{40} = \frac{11}{40}.$$

EXERCISE 19

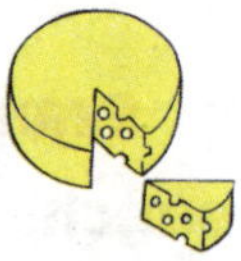

1. Anju's mummy bought two pieces of cheese, one weighing $\frac{7}{8}$ of a kg and the other weighing $\frac{6}{8}$ of a kg. How many kilograms did the two weigh together?

2. Bob cut a pastry into 6 pieces. There were 6 sixths. He and his friends ate 5 sixths of the pastry. How many sixths were left ?

3. Mother bought two chickens. One weighed $1\frac{3}{4}$ kg and the other $1\frac{1}{2}$ kg. How much did they both weigh together?

4. A family eats $1\frac{2}{3}$ breads at a meal. Will 3 breads be enough for 2 meals?

   [**Hint.** $1\frac{2}{3}+1\frac{2}{3}=2+\frac{4}{3}=3\frac{1}{3}$. So more than 3 breads are required for 2 meals].

5. Mary painted $\frac{1}{2}$ of the table. Jane painted $\frac{1}{4}$ of the table. How much of the table did the girls paint?

6. John lives at $1\frac{7}{10}$ kilometres from school. Paul lives at $1\frac{2}{5}$ kilometres. How much farther from school does John live than Paul?

7. Babita has $2\frac{1}{3}$ m of blue ribbon and $\frac{1}{2}$ m of red ribbon. How much blue and red ribbon does she have?

8. A scooter tank has $2\frac{1}{3}$ litres of petrol. Pramod rides around for 3 hours and sees that the petrol level has come down to $\frac{1}{2}$ litre. How much petrol does he use riding around?

9. David lives 1 kilometre from town. Paul lives $\frac{5}{6}$ of a kilometre from town. How much closer to town does Paul live than David ?

10. Dilip spent $\frac{3}{8}$ hours studying English and Mathematics. He spent $\frac{1}{12}$ of an hour studying English. How long did he spend studying Mathematics?

11. A drum full of wheat weighs $84\frac{1}{2}$ kg. If the drum itself weighs $12\frac{1}{5}$ kg, determine the weight of the wheat.

12. Preeti completed $\frac{1}{10}$th of her work on Sunday, $\frac{1}{5}$th of the work on Monday and $\frac{1}{2}$ of it on Tuesday. What part of the work remained yet to be completed ?

7 MULTIPLICATION AND DIVISION OF FRACTIONS

MULTIPLICATION OF FRACTIONS

Example 1. **Ajay wants to use $\frac{1}{2}$ of a page for writing his story. He uses $\frac{1}{3}$ of this $\frac{1}{2}$ page to make pictures. How much of the whole page has been used by the pictures ?**

Solution.

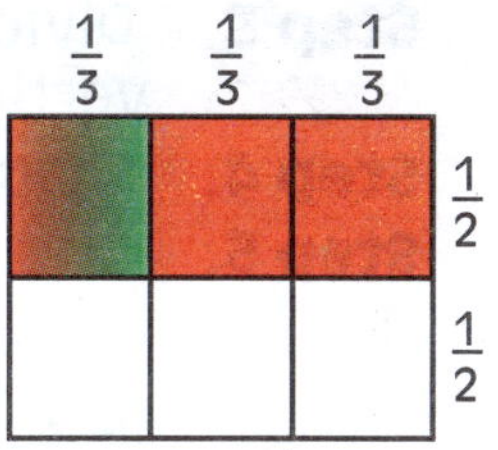

Step 1. Colour the $\frac{1}{2}$ page, used for writing the story, red.

Step 2. Colour $\frac{1}{3}$ of this $\frac{1}{2}$ page, used for pictures, green.

Which part of the whole page is shaded twice ? That part is $\frac{1}{6}$ of the whole and also shows $\frac{1}{2} \times \frac{1}{3}$.

Therefore, from the above model,

$\frac{1}{2} \times \frac{1}{3} = \frac{1}{6}$ ← Number of parts shaded twice (numerator); Total number of parts (denominator)

Example 2. **Evaluate $\frac{2}{3} \times \frac{4}{5}$ with the help of a model.**

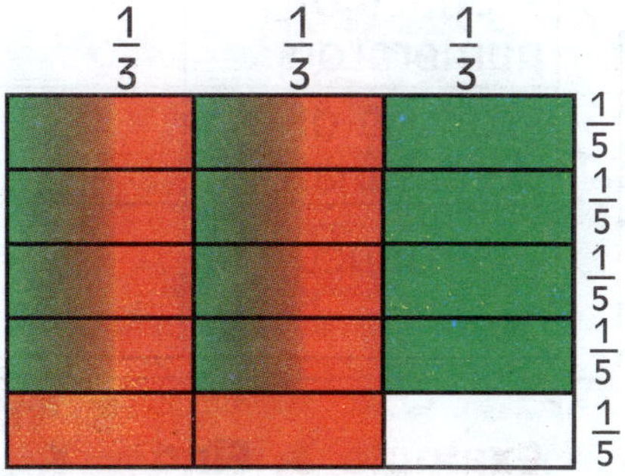

Solution. To mark $\frac{2}{3} \times \frac{4}{5}$ on the rectangle :

Step 1. To mark $\frac{4}{5}$, divide the rectangle into 5 equal parts and shade $\frac{4}{5}$ of it, say, green.

Step 2. To mark $\frac{2}{3}$, divide the same rectangle into 3 equal parts and shade $\frac{2}{3}$ of it, say, red.

Step 3. Identify the part that is shaded twice. That shows $\frac{2}{3} \times \frac{4}{5}$.

Therefore, $\frac{2}{3}\times\frac{4}{5}=\frac{8}{15}$. Thus, we can say that :

Product of fractions $=\frac{2}{3}\times\frac{4}{5}=\frac{2\times4}{3\times5}=\frac{8}{15}=\frac{\text{Product of numerators}}{\text{Product of denominators}}$.

ACTIVITY

Given a square, you can follow the given steps to find $\frac{5}{8}\times\frac{3}{4}$.

Step 1. Divide the square into fourths by drawing horizontal lines.
Step 2. Shade three-fourths of the square red.
Step 3. Divide the square into eighths by drawings vertical lines.
Step 4. Shade five-eighths of the square blue.
Step 5. Identify the part that is shaded twice.

$\frac{5}{8}\times\frac{3}{4}=$ ____________

Complete the table.

	$\frac{2}{7}\times\frac{3}{5}$	$\frac{1}{2}\times\frac{5}{6}$	$\frac{7}{9}\times\frac{2}{3}$	$\frac{5}{11}\times\frac{5}{11}$	$\frac{3}{4}\times\frac{3}{4}$
Product of numerators	6				
Product of denominators	35				
Product	$\frac{6}{35}$				

Example 3. **Find** $\frac{3}{8}\times\frac{12}{21}$.

Solution.

$\frac{3}{8}\times\frac{12}{21}=\frac{3\times12}{8\times21}=\frac{\not{36}\ \not{12}\ 3}{\not{168}\ \not{56}\ 14}=\frac{3}{14}$.

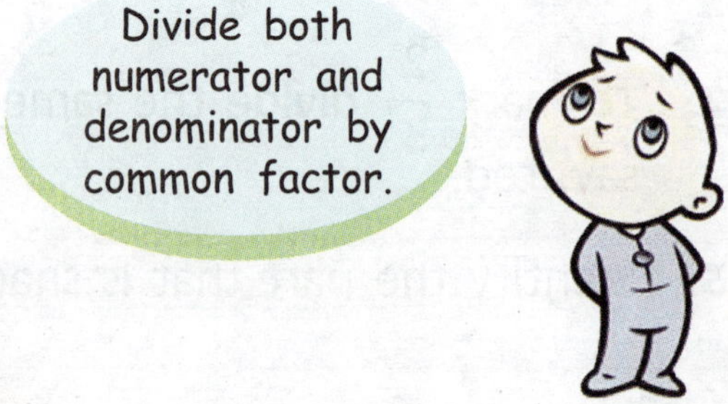

In actual practice the work is shortened as under:

$$\frac{3}{8}\times\frac{12}{21}=\frac{\overset{1}{\cancel{3}}}{\underset{2}{\cancel{8}}}\times\frac{\overset{3}{\cancel{12}}}{\underset{7}{\cancel{21}}}=\frac{1\times3}{2\times7}=\frac{\mathbf{3}}{\mathbf{14}}.$$

You may omit this step and write directly $\frac{3}{14}$ by working out mentally that $1 \times 3 = 3$ and $2 \times 7 = 14$.

Example 4. Find $\frac{42}{15}\times\frac{3}{7}\times\frac{5}{36}$.

Solution. $\frac{\overset{\overset{1}{\cancel{6}}}{\cancel{42}}}{\underset{\underset{1}{\cancel{3}}}{\cancel{15}}}\times\frac{\overset{1}{\cancel{3}}}{\underset{1}{\cancel{7}}}\times\frac{\overset{1}{\cancel{5}}}{\underset{6}{\cancel{36}}}=\frac{1}{6}.$

> **Note:** You can divide **any** numerator and **any** denominator by a common factor.

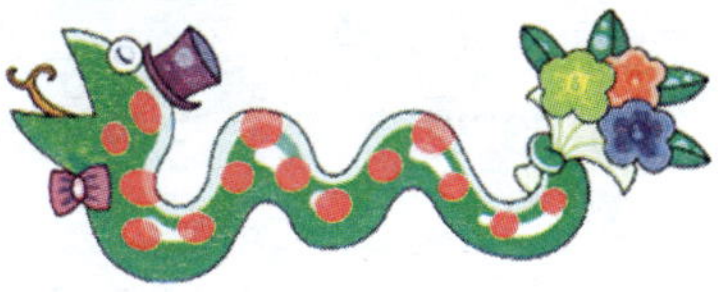

Example 5 : Find $2\frac{3}{5}\times1\frac{1}{4}$.

Solution. First convert the mixed numbers to improper fractions. Then multiply

$$2\frac{3}{5}\times1\frac{1}{4}=\frac{13}{\underset{1}{\cancel{5}}}\times\frac{\overset{1}{\cancel{5}}}{4}=\frac{13}{4}=\mathbf{3\frac{1}{4}}.$$

EXERCISE 20

Copy and multiply :

(a) $\frac{2}{3}\times\frac{3}{4}$ (b) $\frac{7}{8}\times\frac{16}{21}$ (c) $\frac{1}{2}\times\frac{14}{3}$ (d) $\frac{1}{4}\times\frac{3}{5}$ (e) $\frac{1}{2}\times\frac{4}{5}$

(f) $\frac{3}{5}\times\frac{12}{15}$ (g) $\frac{2}{3}\times\frac{9}{10}$ (h) $\frac{17}{63}\times\frac{2}{5}$ (i) $\frac{8}{27}\times\frac{15}{16}$ (j) $\frac{5}{18}\times\frac{54}{45}$

Multiply :

2. (a) $\frac{4}{5}\times\frac{2}{3}\times\frac{25}{16}$ (b) $\frac{7}{11}\times\frac{33}{49}\times\frac{14}{44}$ (c) $\frac{33}{16}\times\frac{24}{55}\times\frac{8}{9}$

3. (a) $1\frac{7}{8}\times3\frac{1}{5}$ (b) $9\frac{1}{11}\times10\frac{1}{12}$ (c) $7\frac{1}{2}\times3\frac{1}{3}$ (d) $2\frac{1}{7}\times3\frac{1}{2}$

(e) $3\frac{11}{15}\times1\frac{42}{63}$ (f) $1\frac{2}{7}\times5\frac{5}{6}$ (g) $4\frac{3}{4}\times1\frac{7}{57}$ (h) $12\frac{1}{2}\times3\frac{1}{5}$

4. Simplify:

(a) $2\frac{3}{4}\times4\frac{4}{5}\times6\frac{3}{8}$ (b) $1\frac{2}{3}\times\frac{7}{8}\times3\frac{1}{5}$ (c) $2\frac{1}{2}\times3\frac{2}{3}\times\frac{6}{11}$

Multiplication of a Fraction by a Whole number

What will be $4 \times \frac{1}{5}$?

$$4 \times \frac{1}{5} = \frac{1}{5} + \frac{1}{5} + \frac{1}{5} + \frac{1}{5} = \frac{4}{5}$$

It can be shown on a number line as

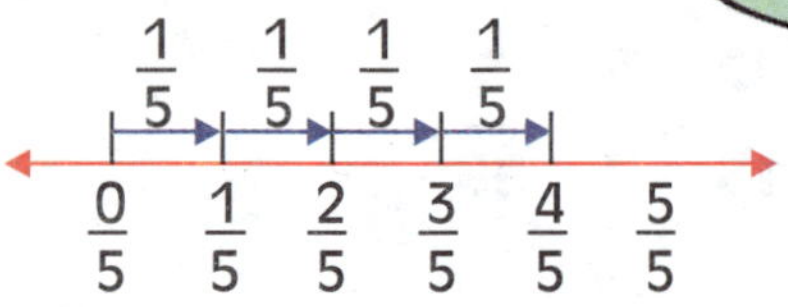

The working can be shown as $4 \times \frac{1}{5} = \frac{4}{1} \times \frac{1}{5} = \frac{4 \times 1}{1 \times 5} = \mathbf{\frac{4}{5}}$.

Similarly, $5 \times \frac{3}{8} = \frac{5}{1} \times \frac{3}{8} = \frac{5 \times 3}{1 \times 8} = \frac{15}{8} = \mathbf{1\frac{7}{8}}$.

Pictorially, it can be shown as $5 \times \frac{3}{8} = \frac{3}{8} + \frac{3}{8} + \frac{3}{8} + \frac{3}{8} + \frac{3}{8} = \frac{15}{8} = \mathbf{1\frac{7}{8}}$.

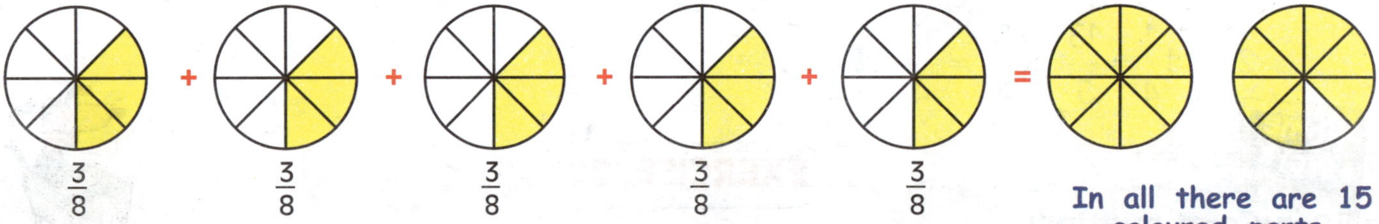

In all there are 15 coloured parts.

EXERCISE 21

Write the product in the simplest form:

1. $8 \times \frac{5}{6}$ **2.** $4 \times \frac{1}{4}$ **3.** $7 \times \frac{3}{5}$ **4.** $6 \times 1\frac{7}{8}$

5. $10 \times 1\frac{2}{15}$ **6.** $34 \times \frac{5}{51}$ **7.** $22 \times \frac{13}{33}$ **8.** $75 \times \frac{19}{100}$

9. $12 \times \frac{7}{16}$ **10.** $18 \times \frac{4}{27}$ **11.** $90 \times \frac{14}{105}$ **12.** $110 \times \frac{21}{66}$

WORD PROBLEMS

Example 6. Mrs Consul had $\frac{5}{9}$ of a sheet of chart paper. She used $\frac{3}{5}$ of it to make some cards. How much chart paper did she use ?

Solution. Chart paper used $= \frac{3}{5}$ of $\frac{5}{9} = \frac{\cancel{3}^{1}}{\cancel{5}_{1}} \times \frac{\cancel{5}^{1}}{\cancel{9}_{3}} = \frac{1}{3}$.

She used $\frac{1}{3}$ of the sheet of chart paper to make cards.

EXERCISE 22

1. Three-fourths of a class are in the drawing club. $\frac{2}{3}$ of them are taking part in a drawing competition. What fraction of the class is taking part in the competition.

2. Ayushree took $\frac{5}{8}l$ of orange juice to school. She drank $\frac{1}{10}$ of it in school. How many litres of orange juice did she drink ? How much was left ?

3. David had $\frac{8}{11}$ hour of free time. He spent $\frac{3}{4}$ of his time cleaning his room. How much time did he spend cleaning his room ?

4. If 1 litre of milk cost Rs $21\frac{1}{2}$, what is the cost of $6\frac{1}{4}$ litres of milk ?
   [Hint : Quick and as way. Total cost $= 6 \times$ Rs $21\frac{1}{2} + \frac{1}{4} \times$ Rs $21\frac{1}{2}$]

5. 15 students contributed Rs $30\frac{3}{5}$ each for a picnic. What was the total contribution ?

6. Ravi is $1\frac{1}{4}$m tall. Dinesh is $1\frac{1}{5}$ times as tall as Ravi. What is Dinesh's height ?

7. Anjali bought $\frac{6}{7}$ kg of flour. She used $\frac{2}{3}$ of it to make puris. How much flour did she use ?

8. One shirt requires $2\frac{3}{4}$ m of cloth. How much cloth will be needed for 12 such shirts ?

DIVISION OF FRACTIONS

To understand division of fractions, we should know what reciprocal of a fraction means.

Reciprocal of a Fraction :

Study the following example :

$$1\frac{1}{2} \times \frac{2}{3} = \frac{\cancel{3}^{1}}{\cancel{2}_{1}} \times \frac{\cancel{2}^{1}}{\cancel{3}_{1}} = \frac{1}{1} = 1$$

The numbers $\frac{3}{2}$ and $\frac{2}{3}$ are alike in the sense that they consist of the same numerals 2 and 3. If you compare their numerators and denominators, you will find that numerator of one is the denominator of the other.

The product of such numbers is always 1 and they are called reciprocals. Reciprocals are number pairs that have the product 1.

Examples.

$\frac{4}{5}$ and $\frac{5}{4}$

$\frac{4}{5} \times \frac{5}{4} = \frac{20}{20} = 1$

6 and $\frac{1}{6}$

$\frac{6}{1} \times \frac{1}{6} = \frac{6}{6} = 1$

$\frac{25}{4}$ and $\frac{4}{25}$

$\frac{\cancel{25}^{1}}{\cancel{4}_{1}} \times \frac{\cancel{4}^{1}}{\cancel{25}_{1}} = 1$

Skill Practice — **Class Work**

Name the reciprocals of :

Number	Reciprocal	Number	Reciprocal
$\frac{2}{3}$		$\frac{1}{7}$	
$\frac{5}{6}$		$3\frac{1}{7}$	
4		$1\frac{2}{9}$	
6		$\frac{1}{9}$	

DIVIDING A WHOLE NUMBER BY A FRACTION

$3 \div \frac{1}{5}$ is the same as finding out how many one-fifths are there in 3 wholes.

Pictorially, it can be shown as :

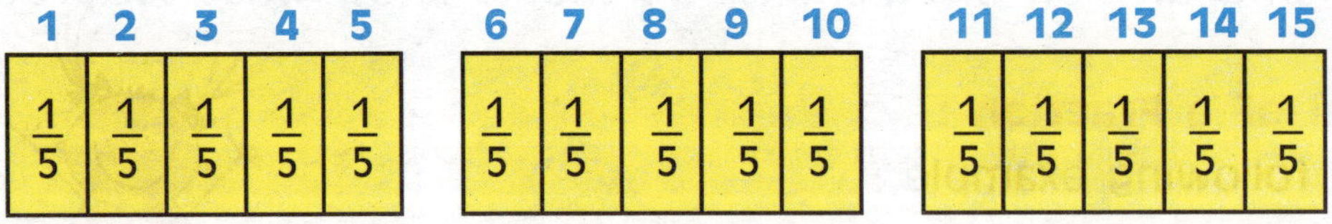

Counting the number of $\frac{1}{5}$'s in 3 wholes, we get

$3 \div \frac{1}{5} = 15$, which can also be obtained by working out as $3 \times 5 = 15$

Let us take another example. Suppose we have to find $6 \div \frac{3}{4}$.

$6 \div \frac{3}{4}$ is again the same as finding out how many three-fourths are there in 6, which can be shown pictorially as:

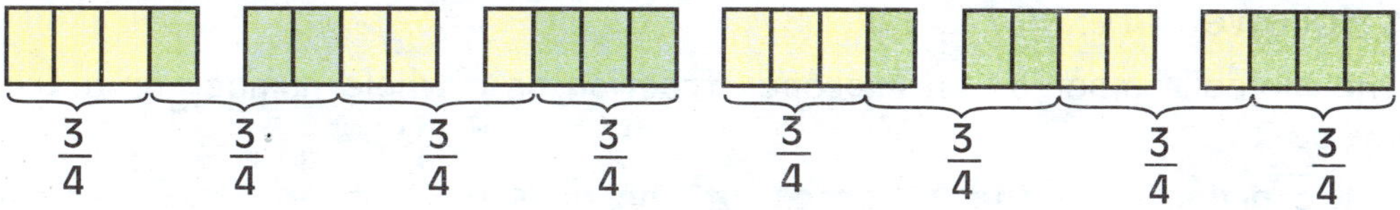

Counting, you get 8 three-fourths in 4.

So, $6 \div \frac{3}{4} = 8$, which can also be obtained by working out as $\overset{2}{\cancel{6}} \times \frac{4}{\cancel{3}_1} = 8$.

Thus,

changes to

$$3 \div \frac{1}{5} = 3 \times 5 = \mathbf{15}$$

Reciprocal

changes to

$$6 \div \frac{3}{4} = 6 \times \frac{4}{3} = \mathbf{8}$$

Reciprocal

To divide fractions, multiply the first fraction by the reciprocal of the second fraction.

DIVIDING A FRACTION BY A WHOLE NUMBER

In a birthday party, a cake was divided into 3 equal parts. One part was eaten by the birthday boy and two-thirds of the cake was distributed equally among 5 of his friends.

What part of the whole cake did each friend get ?

Each child's share $= \frac{2}{3} \div 5$

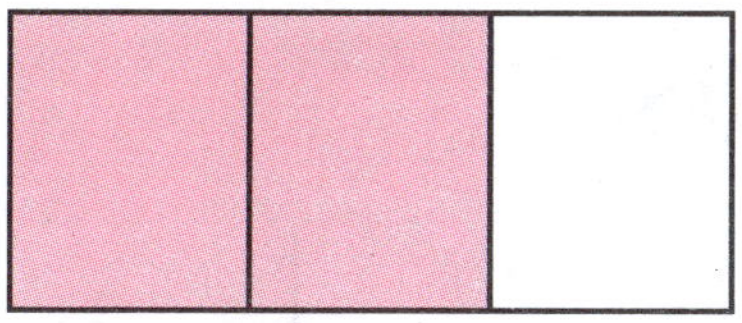

The shaded part shows $\frac{2}{3}$ of the cake.

Each child's share

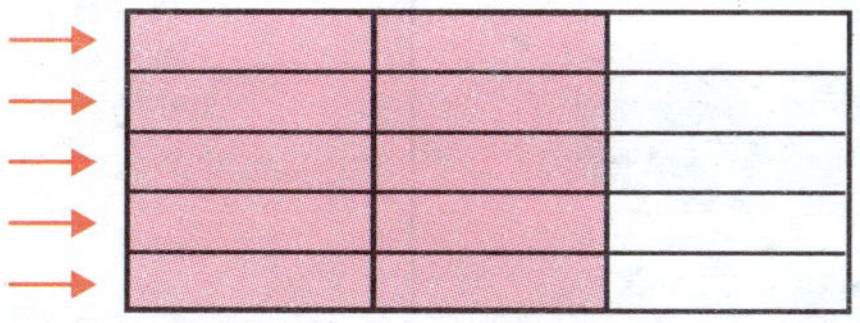

To find each child's share, divide the above rectangle into 5 equal parts.

Each small rectangle shows $\frac{1}{15}$ of the whole.

Therefore, each child's share $= \frac{2}{15}$ of the whole cake.

This can also be obtained as :

$$\frac{2}{3} \div 5 = \frac{2}{3} \times \frac{1}{5} = \frac{2 \times 1}{3 \times 5} = \frac{2}{15}$$

By now, you must have completely understood the method of dividing fractions.

The **steps** followed are:

1. Write the mixed number as an improper fraction or a whole number as a fraction as the case may be.
2. Multiply the dividend by the reciprocal of the divisor.
3. Write the product in the simplest form.
4. Convert to a mixed number, if the product is an improper fraction.

Example. Divide:

(a) $\frac{5}{12}$ by $\frac{4}{15}$ **(b)** 35 by $\frac{7}{3}$ **(c)** $\frac{8}{15}$ by 4 **(d)** $2\frac{1}{2}$ by $2\frac{1}{4}$

Solution. (a) $\frac{5}{12}\div\frac{4}{15}=\frac{5}{\cancel{12}_{4}}\times\frac{\cancel{15}^{5}}{4}=\frac{5\times5}{4\times4}=\frac{25}{16}=\mathbf{1\frac{9}{16}}$.

(b) $35\div\frac{7}{3}=\cancel{35}^{5}\times\frac{3}{\cancel{7}_{1}}=\frac{5\times3}{1}=\frac{15}{1}=\mathbf{15}$.

(c) $\frac{8}{15}\div4=\frac{\cancel{8}^{2}}{15}\times\frac{1}{\cancel{4}_{1}}=\mathbf{\frac{2}{15}}$.

(d) $2\frac{1}{2}\div2\frac{1}{4}=\frac{5}{2}\div\frac{9}{4}=\frac{5}{\cancel{2}_{1}}\times\frac{\cancel{4}^{2}}{9}=\frac{5\times2}{9}=\frac{10}{9}=\mathbf{1\frac{1}{9}}$. (Change to improper fractions and then divide)

Divide:

1. $4\div\frac{1}{8}$	**2.** $\frac{3}{4}\div9$	**3.** $\frac{4}{9}\div\frac{1}{6}$	**4.** $2\frac{1}{5}\div1\frac{1}{4}$

EXERCISE 23

Find the quotient in the simplest form

1. $3\div\frac{1}{4}$ **2.** $2\div\frac{1}{5}$ **3.** $8\div\frac{1}{2}$ **4.** $9\div\frac{3}{2}$ **5.** $\frac{7}{9}\div7$

6. $\frac{8}{15}\div\frac{4}{3}$ **7.** $\frac{3}{8}\div4$ **8.** $\frac{6}{7}\div\frac{5}{6}$ **9.** $\frac{6}{5}\div\frac{6}{5}$ **10.** $\frac{1}{2}\div\frac{9}{4}$

11. $\frac{9}{10} \div \frac{4}{10}$ 12. $\frac{9}{28} \div \frac{14}{3}$ 13. $3\frac{3}{4} \div 1\frac{1}{4}$ 14. $\frac{2}{3} \div 3\frac{1}{5}$ 15. $4\frac{1}{2} \div 6$

16. $1 \div 4\frac{3}{7}$ 17. $8\frac{4}{7} \div \frac{4}{21}$ 18. $1\frac{7}{15} \div 11$ 19. $0 \div 6\frac{1}{11}$ 20. $10 \div 3\frac{1}{3}$

21. $2\frac{2}{3} \div 1\frac{2}{3}$ 22. $1\frac{7}{8} \div 1\frac{2}{3}$ 23. $3\frac{4}{5} \div \frac{3}{5}$ 24. $16 \div 2\frac{2}{15}$ 25. $5\frac{1}{4} \div 2\frac{4}{5}$

26. $1\frac{11}{15} \div 1\frac{4}{35}$ 27. $3\frac{3}{17} \div 2\frac{6}{51}$ 28. $3\frac{11}{38} \div 1\frac{18}{57}$

WORD PROBLEMS

Example : Mrs. Thomas cut a 20 metre rope into pieces, each $2\frac{1}{2}$ m long. How many pieces could she cut ?

Solution. Number of pieces $= 20 \div 2\frac{1}{2} = 20 \div \frac{5}{2} = \cancel{20}^{4} \times \frac{2}{\cancel{5}_{1}} = \mathbf{8}$.

EXERCISE 24

1. Rita gave $\frac{2}{3}$ of her chocolates equally to 6 of her friends. What fraction of Rita's chocolates did each friend get ?

2. A teapot contains $\frac{4}{7}$ l of tea. The tea is poured equally in to 8 cups. How much tea is there in each cup ?

3. How many pieces each $2\frac{3}{4}$ m long may be cut from a string $16\frac{1}{2}$ m long ?

4. Manasi did her homework in $3\frac{1}{5}$ hours. She spent the same amount of time in each of 4 subjects. What time did she spend on each subject ?

5. Shubham practises on his guitar $6\frac{1}{4}$ hours a week. If he practises $1\frac{1}{4}$ hours each day, how many days does he practise each week ?

6. Shubra feeds her chickens $3\frac{1}{7}$ kg of grain each day. At this rate, how many days will a 22 kg bag of grain last ?

7. Some students hiked $10\frac{1}{2}$ km in $1\frac{3}{4}$ hours. At the same rate, how far can they hike in one hour ?

8. A line segment $6\frac{1}{2}$ cm long is divided into 5 equal parts. What will be the length of each part.

9. Naina wanted some sheets of paper. She buys a big sheet of paper and cuts it into pieces, each of which is $\frac{1}{18}$ of the full sheet. How many small sheets can she get from the full sheet of paper ?

[**Hint :** Size of one sheet $= \frac{1}{18}$ of the full sheet $\therefore$ Number of sheets $= 1 \div \frac{1}{18}$]

10. Two-thirds of a tank can be filled in $17\frac{1}{3}$ minutes. How many minutes will be required to fill the whole tank ?

FRACTIONAL PART OF A WHOLE NUMBER

Example 1 : $\frac{1}{4}$ of 20 beads in a bracelet were green. What is the number of green beads in the bracelet?

Solution. $\frac{1}{4}$ means one part out of 4 equal parts. Divide 20 into 4 equal parts.

Each part has 20 ÷ 4 beads = 5 beads

$\frac{1}{4}$ of 20 is the same as $\frac{1}{4} \times 20$.

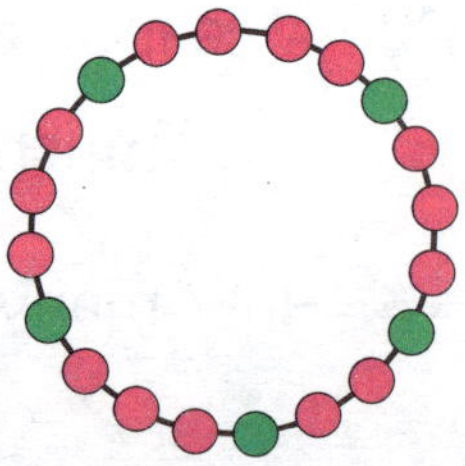

Therefore, $\frac{1}{4}$ of 20 $= \frac{1}{\cancel{4}_{1}} \times \cancel{20}^{5} = 5$

Hence, there are 5 green beads in the bracelet.

Example 2 : There are 28 students in a class. $\frac{2}{7}$ of the students are girls. How many girls are there in the class ?

Solution. $\frac{2}{7}$ means 2 parts out of 7 equal parts.

Divide 28 into 7 equal parts. Each part has 28 ÷ 7 students = 4 students.

1 part is 4 students, therefore, 2 parts is 2 × 4 = 8 students.

$\frac{2}{7}$ of 28 students = 8 girls

You can also work out as under:

$$\frac{2}{7} \text{ of } 28 = \frac{2}{\cancel{7}_{1}} \times \overset{4}{\cancel{28}} = \frac{2 \times 4}{1} = \mathbf{8.}$$

Example 3 : Find : (a) $\frac{3}{4}$ of 1 rupee (b) $\frac{2}{5}$ of a non-leap year.

Solution. (a) $\frac{3}{4}$ of 1 rupee $= \frac{3}{4}$ of 100 paise $= \frac{3}{\cancel{4}_{1}} \times \overset{25}{\cancel{100}} = \frac{3 \times 25}{1} = 75 \text{ paise}$

(b) $\frac{2}{5}$ of a non-leap year $= \frac{2}{5} \times 365 \text{ days}$

$$= \frac{2}{\cancel{5}_{1}} \times \overset{73}{\cancel{365}} = \frac{2 \times 73}{1} = \mathbf{146 \text{ days}}.$$

Thus, if "of" occurs between two fractions replace it by ×.

Find the following :

1. $\frac{1}{3}$ of 12	**2.** $\frac{3}{7}$ of 56	**3.** $\frac{3}{5}$ of 20 kg	**4.** $\frac{1}{3}$ of 30 days

EXERCISE 25

1. Find :

(a) $\frac{1}{8}$ of 64 (b) $\frac{1}{9}$ of 81 (c) $\frac{2}{5}$ of 35 (d) $\frac{2}{5}$ of 30

(e) $\frac{4}{11}$ of 99 (f) $\frac{1}{18}$ of 162 (g) $\frac{1}{12}$ of 108 (h) $\frac{1}{12}$ of 3000

2. Fill in the blanks:

(a) $\frac{2}{3}$ of one day = ______ hours

(b) $\frac{1}{6}$ of an hour = ______ minutes

(c) $\frac{4}{5}$ of 10 km = ______ km

(d) $\frac{2}{7}$ of a week = ______ days

(e) $\frac{7}{12}$ of a minute = ______ seconds

(f) $\frac{11}{20}$ of a metre = ______ centimetre

3. There are 50 apples in a box. $\frac{2}{5}$ of the apple are rotten. How many apples are rotten ?

4. Manish has 48 stamps. He put $\frac{5}{6}$ of the stamps in his album and gave the rest to his friend. How many stamps did Manish put in his album ? How many did he give to his friend ?

5. Of 60 books that Ayushree had, $\frac{2}{3}$ were story books. How many story books did she have ? How many other books did she have ?

6. The distance between Delhi and Dehradun is 280 km. Rina travelled $\frac{3}{4}$ of the distance by train and the rest by car. How many kilometres did Rina travel by train ?

MENTAL MATHS–5

1. Arrange in as ascending : $\frac{1}{2}, \frac{1}{5}, \frac{1}{3}, \frac{1}{8}, \frac{1}{6}, \frac{1}{4}$

2. Complete $\frac{2}{7} = \frac{6}{\square} = \frac{\square}{35} = \frac{12}{\square}$

3. What is a half of a half of 84 ?

4. Of a mushroom $\frac{19}{20}$ is water; what fraction of it is solid matter ?

5. Which is greater : $\frac{2}{3}$ or $\frac{3}{4}$?

6. $\frac{5}{6}$ of 60 + fifth multiple of 5 = ______________ .

7. How many slices of a cake are there, if it is divided into fifths ?

8. Write the value of $18\frac{2}{5} - 5\frac{1}{5}$?

9. The difference between $\frac{2}{3}\times\frac{4}{5}$ and $\frac{2\times4}{3\times5}$ is 1. True or False ?

10. Simplify : $\frac{8}{21}\div\frac{4}{7}$.

11. $7\frac{1}{3}\times6$ = ☐

12. Simplify : $\frac{1}{2}\times\frac{2}{3}\times\frac{3}{4}$.

13. What is $1\div2\frac{1}{2}$?

14. What one-third of one-half of 30 ?

15. Write the reciprocal of $\left(\frac{1}{2}+\frac{1}{4}\right)$.

16. Simplify $\dfrac{\frac{2}{5}\times\frac{4}{5}}{\frac{12}{5}}$.

17. In a class of 35, $\frac{2}{7}$ were boys. What was the number of girls ?

18. What mixed number does the point P represent ?

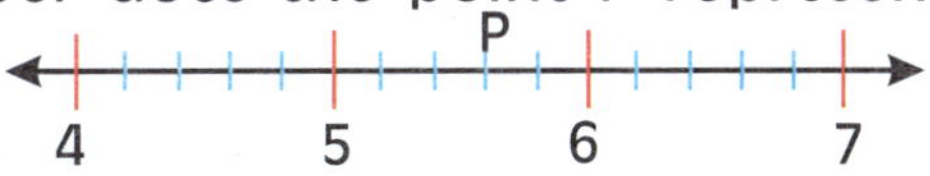

KNOW YOUR NUMBERS

THE NUMBER THREE

3 is a prime number
1 + 2 = 3

My telephone number is three, triple eight, five, 38885
Triple, trio,—— indicate the number three. It is a super special number. Tri___ means three. Triangles have three sides, tripods have three legs and the dinosaur triceratops had three horns-two on its head and a little one on its nose.

The Indian National Flag is a tricolour because it has three colours.
A **hat-trick** is success achieved in three consecutive attempts. If any one player scores three goals one after another he is said to have achieved a hat-trick. If a bowler takes three wickets off consecutive balls in cricket, it is a hat-trick.
Oaths are generally repeated three times.
A three-legged race is run by two people each with a leg tied to the partner.

Most colours can be mixed from just three primary colours. For example, the colours you see on a television screen are mixtures of red, green and blue light.

8 DECIMALS

The cost of 1 packet of sweets, your weight, height of your sister, time taken for a 100 m dash etc. are quantities which are expressed as decimal numbers

Rs 112.60

1.35 m

100m race
Time : 10.51sec.

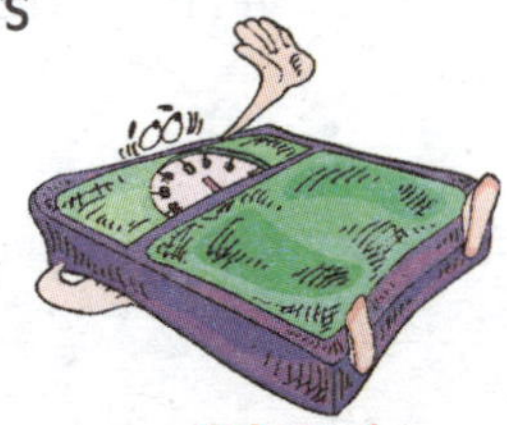

34.7 kg

You have studied the meaning of decimals, the way to write them and also their addition and subtraction in class IV. We will first revise the same before proceeding further.

MEANING OF DECIMALS

A decimal is another way of expressing a fraction.

TENTHS

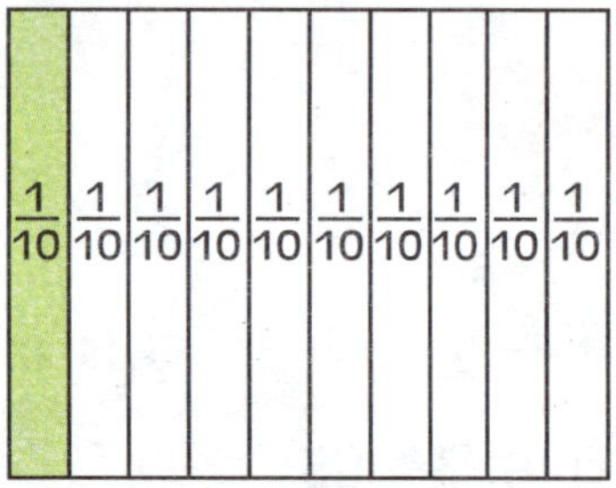

The fraction $\frac{1}{10}$, as a decimal is written as **0.1** and read as **zero point one.** Thus, the following shaded parts can be read as :

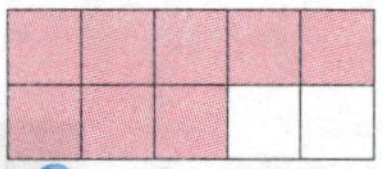

$\frac{8}{10}$ or 8 tenths

0.8 or zero point eight

$\frac{7}{10}$ or 7 tenths

0.7 or zero point seven

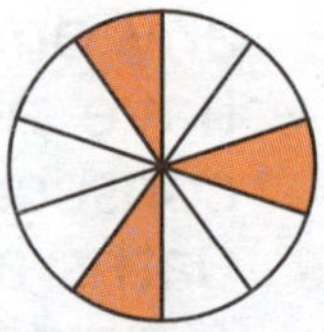

$\frac{3}{10}$ or 3 tenths

0.3 or zero point three

> The dot or the point between the two digits is called the decimal point.

Whole numbers and decimal numbers can also be combined as shown on next page :

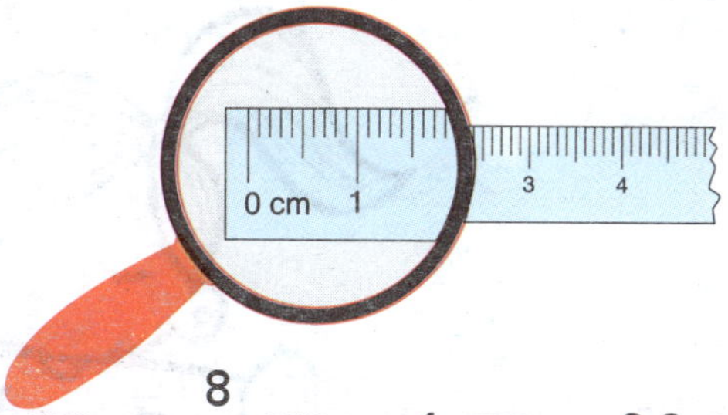

The picture given alongside shows on a ruler, a length of 1 cm + 8 parts out of 1 cm. Since each cm is further divided into 10 equal parts (as can be seen), the above given number can be written as 1.8.

$1 \text{ cm} + \frac{8}{10} \text{ cm} = 1 \text{ cm} + 0.8 \text{ cm} = 1.8 \text{ cm}$. **1.8** is also a decimal number and is read as **one point eight.**

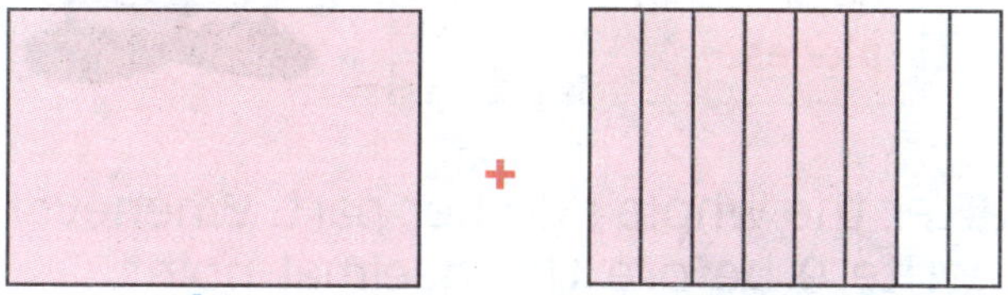

1 $\quad$ $\frac{8}{10}$ or 0.8

The numeral 1.8 names the same number as $\frac{18}{10}$ or $1\frac{8}{10}$ and represents one and eight-tenths.

HUNDREDTHS

Let us now divide a whole into 100 equal parts or one-tenth into 10 equal parts.

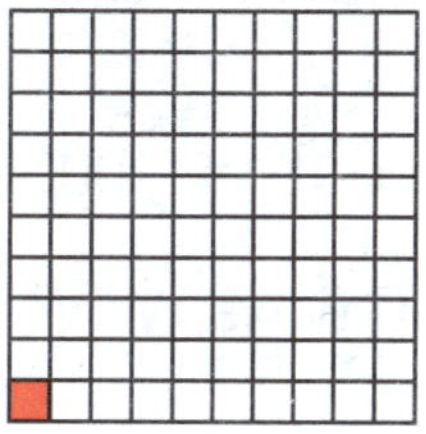

Each part out of 100 equal parts is $\frac{1}{100}$ or 0.01

0.01 is 1 hundredth

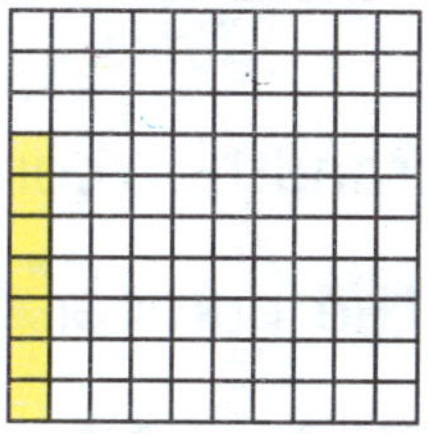

7 parts out of 100 equal parts is $\frac{7}{100}$ or 0.07

0.07 is 7 hundredths

73 parts out of 100 equal parts is $\frac{73}{100}$ or 0.73

0.73 is 73 hundredths

We read 0·01 as **zero point zero one**, 0·07 as **zero point zero seven** and 0·73 as **zero point seven three.**

Hey! It is wrong to read 0·73 as zero point seventy three

0.37 is 37 hundredths or 3 tenths 7 hundredths.

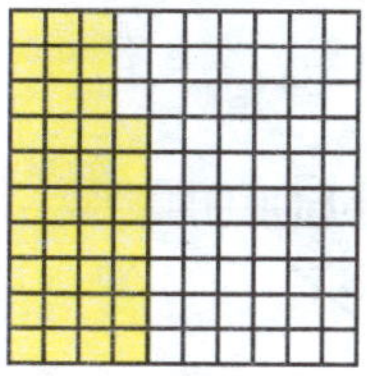

37 hundredths

=

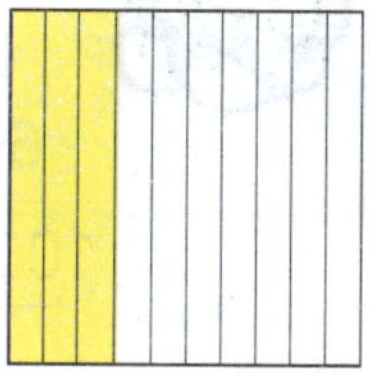

3 tenths

+

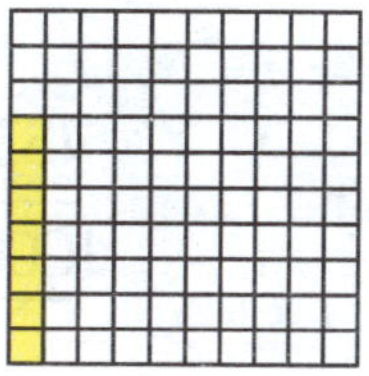

7 hundredths

$$0.37 = \frac{37}{100} = \frac{30}{100} + \frac{7}{100}$$

$$= \frac{3}{10} + \frac{7}{100}$$

Similarly, $6\frac{79}{100} = 6.79 = 6 + \frac{7}{10} + \frac{9}{100}$

$$5\frac{3}{100} = 5.03 = 5 + \frac{0}{10} + \frac{3}{100}$$

Number of digits after the decimal point is called the number of decimal places.

So, a decimal number has two parts– the whole number part and the decimal part separated by a decimal point.

Whole number part greater than one → 6.79 ← Decimal part less than one

↑ Decimal point

In the above number there are 2 decimal places after the whole number part. Whenever there is no whole number part, do not forget to write 0 before the decimal point.

e.g. .65 is always written as **0**.65

THOUSANDTHS

If we divide one hundreth further into 10 equal parts, the whole gets divided into thousand equal parts. Then each part of the whole represents one thousandth.

Thus $\frac{1}{1000}$ **= 0.001** or **one-thousandth** read as **zero point zero zero one.**

The decimal $0.895 = \frac{895}{1000}$ (895 thousandths) represents 895 parts out of 1000 parts.

Ankit and Sneha are both reading the number 0.895 in different ways. Which is the correct one?

I read 0.895 as zero point eight nine five

I read 0.895 as zero point eight hundred ninety five

Also, $0.895 = \frac{8}{10} + \frac{9}{100} + \frac{5}{1000}$

39.107 is thirty nine and one hundred seven thousandths and is read as **thirty nine point one zero seven.**

$39.107 = 39 + \frac{1}{10} + \frac{0}{100} + \frac{7}{1000}.$

Notice the following pattern.

$\frac{7}{10} = 0{\cdot}7$ — one digit

$\frac{9}{100} = 0{\cdot}09$ — Two digits

$\frac{453}{1000} = 0{\cdot}453$ — Three digits

$\frac{69}{10000} = 0{\cdot}0069$ — Four digits

The number of digits after the decimal point in the decimal numeral is equal to the number of zeros after 1 in the denominator of the corresponding common fraction.

Fill in the blanks.

	Fraction	Decimal Number	Read as
1.	$\frac{7}{10}$	0.7	zero point seven
2.	$\frac{65}{100}$		
3.		0.435	
4.	$\frac{237}{100}$		
5.			Fifteen point three
6.	$145\frac{695}{1000}$		

PLACE VALUE AND DECIMALS

The place value chart shown below can also be used to understand decimals. The place value of each digit in the number 5555.555 is shown below.

digits →	5	5	5	5	•	5	5	5
place value →	1000	100	10	1		$\frac{1}{10}$	$\frac{1}{100}$	$\frac{1}{1000}$
	thousands	hundreds	tens	ones		tenths	hundredths	thousandths

Notice that each place has a value one tenth $\left(\frac{1}{10}\right)$ that of the place to its left. The first place to the right of the decimal point is the tenths' place, the second place to the right is the hundredths' place and so on. The last place after the decimal point tells us how to name the decimal part. Here it is 555 thousandths.

Using the place value chart, we can write the expanded form of any decimal number. Take the number 6785.093.

6785.093 = 6 thousands + 7 hundreds + 8 tens + 5 ones + 0 tenths + 9 hundredths + 3 thousandths

$$= 6 \times 1000 + 7 \times 100 + 8 \times 10 + 5 \times 1 + 0 \times \frac{1}{10} + 9 \times \frac{1}{100} + 3 \times \frac{1}{1000}$$

or

$$6785\frac{93}{1000} = 6000 + 700 + 80 + 5 + \frac{9}{100} + \frac{3}{1000}$$ **(Fractional form)**

$$= 6000 + 700 + 80 + 5 + 0.09 + 0.003$$ **(Decimal form)**

I. Write the value of the circled digits :

1. 1 2 . 3 ③ — 3 hundredths

2. 4 5 . ⑥ 4 — ______

3. 9 . 8 0 ⑤ — ______

4. 2 ① 9 . 6 3 — ______

5. 1 ⑤ . 0 3 5 — ______

6. 8 9 . 4 ⑧ 3 — ______

II. Write in decimal form :

1. $400 + 20 + 5 + \frac{3}{10} + \frac{4}{100} + \frac{7}{1000} =$ ______________________

2. 8000 + 600 + 7 + 0.3 + 0.02 + 0.001 = ______________________

III. Write in expanded form :

1. 178.345 (Decimal expansion) = ______________________

2. 26.809 (Fractional expansion) = ______________________

CONVERTING FRACTIONS TO DECIMALS AND VICE VERSA

Fractions whose denominators are 10,100 or 1,000 can be easily converted to decimals by putting the decimal point in the numerator accordingly. Thus,

$\frac{5}{10} = 0.5$

$\frac{32}{10} = 3.2$

$\frac{143}{10} = 14.3$

If, Denominator = 10

Number of decimal places = 1

$\frac{5}{100} = 0.05$

$\frac{32}{100} = 0.32$

$\frac{143}{100} = 1.43$

Denominator = 100

Number of decimal places = 2

$\frac{5}{1000} = 0.005$

$\frac{32}{1000} = 0.032$

$\frac{143}{1000} = 0.143$

Denominator = 1000

Number of decimal places = 3

For fractions which can be converted to equivalent fraction having denominators 10 or multiples of 10, we apply the same method as above.

$\frac{1}{2} = \frac{5}{10} = 0.5$ (× 5 above and below) $\qquad \frac{1}{4} = \frac{25}{100} = 0.25$ (× 25 above and below) $\qquad \frac{4}{5} = \frac{8}{10} = 0.8$ (× 2 above and below) $\qquad \frac{5}{8} = \frac{625}{1000} = 0.625$ (× 125 above and below)

Similarly, $2\frac{3}{5} = 2 + \frac{3}{5} = 2 + \frac{6}{10} = 2 + 0.6 = \mathbf{2.6.}$

$$4\frac{8}{25} = 4 + \frac{8}{25} = 4 + \frac{32}{100} = 4 + 0.32 = \mathbf{4.32.}$$

Decimals can also be converted to fractions as :

$0.8 = \frac{8}{10} = \frac{4}{5} \qquad 0.36 = \frac{36}{100} = \frac{9}{25} \qquad 2.25 = \frac{225}{100} = \frac{9}{4}$

Reduce to simplest form — Reduce to simplest form — Reduce to simplest form

Decimal Number

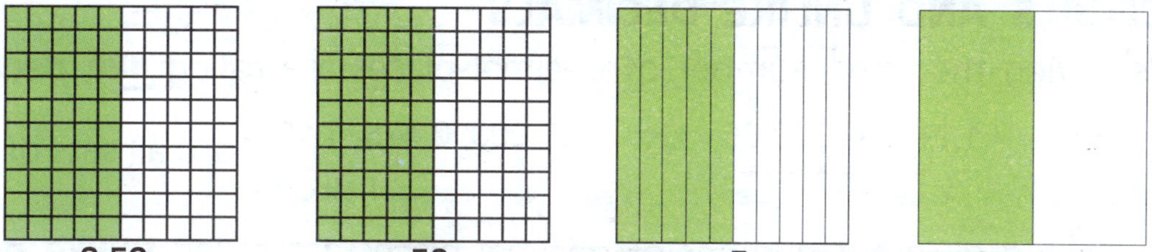

Fraction in simplest form

0.50 $\qquad \frac{50}{100} \qquad \frac{5}{10} \qquad \frac{1}{2}$

Write as decimals :

1. $\frac{9}{100} =$ _____	**2.** $\frac{41}{10} =$ _____	**3.** $\frac{27}{1000} =$ _____
4. $\frac{273}{100} =$ _____	**5.** $\frac{153}{10} =$ _____	**6.** $\frac{1704}{1000} =$ _____
7. $\frac{3}{4} =$ _____	**8.** $\frac{4}{5} =$ _____	**9.** $\frac{7}{20} =$ _____
10. $\frac{3}{8} =$ _____	**11.** $\frac{2}{25} =$ _____	**12.** $\frac{19}{50} =$ _____

Write as fractions in simplest form :

13. 0.8 = _____	**14.** 0.45 = _____	**15.** 1.2 = _____
16. 3.5 = _____	**17.** 2.7 = _____	**18.** 3.4 = _____

EQUIVALENT DECIMALS

Decimals that name the same amount are called equivalent decimals.

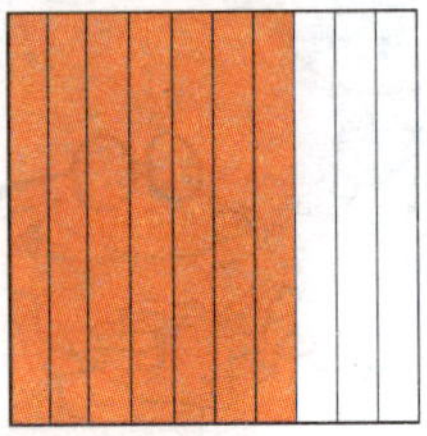

7 tenths = 0.7

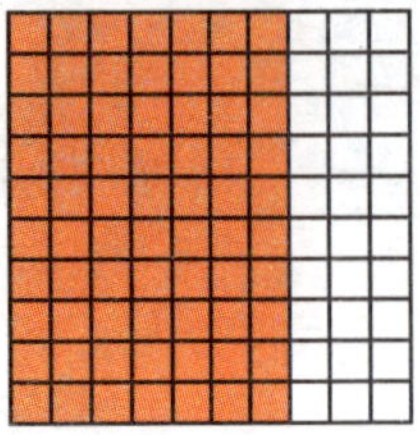

7 hundredths = 0.70

As can be seen, these two pictures name the same amount, so 0.7 = 0.70

7 tenths = 70 hundredths.

$$\frac{7}{10} = \frac{70}{100}$$

Thus, 0.7 = 0.70 = 0.700 = 0.7000 = ...

This shows that, **writing zeros at the right-hand end of a decimal number does not change its value.**

1.8 = 1.80 = 1.800 ; 45.21 = 45.210 = 45.2100 etc.

LIKE DECIMALS AND UNLIKE DECIMALS

Decimals having the same number of decimal places are called **like decimals**.

2.5, 0.8, 9.1 (1 decimal place) ; 1.21, 35.83, 0.06 (2 decimal places) ; 135.008, 0.475, 1.105 (3 decimal places) are all sets of like decimals.

Decimals having different number of decimal places are called **unlike decimals.**

1.2, 4.03, 0.895 are all unlike decimals.

Unlike decimals can be converted to like decimals by finding their equivalent decimals.

Example : Convert 4.8, 2.25 and 0.893 to like decimals.

Solution. The greatest number of decimal places is 3, so we convert all of them to equivalent decimals with 3 decimal places.

4.8 → 4.800, 2.25 → 2.250, 0.893.

Thus, 4.8, 2.25, 0.893 when converted to like decimals become 4.800, 2.250, 0.893.

Convert these unlike decimals to like decimals:

1. 6.8, 7.83, 12.040 ____________________
2. 0.9, 12.825, 1.5, 6.83 ____________________
3. 6.5, 6.08, 60.93, 61.405 ____________________

COMPARING AND ORDERING DECIMALS

To compare decimal numbers:

1. Convert to like decimals.
2. First compare the whole number parts. The number with the greater whole number part is greater.

3. If the whole number parts are the same, compare the tenths digits. The decimal number having greater tenths digit names a greater number.
4. If the tenths digits are the same, compare the hundredths digits and so on.

Example : Compare 4.712 and 4.73.

Solution. The numbers are 4.7 1 2
4.7 3 0

Comparing the digits from the left, you see that the hundredths digits differ.

Since 1 hundredths $<$ 3 hundredths, therefore $4.712 < 4.730$.

Example : Arrange 12.113, 12.114, 21.311 in decreasing order.

Solution. Comparing the whole number parts, you see that 21.311 is the greatest.

Now compare 12.1 1 3
12.1 1 4

The whole number parts being the same, we start comparing from the tenths digit. The thousandths digits differ. Since 3 thousandths $<$ 4 thousandths, so $12.114 > 12.113$.

Thus, the given numbers in decreasing order are: 21.311, 12.114, 12.113.

Skill Practice — **Class Work**

1. Compare by writing $>$, $<$ or $=$ in ◯

(a) 5.62 ◯ 5.26 (b) 53.4 ◯ 53.40 (c) 4.153 ◯ 4.2

(d) 0.008 ◯ 0.0008 (e) 0.7007 ◯ 10.1 (f) 0.6 ◯ 0.006

(g) 9.9 ◯ 10.1 (h) 4.625 ◯ 4.713 (i) 0.09 ◯ 0.11

Home Work

EXERCISE 26

1. Write in increasing order :

(a) 5.13, 5.31, 5.3 (b) 0.72, 7, 7.2 (c) 10.002, 0.103, 1.03

(d) 60.060, 60.000, 60.60 (e) 6.303, 6.033, 5.330, 5.331

(f) 0.009, 0.901, 0.109, 0.910 (g) 0.343, 0.434, 0.334, 0.443

2. Arrange these numbers in order, largest to smallest

15.69, 15.49, 15.7, 15.87, 16.02, 15.78, 15.99, 15.1

3. Labels showing the distances of the various parts of the given plant from the ground have blown off. Which label goes in which space?

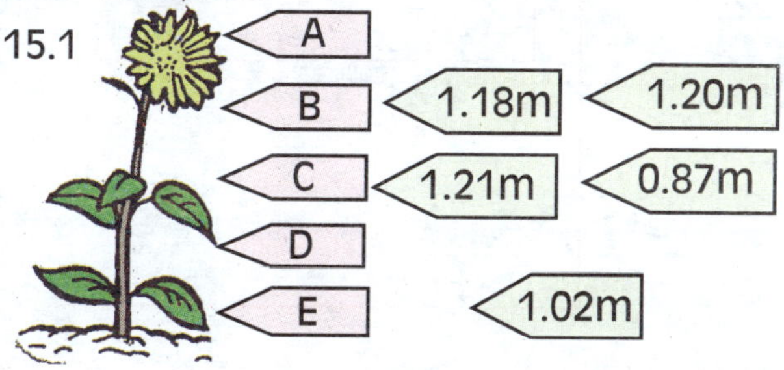

4. The land area of various continents in terms of the land area of the earth are as under: List the continents in the decreasing order of their land areas.

   Africa – 0.205, North America – 0.148, Europe – 0.157, Asia – 0.214, Antartica – 0.097, South America – 0.121.

5. In 2004 Olympic games, the time taken for 100 m dash by some of the atheletes were: Justin Gatlin 9.855s, Asafa Powell 9.945s, Shaw Garford 9.895s, Francis Obikwch is 9.865s. Arrange the names in increasing order of time taken. Who won the race ?

6. **State whether the following statements are true or false.**

   (a) 11.45 > 11.42 **(b)** 9.7 < 9.69 **(c)** 0.252 < 0.343

   (d) 3.305 > 3.503 **(e)** 11.999 > 12 **(f)** 2.35 > 3.05

ADDITION AND SUBTRACTION OF DECIMALS

Follow the following steps to carry out:

1. *Change to like decimals.*
2. *Line up the decimal points, i.e., the addends are placed one below the other such that the tens digit is below tens, ones is below ones, decimal points below the decimal point, tenths below the tenths and so on.*
3. *Add as in case of whole numbers. Carry wherever necessary.*
4. *Place the decimal point in the sum directly below the decimal point in the addends.*

Example 1. Add: 5.83, 680.9 and 74.649.

Solution. Converting to like decimals:

1	1	2				
		5	.	8	3	0
6	8	0	.	9	0	0
	7	4	.	6	4	9
7	6	1	.	3	7	9

Line up the decimals

Add the thousandths: 0 + 0 + 9 = 9 thousandths

Add the hundredths: 3 + 0 + 4 = 7 hundredths

Add the tenths : 8 + 9 + 6 = 23 tenths

23 tenths = 2 ones + 3 tenths

write 3 in the tenths place

Carry over 2 to ones place

Add the ones: 2 + 5 + 0 + 4 = 11 ones

11 ones = 1 tens + 1 ones.

write 1 in the ones place.

Carry over 1 to tens place.

Add the tens : 1+ 8 + 7 = 16 tens

16 tens = 1 hundreds + 6 tens

Write 6 in the tens place.

Carry over 1 to hundreds place

Add the hundreds : 1 + 6 = 7 hundreds

Example 2 : Add : 9.254, 3.45, 8, 17.8.

Solution. Convert to like decimals : 9.254, 3.45 → 3.450, 8 → 8.000, 17.8 → 17.800

Now, line up the decimals and add :

```
  2 1 1
   9.254
   3.450
   8.000
+ 17.800
--------
  38.504
--------
```

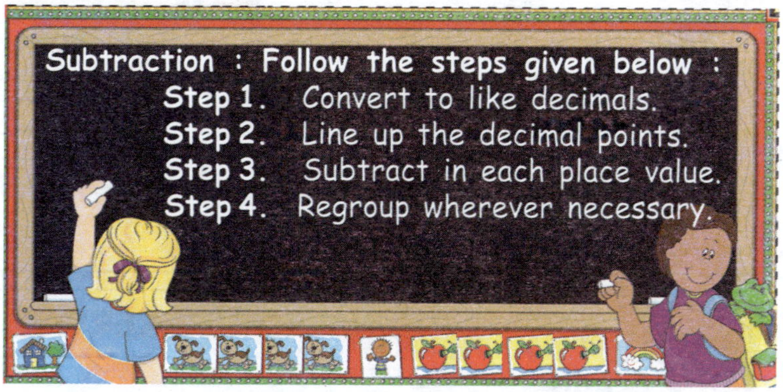

A whole number can be expressed as decimal number as : 8 = 8.0 or 8.00

Example 3: Subtract 7.385 from 16.03

Solution. Convert to like decimals : 7.385, 16.03 → 16.030 Line up the decimal points.

```
       2 10           9 12           15 9 12
                    5  10 2 10       5 10 2 10
  16.0 3 0        1 6 . 0 3 0      1 6 . 0 3 0
-  7.3 8 5       - 7 . 3 8 5      - 7 . 3 8 5
  --------        -----------      -----------
         5               4 5          8 . 6 4 5
```

Subtract the thousandths :
You cannot subtract 5 thousandths from 0 thousandths, so borrow 1 from hundredths.
1 hundredths = 10 thousandths

3 hundredths → 2 hundredths
0 thousandths → 10 thousandths

Now **subtract thousandths**,
10 − 5 = 5 thousandths

Subtract the hundredths :
You cannot subtract 8 hundredths from 2 hundredths, so borrow 1 from Tenths.
There are no tenths, so use 1 ones

6 ones → 5 ones
0 tenths → 10 tenths → 9 tenths
2 hundredths → 12 hundredths

1 = 10 tenths = 9 tenths + 10 hundredths

Now **subtract the hundredths:**
12 − 8 = 4 hundredths.

Subtract the tenths:
9 − 3 = 6 tenths

Subtract the ones:
Since 5 < 7, so borrow 1 from tens.

5 ones → 15 ones
1 tens → 0 tens

Now subtract ones:
15 − 7 = 8 ones.

Example 4: Subtract 14.256 from 27.83.

Solution.

```
       12
      7 2 10
  2 7 . 8 3 0
- 1 4 . 2 5 6
-------------
  1 3 . 5 7 4
-------------
```

Skill Practice

Class Work

Add:

1. $\begin{array}{r} 6.2 \\ +\ 14.858 \\ +\ 25.34 \\ \hline \end{array}$

2. $\begin{array}{r} 0.525 \\ +\ 8.01 \\ +\ 125.3 \\ \hline \end{array}$

Subtract:

3. $\begin{array}{r} 12.83 \\ -\ 6.954 \\ \hline \end{array}$

4. $\begin{array}{r} 80.042 \\ -\ 25.34 \\ \hline \end{array}$

Home Work

EXERCISE 27

Add:

1. 12.67, 4.34
2. 26.83, 18.45
3. 5.357, 11.303, 27.152
4. 425.93, 85.65, 12.96
5. 83.237, 12.125, 37.489, 52.675
6. 6.9, 34.05, 9.468
7. 12.08, 4, 9.1, 8.251
8. 164.3, 57.8, 240.03, 82, 179.9
9. 35.002, 8.2, 19.5, 2.783
10. 46.4, 19.83, 12.054, 24.43
11. 175.8, 200.03, 412.358, 18.9
12. 0.017, 0.1234, 0.9, 0.819, 0.05
13. Raju buys a game for Rs 135.45. He already has one that costs Rs 85.58. How much has he spent on both the games ?

14. A tailor needs 52.18 cm of white lace for Richa's skirt and 75.85 cm of the same lace for the top. How much lace does he need in all ?
15. Dev saved these amounts of money from his pocket money four months in a row: Rs 76.82, Rs 45.62, Rs 63.25 and Rs 38.18. How much did he save in all?
16. Find the perimeter of the adjoining figure ?

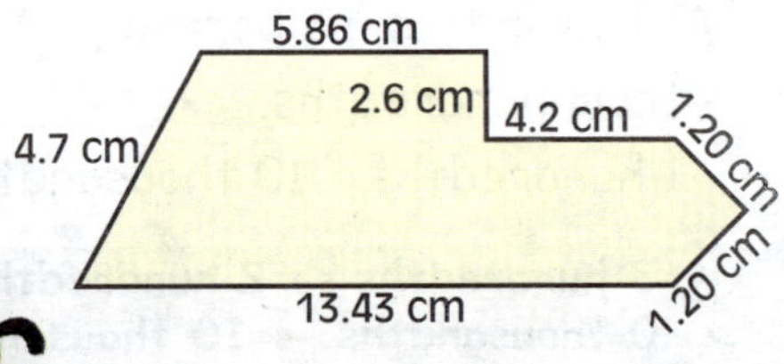

Home Work

EXERCISE 28

Subtract:

1. 6.253 – 1.678
2. 19.250 – 5.135
3. 11.2 – 3.243
4. 312.8 – 59.98
5. 160 – 39.284
6. 12.43 – 8.425
7. A roll of fabric has 85.63 metres on it. If 43.76 metres are sold, how much is left?
8. Sharon buys two bottles of perfume; one contains 52.5 ml and the other contains 156.54 ml. How much more perfume is in the larger of the two bottles ?

9. Sandhya and Mridula together saved Rs 235.75. Sandhya saved Rs 115.83. How much did Mridula save ?

10. What should be added to 1.483 to get 5.25 ?
11. I purchased some books worth Rs 89.37 and gave the shopkeeper a 100-rupee note. How much change did I receive back ?
12. Sonu runs 153.35 m in 50 seconds and John runs 92.87m in the same time. How much farther does Sonu run than John ?

Simplify :

13. Rs. 3.598 + 1.72 + 4.037 – 5.109

14. 639.209 + 17.813 + 6.275 – 3.468

MENTAL MATHS–6

1. Write in figures : **(a)** twelve point seven **(b)** three hundred seventy point five two.
2. What is the mixed number equivalent in its simplest form of the decimal number 5.6.
3. What is one hundredth of the place value of 3 in 5 3961 ?
4. What number is halfway between seven point four and nine point four on the number line?

5. Write in decimal form **(a)** $4\frac{7}{10}$ **(b)** $9\frac{35}{100}$ **(c)** $84\ \frac{1}{100}$.
6. Which is greater 0.25 or $\frac{1}{3}$?
7. What is the sum of 99.999 + 0.001.
8. Express seven thousandth as a numeral.
9. What is the difference 50 – 0.01 ? Express in words.

KNOW YOUR NUMBERS

THE NUMBER FOUR

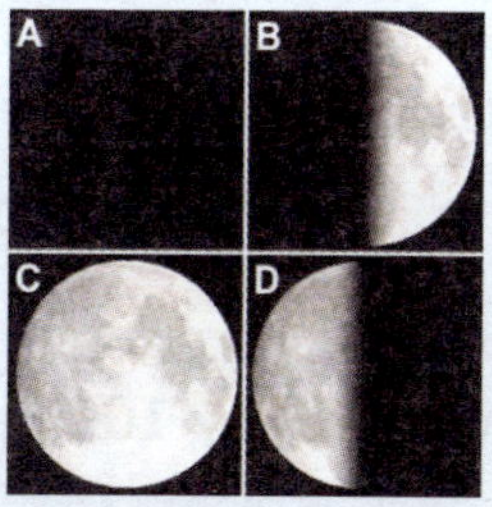

The word *four* has four letters. In the English language there is no other number whose number of letters is equal to its value. Many things are arranged in fours. There are four wings on a bee, and four phases of the moon. The four seasons are spring, summer, autumn and winter.

Quad – means four. A quadrilateral is a figure with four sides. A quadruped is a four-footed animal like a bear, or almost any animal.

Only four colours are all that are needed to show adjacent countries on a map. No one has ever found a map that needs more than four colours.

QUADRILATERAL

MULTIPLICATION AND DIVISION OF DECIMALS

MULTIPLICATION BY 10, 100, 1000

Study the following examples.

1. $0.4 \times 10 = \frac{4}{10} \times 10 = 4$

Ones	Tenths
	4
4	

× 10

4 tenths × 10 = 4 ones

2. $6.3 \times 10 = \frac{63}{10} \times 10 = 63$

Tens	Ones	Tenths
	6	3
6	3	

× 10

6 ones 3 tenths × 10 = 6 tens 3 ones

Multiplying a decimal by 10 moves the decimal point 1 place to right. 4.75 × 10 = 47.5

3. $6.43 \times 100 = \frac{643}{100} \times 100 = 643$

Hundreds	Tens	Ones	Tenths	Hundredths
		6	4	3
6	4	3		

× 100

6 ones 4 tenths 3 hundredths × 100 = 6 hundreds 4 tens 3 ones

4. $0.8 \times 100 = \frac{8}{10} \times 100 = 8 \times 10 = 80$

Tens	Ones	Tenths
		8
8		

× 100

8 tenths × 100 = 8 tens

Multiplying a decimal by 100 moves the decimal point 2 places to right.

5. $0.005 \times 1000 = \frac{5}{1000} \times 1000 = 5$

Ones	Tenths	Hundreths	Thousandths
			5
5			

5 thousandths × 1000 = 5 ones

6. $2.1 \times 1000 = \frac{21}{10} \times 1000 = 2100$

7. $6.93 \times 1000 = 6930$

Multiplying a decimal by 1000 moves the decimal point 3 places to the right.
18.9 × 1000 = 18900

Note: A number without a decimal point has the decimal point always to the extreme right. 584 means 584.0.

Class Work

1. Multiply.

Decimal	× 10	× 100	× 1000
2.7			
3.06			
0.519			
0.23			
18.25			
3.007			
16.73			
148.954			
24.06			
100.035			

Home Work

EXERCISE 29

I. Write the missing numbers in the boxes:

1. 0.07 × ☐ = 0.7 **2.** 4.354 × ☐ = 435.4

3. 67.2 × ☐ = 6720 **4.** 15.54 × ☐ = 15540

5. 4.625 × ☐ = 4625 **6.** 2.3 × ☐ = 2300

7. 0.08 × ☐ = 8 **8.** 6.3 × ☐ = 63

II. Multiply each number on the left by 10 and draw a line to join it to the correct answer on the right.

1.8	42.5
4.25	5193.5
19.483	18
519.35	8.5
0.85	194.83

III. Write the one step operation you would use to change these numbers.

1. 4.5 to 45	× 10	**2.** 6.25 to 62.5	
3. 4.03 to 40.3		**4.** 0.01 to 1	
5. 6.7 to 670		**6.** 12.875 to 12875	
7. 4.2 to 4200		**8.** 0.85 to 850	

MULTIPLICATION BY A COUNTING NUMBER

Example 1. Manav bought 7 stickers at Rs. 3.56 each. How much did he pay for the stickers ?

Solution. The amount paid by Manav = Rs 3.56 × 7

$$= \text{Rs}\,\frac{356}{100} \times 7 = \text{Rs}\,\frac{356 \times 7}{100} = \text{Rs}\,\frac{2492}{100} = \text{Rs}\,24.92.$$

The work can also be set as under :

Step 1. Multiply as you would multiply counting numbers.

```
  356
×   7
─────
 2492
```

Step 2. Place the decimal point in the product. The number of decimal places in the product is equal to the total number of decimal places of the factors.

```
  3.56   ← 2 decimal places
×    7   ← 0 decimal places
─────
 24.92   ← 2 decimal places
             (0 + 2)
```

Example 2. Multiply 3.098 by 47

Solution.

```
   3.098   ← 3 decimal places
  ×   47   ← 0 decimal places
  ──────
  21 686
  123 92
  ──────
 145.606   ← 3 decimal places → (0 + 3)
```

Multiply

1. 6.23 × 2	**2.** 12.853 × 5	**3.** 4.16 × 12	**4.** 6.095 × 45

EXERCISE 30

Multiply

1. 0.53 by 2 **2.** 0.85 by 5 **3.** 0.08 by 3 **4.** 0.9 by 4

5. 4.67 by 3 **6.** 7.42 by 5 **7.** 5.93 by 8 **8.** 28.07 by 6

9. 85.12 by 7 **10.** 8.25 by 16 **11.** 19.83 by 25 **12.** 25.8 by 11

13. 615.3 by 78 **14.** 25.035 by 15 **15.** 62.131 $\times$ 42 **16.** 723.04 $\times$ 89

17. If 5867 $\times$ 6 = 35202, then find the product of each of the following without actually performing multiplication.

(a) 5.867 $\times$ 6 (b) 58.67 $\times$ 6 (c) 586.7 $\times$ 6 (d) 0.5867 $\times$ 6

18. If 1 litre of petrol costs Rs 45.34, find the cost of 25 litres.

19. The weight of a pineapple is 1.46 kg. What is the weight of 18 such pineapples.

20. The cost of a pair of shoes is Rs 723.95. What will be the cost of 38 such pairs of shoes ?

MULTIPLICATION OF A DECIMAL NUMBER BY A DECIMAL NUMBER

Look at this example:

$$18.9 \times 0.4 = \frac{189}{10} \times \frac{4}{10} = \frac{189 \times 4}{100} = \frac{756}{100} = \mathbf{7.56}$$

The above work can also be set as :

189 × 4 756 Multiply as you would multiply counting numbers	18.9 ← 1 decimal place × 0.4 ← 1 decimal place 7.56 ← 2 decimal places (1 + 1) The total number of decimal places in the product is equal to the sum of the decimal places of the factors.

Example 1. Multiply

(a) 0.6 × 0.3 **(b) 0.008 × 0.05**

Solution. (a) 0.6 × 0.3 = **0.18**

(0.6: 1 d.p.; 0.3: 1 d.p.; 0.18: 2 d.p.)

(b) 0.008 × 0.05 = 0.00040 = **0.0004**

(0.008: 3 d.p.; 0.05: 2 d.p.; 0.00040: 5 d.p.)

8 × 5 = 40
The product has 5 decimal places, so add zeros to the left of the product.

Example 2. Multiply 5.48 by 0.17.

Solution.

```
   5.48  ← 2 decimal places
 × 0.17  ← 2 decimal places
   3836
   548
 0.9316  ← 4 decimal places (2 + 2)
```

Example 3. Multiply 6.013 by 2.95

Solution.

```
      6.013   ← 3 decimal places
    ×  2.95   ← 2 decimal places
      30065
     54117
   12026
  17.73835    ← 5 decimal places (3 + 2)
```

Example 4. Multiply

(a) 4.6 by 20 (b) 28.73 by 600

Solution. (a) $4.6 \times 20 = 4.6 \times 2 \times 10 = 9.2 \times 10 =$ **92.**

(b) $28.73 \times 600 = 28.73 \times 6 \times 100 = 172.38 \times 100 =$ **17238.**

Multiply :

1. 0.7×0.2	2. 4.04×2.8	3. 12.6×50	4. 0.008×300

EXERCISE 31

Multiply:

1. 0.3×0.8
2. 0.9×0.8
3. 0.007×0.6
4. 1.13×0.7
5. 21.05×0.8
6. 19.2×0.003
7. 0.004×1.3
8. 1.25×15.3
9. 83.7×5.9
10. 89.65×6.8
11. 1.57×0.018
12. 5.78×3.5
13. 2.08×0.15
14. 1.035×16.3
15. 7.79×1.9
16. 23.02×49.3
17. 25.3×16.7
18. $0.2 \times 0.2 \times 0.2$
19. $0.3 \times 0.4 \times 0.02$
20. $0.03 \times 0.03 \times 0.03$
21. $0.02 \times 0.03 \times 0.0001$
22. $0.1 \times 0.02 \times 0.003 \times 0.004$
23. 39.8×400
24. 65.12×7000
25. **If $59 \times 8.7 = 513.3$, find the product in each of the following without multiplication.**

(a) 5.9×8.7 (b) 5.9×87 (c) 0.59×87 (d) 0.59×8.7
(e) 0.59×0.87 (f) 5.9×0.87 (g) 0.059×0.0087.

26. **Multiply 7.349 by 9. Hence without multiplying again, write down the product when 7.349 is multiplied by:**

(a) 90 (b) 900 (c) 9000 (d) 0.9 (e) 0.09 (f) 0.009

27. A jeep can travel 6.8 km in 1 litre of petrol. How far can it travel in 20.5 litres, provided it goes at the same speed.

28. Alok weighs 50.87 kg. His father is 1.5 times heavier than he is. Calculate his father's weight.

DIVISION BY 10, 100, 1000

Study the following examples :

1. $3 \div 10 = 3 \times \frac{1}{10} = \frac{3}{10} = 0.3.$

2. $0.4 \div 10 = \frac{4}{10} \div 10 = \frac{4}{10} \times \frac{1}{10} = \frac{4}{100} = 0.04.$

Ones	Tenths
3	
	3

÷ 10

3 ones ÷ 10 = 3 tenths

Ones	Tenths	Hundredths
	4	
		4

÷ 10

4 tenths ÷ 10 = 4 hundredths

Dividing a decimal by 10 moves the decimal point one place to the left.
e.g. **0.08 ÷ 10 = 0.008**

3. $6 \div 100 = 6 \times \frac{1}{100} = \frac{6}{100} = 0.06$

Ones	Tenths	Hundredths
6		
		6

÷ 100

6 ones ÷ 100 = 6 hundreths

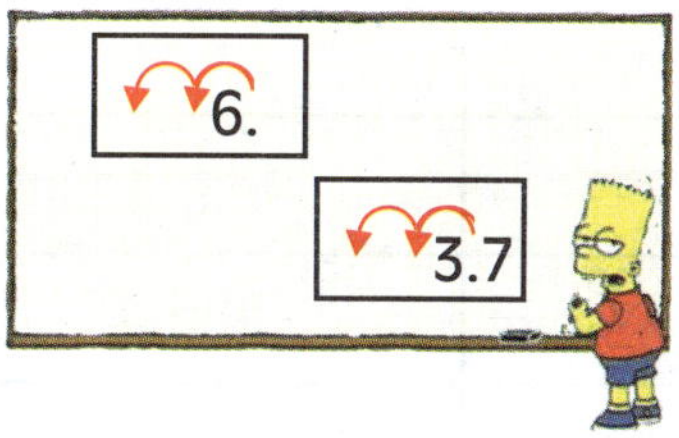

4. $3.7 \div 100 = \frac{37}{10} \div 100 = \frac{37}{10} \times \frac{1}{100} = \frac{37}{1000} = 0.037$

Ones	Tenths	Hundredths	Thousandths
3	7		
		3	7

÷ 100

3 ones 7tenths ÷ 100 = 3 hundredths 7 thousandths

Dividing a decimal by 100 moves the decimal point 2 places to the left
e.g. **12.57 ÷ 100 = 0.1257**

5. $8 \div 1000 = 8 \times \frac{1}{1000} = \frac{8}{1000} = 0.008$ Move 3 places to the left

Ones	Tenths	Hundredths	Thousandths
8			
			8

÷ 1000

8 ones ÷ 1000 = 8 thousandths

6. $29 \div 1000 = 29 \times \frac{1}{1000} = \frac{29}{1000} = 0.029$

7. $345 \div 1000 = 345 \times \frac{1}{1000} = \frac{345}{1000} = 0.345$

Dividing a decimal by 1000 moves the decimal point 3 places to the left.
e.g. 6.7 ÷ 1000 = 0.0067.

Fill in the blanks:

Number	÷ 10	÷ 100	÷ 1000
5			
18			
389			
0.8			
0.92			
4.532			
87.1			
93.85			
105.293			
60.07			

EXERCISE 32

I. Write the missing numbers in the boxes:

1. 8 ÷ ☐ = 0.8

2. 2.58 ÷ ☐ = 0.258

3. 12.89 ÷ ☐ = 0.1289

4. 0.4 ÷ ☐ = 0.04

5. 62 ÷ ☐ = 0.62

6. 111.2 ÷ ☐ = 0.1112

7. 8.9 ÷ ☐ = .0089

8. 123.4 ÷ ☐ = 1.234

II. Divide each number on the left by 100 and draw a line to join it to the correct answer on the right.

Join

1. 2.4	0.0008
2. 16	4.215
3. 421.5	0.024
4. 645	0.16
5. 0.08	6.45

III. Write the one step operation you would use to change these numbers.

1. 24 to 0.24 $\div 100$

2. 6.2 to 0.62 ☐

3. 87.53 to 8.753 ☐

4. 4 to 0.004 ☐

5. 9.73 to 0.0973 ☐

6. 89 to 0.089 ☐

7. 60 to 0.6 ☐

8. 70.6 to 0.706 ☐

Division by a Counting Number

Example 1. 14.28 ÷ 4.

Compare the two workings shown below.

Step 1. $4\overline{)14.28}$ ← dividend

↑ divisor

1. $\dfrac{14.28}{4}$

Step 2. $4\overline{)14.28}$ (decimal point moved two places to the right)

2. $\dfrac{14.28}{4} \times 100 = \dfrac{1428}{4}$

Step 3. $\begin{array}{r} 357 \\ 4\overline{)1428} \end{array}$

3. $\dfrac{1428}{4} = 357$

Step 4. $\begin{array}{r} 3.57 \\ 4\overline{)14.28} \end{array}$

4. $357 \div 100 = 3.57$

ACTUAL WORKING

$$\begin{array}{r} 3.57 \\ 4\overline{)14.28} \\ 12 \\ \hline 2\,2 \\ -\,2\,0 \\ \hline 28 \\ -\,28 \\ \hline 0 \end{array}$$

Rule : 1. Divide as you would divide counting numbers.

2. Place the decimal point in the quotient **directly above** the decimal point in the dividend.

Short Form

$$\begin{array}{r} 4)14.28 \\ \hline 3.57 \end{array}$$

We give below some more examples.

Example 2. 5.04 ÷ 8

Solution.

```
     0.63
  8)5.04
   -4 8↓
      24
     -24
       0
```

8 > 5, so we put a zero in the quotient.

Ans. 0.63

Check
$8 \times 0.63 = 5.04$

Example 3. 1414.332 ÷ 17

Solution.

```
       83.196
  17)1414.332
    - 136↓
        54
      - 51
        33
      - 17
        163
      - 153
         102
       - 102
           0
```

Ans. 83.196

Example 4. (a) 132.48 ÷ 23

```
       5.76
  23)132.48
    -115
      174
     -161
      138
     -138
        0
```

Ans. 5.76

Check
$23 \times 5.76 = 132.4$

(b) 0.22533 ÷ 259

```
       0.00087
  259)0.22533
     - 2072↓
       1813
     - 1813
          0
```

Ans. 0.00087

If the last remainder is not zero, write as many zeros on the right of the dividend as are necessary to make the last remainder zero. This is illustrated in the following examples.

Example 5. 24.66 ÷ 12

```
       2.055
  12)24.66,0  ← Write one zero
     24
      66
     -60
      60   ← The remainder is not zero
     -60
       0
```

$\therefore$ 24.66 ÷ 12 = **2.055.**

Example 6. 9.2 ÷ 16

```
       0.575
  16)9.2,0,0  ← Write two zeros
    - 80
      120     ← Remainder not zero
    - 112
        80    ← Remainder not zero
      - 80
         0
```

$\therefore$ 9.2 ÷ 16 = **0.575.**

When the divisor is a multiple of 10 or 100 or 1000, etc., then the process of division can be shortened as shown below.

Example 7. **94.24 ÷ 80**

```
       1.178
 80)94.24,0   ← Write one zero
    80
    ----
    14 2
   - 8 0
    ----
     6 24
   - 5 60
    -----
      640
    - 640
    -----
        0
```

> You can divide 9.424 mentally by 8

Short Method

$$= \frac{94.24}{80} = \frac{94.24}{10\times 8} = \frac{9.424}{8} = 1.178$$

Note: $94.24 \div 10 = 9.424$

Divide :

1. $6\overline{)20.22}$
2. $8\overline{)42.56}$
3. $3\overline{)2.67}$
4. 18.78 ÷ 12
5. $15\overline{)35.55}$

EXERCISE 33

Divide and check your answer using multiplication:

1. $8\overline{)37.04}$	**2.** $7\overline{)41.3}$	**3.** $3\overline{)17.1}$	**4.** $2\overline{)51.16}$
5. $8\overline{)34.4}$	**6.** $6\overline{)3.42}$	**7.** $9\overline{)0.288}$	**8.** $8\overline{)0.0344}$
9. $8\overline{)0.00344}$	**10.** $7\overline{)14.805}$	**11.** $11\overline{)21.01}$	**12.** $23\overline{)151.8}$
13. $29\overline{)5.887}$	**14.** $84\overline{)109.2}$	**15.** $14\overline{)7.868}$	**16.** $5\overline{)22.6}$
17. $4\overline{)8.7}$	**18.** $8\overline{)122.6}$	**19.** $15\overline{)88.11}$	**20.** $35\overline{)262.85}$

Divide

21. 18 ÷ 20	**22.** 6.3 ÷ 70	**23.** 0.82 ÷ 40	**24.** 120 ÷ 400
25. 84 ÷ 400	**26.** 6.4 ÷ 200	**27.** 21 ÷ 3000	**28.** 120 ÷ 6000
29. 343 ÷ 7000	**30.** 724.36 ÷ 80	**31.** 540.96 ÷ 400	**32.** 780.6 ÷ 5000

33. 8.75 litres of milk is distributed equally among 35 students. How much milk will each student get ?

34. The cost of 15 chocolates is Rs 143.10. What is the cost of 1 chocolate ?

DIVISION OF A DECIMAL NUMBER BY A DECIMAL NUMBER

You have learnt how to divide a decimal number by a whole number. To divide a decimal number by a decimal number, we make the divisor a whole number. Look at the following example.

$5.04 \div 0.8 = ?$

1. **Multiply the divisor by a multiple of 10 to make it a whole number.**
   Thus, $0.8 \times 10 = 8$
2. **Multiply the dividend by the same multiple of 10.**
   Thus, $5.04 \times 10 = 50.4$

$$\begin{array}{r} 6.3 \\ 8\overline{)50.4} \\ -48 \\ \hline 2\,4 \\ -\,2\,4 \\ \hline 0 \end{array}$$

Now, divide as usual $5.04 \div 8 = \frac{5.04}{8} = 6.3.$

The work can be set as :

$$5.04 \div 8 = \frac{5.04}{0.8} = \frac{5.04 \times \boxed{10}}{0.8 \times \boxed{10}}$$

$$= \frac{50.4}{8} = 6.3$$

> Multiplying both the divisor and the dividend by 10 means multiplication by $\frac{10}{10} = 1$.

Example 1. Divide (a) 6.265 ÷ 0.07 (b) 1.5 ÷ 0.003

Solution. (a) $6.265 \div 0.07 = \frac{6.265}{0.07} = \frac{6.265 \times \boxed{100}}{0.07 \times \boxed{100}} = \frac{626.5}{7} = \mathbf{89.5}.$

(b) $1.5 \div 0.003 = \frac{1.5}{0.003} = \frac{1.5 \times \boxed{1000}}{0.003 \times \boxed{1000}} = \frac{1500}{3} = \mathbf{500}.$

Example 2. Divide 2.625 by 2.1

> **Think !**
> 1 decimal place in **divisor**, so shift decimal point both in divisor and dividend 1 place to the right.

Solution. $2.625 \div 2.1 = \frac{2.625}{2.1} = \frac{2.625}{2.1} = \frac{26.25}{21}$ (1 place, 1 place)

$$\begin{array}{r} 1.2\,5 \\ 21\overline{)26.2\,5} \\ -21 \\ \hline 5\,2 \\ -\,4\,2 \\ \hline 1\,0\,5 \\ -1\,0\,5 \\ \hline 0 \end{array}$$

$\therefore\ 2.625 \div 2.1 = \mathbf{1.25.}$

Example 3: **20.169 ÷ 0.83**

Solution.

$$= \frac{20.169}{0.83} \quad \text{(2 places)}$$

$$= \frac{2016.9}{83}$$

$$= \mathbf{24.3.}$$

```
        2 4.3
  83)2 0 1 6.9
    -1 6 6
       3 5 6
      -3 3 2
         2 4 9
        -2 4 9
             0
```

Example 4. **Divide 3.154 by 1.25.**

Solution.

1. 25)3.154 ⟶ 125)315.4 ⟶

```
        2.5232
  125)315.4000
     - 250
        65 4
      - 62 5
         2 90
       - 2 50
           400
         - 375
           250
         - 250
             0
```

Shift the decimal point 2 places to the right both in the divisor and the dividend.

3 zeros have been written to the right of 4. These zeros do not change the value of the number.

$\therefore$ 3.154 ÷ 1.25 = **2.5232.**

Example 5. **Divide (a) 30 ÷ 0.60** **(b) 91 ÷ 0.875**

Solution.

(a) $30 \div 0.60 = \dfrac{30}{0.60} = \dfrac{30 \times 100}{0.60 \times 100} = \dfrac{3000}{60} = \mathbf{50}.$

(b) $91 \div 0.875 = \dfrac{91}{0.875} = \dfrac{91 \times 1000}{0.875 \times 1000} = \dfrac{91000}{875} = \mathbf{104}.$

```
          104
  875)91000
     -875
       3500
     - 3500
          0
```

Divide :

1. 2.4 ÷ 0.8

2. 8.1 ÷ 0.3

3. 2.4 ÷ 0.02

4. 25 ÷ 0.5.

EXERCISE 34

Divide:

1. 3.5 ÷ 0.5 **2.** 3.2 ÷ 0.8 **3.** 7.2 ÷ 0.9 **4.** 32.4 ÷ 0.6

5. 32.4 ÷ 0.06 **6.** 1.92 ÷ 1.2 **7.** 0.819 ÷ 1.3 **8.** 36 ÷ 0.06

9. 4.20 ÷ 1.05 **10.** 0.625 ÷ 0.005 **11.** 8.01 ÷ 0.009 **12.** 78.75 ÷ 0.125

13. 13.0563 ÷ 0.0009 **14.** 66.468 ÷ 76.4 **15.** 84.826 ÷ 5.81 **16.** 7.55 ÷ 0.0005

17. 25 ÷ 0.50 **18.** 18 ÷ 0.45 **19.** 31 ÷ 0.775 **20.** 64 ÷ 0.004

21. A pamphlet containing 54 sheets of paper was 1.458 cm thick. What was the thickness of each sheet of paper?

22. Ravi worked 26 hours last week and earned Rs 63.18. What was his hourly rate of wage?

23. How many pieces of rope, each 1.2 m long can be cut from a rope which measures 45.6 m?

24. A pile of books is 54.4 cm high. If each book is 3.2 cm thick, how many books are there ?

25. A dress requires 2.9 m of cloth. How many such dresses can be made from a piece of cloth measuring 98.6 m?

CHANGING COMMON FRACTIONS OF DECIMAL FRACTIONS

Example 1. Change $\frac{3}{4}$ to a decimal fraction.

Solution. To convert $\frac{3}{4}$ to a decimal, we divide the numerator 3 by the denominator 4 as follows :

```
    0.7 5
 4) 3.0 0
  - 2 8
    ----
      2 0
    - 2 0
    -----
        0
```

Put decimal point after the whole number and add 2 zeros to ensure complete division.

Ans. $\frac{3}{4} = 0.75$.

Example 2. Express $\frac{11}{8}$ as a decimal.

Solution. To convert $\frac{11}{8}$ to a decimal, divide the numerator 11 by the denominator 8.

```
      1.375
  8) 1 1.000
  -    8
     ----
       3 0
     - 2 4
     -----
         60
      -  56
      -----
          4 0
        - 4 0
        -----
            0
```

Put decimal point after the whole number and add 3 zeros to ensure complete division.

Ans. : $\frac{11}{8} = 1.375$.

Skill Practice

Class Work

Change each fraction given below to decimal.

1. $\frac{4}{5}$	**2.** $\frac{3}{8}$	**3.** $\frac{7}{4}$	**4.** $\frac{11}{20}$	**5.** $\frac{85}{16}$

USING DECIMALS IN METRIC MEASURES

LENGTH

cm to m :

You know that 1 m = 100 cm, therefore $1\text{cm} = \frac{1}{100}\text{m} = 0.01\text{m}.$

(a) $8\text{cm} = \frac{8}{100}\text{m} = \mathbf{0.08\,m}$

(b) $27\text{cm} = \frac{27}{100}\text{m} = \mathbf{0.27\,m}$

(c) 12 m 86 cm = **12.86 m**

(d) 47 m 20 cm = **47.20 m or 47.2m**

m to cm

(a) 6.83 m = 6 m + 0.83 m = (6×100) cm + (0.83×100) cm

= 600 cm + 83 cm = **683 cm**

(b) 23.06 m = 23 m + 0.06 m = (23×100) cm + $(0.06\text{ cm} \times 100)$ cm

= 2300 cm + 6 cm = **2306 cm**

m to km :

1 km = 1000 m, therefore, $1\text{m} = \frac{1}{1000}\text{km} = 0.001\text{km}.$

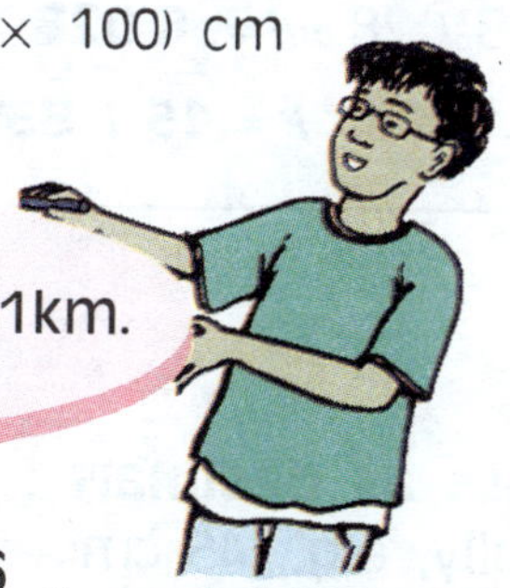

(a) $5\text{m} = \frac{5}{1000}\text{km} = \mathbf{0.005\,km.}$

(b) $26\text{m} = \frac{26}{1000}\text{km} = \mathbf{0.026\,km.}$

(c) $456\text{m} = \frac{456}{1000}\text{km} = \mathbf{0.456\,km.}$

(d) $5\text{km}\,2\text{m} = 5\text{km} + \frac{2}{1000}\text{km} = 5\text{km} + 0.002\text{km} = \mathbf{5.002\,km}.$

km to m :

(a) 25.052 km = 25 km + 0.052 km = (25 × 1000) m + 0.052 × 1000 m

= 25000 m + 52 m = **25052 m**

(b) 44.007 km = 44 km 7 m = 4400 m + 7 m = **44007 m**

(c) 22.625 km = 22 km 625 m = 22000 m + 625 m = **22625 m.**

(d) $123\,\text{km}\,789\,\text{m} = 123\,\text{km} + \frac{789}{1000}\,\text{km} = 123\,\text{km} + 0.789\,\text{km} = \mathbf{123.789\,km}$.

(e) $11\,\text{km}\,35\,\text{m} = 11\,\text{km} + \frac{35}{1000}\,\text{km} = 11\,\text{km} + 0.035\,\text{km} = \mathbf{11.035\,km}$.

WEIGHT : g to kg Since 1 kg = 1000 g, therefore, $1\text{g} = \frac{1}{1000}\text{kg} = 0.001\text{kg}$

Rule : To change g to kg, divide by 1000.

(a) $50\,\text{g} = \frac{50}{1000}\text{kg} = \mathbf{0.050\,kg}$ **(b)** $750\,\text{g} = \frac{750}{1000}\text{kg} = \mathbf{0.750\,kg}$ or 0.75 kg.

(c) 4 kg 255g = **4.255 kg** **(d)** 13 kg 56g = **13.056 kg** **(e)** 24 kg 5 g = **24.005 kg**

(f) 12.280 kg = 12 kg + 0.280 kg = 12 kg + (0.280 × 1000)g

= 12 kg + 280 g = **12 kg 280 g.**

(g) 5.065 kg = **5 kg 65 g** **(h)** 15.008 kg = **15 kg 8 g.**

CAPACITY ***ml* to *l***

Since 1 *l* = 1000 *ml*, therefore, $1ml = \frac{1}{1000}l = 0.001l$

(a) $25\,ml = \frac{25}{1000}l = \mathbf{0.025}\,l$ **(b)** $483\,ml = \frac{483}{1000}l = \mathbf{0.483}\,l$ **(c)** 12 *l* 5 *ml* = **12.005** *l*

(d) 8 *l* 28 *ml* = **8.028** *l* **(e)** 42 *l* 765 *ml* = **42.765** *l* **(f)** 18.006 *l* = **18 *l* 6 *ml***

(g) 15.083 *l* = **15 *l* 83 *ml*** **(h)** 29.425 *l* = **29 *l* 425 *ml*.**

Express as decimals : It should be possible for you to multiply or divide mentally, e.g., 45 cm ÷ 100 = 0.45 m.

1. 45 cm	0.45 m	**2.** 6 cm	m
3. 296 cm	m	**4.** 275 m	km
5. 43 m	km	**6.** 9 m	km

7. 4 m 5 cm	m	8. 19 m 46 cm	m
9. 2 km 525 m	Km	10. 18 km 75 m	km
11. 600 km 5 m	Km	12. 16 m 8 cm	m
13. 89 g	0.089 kg	14. 925 g	kg
15. 3 kg 6 g	kg	16. 21 kg 54 g	kg
17. 14 kg 425 g	kg	18. 1 kg 700 g	kg
19. 425 *ml*	*l*	20. 65 *ml*	*l*
21. 1 *l* 125 *ml*	*l*	22. 5 *l* 5 *ml*	5.005 *l*
23. 10 *l* 25 *ml*	*l*	24. 16 *l* 720 *ml*	*l*

25. Add 2.4 litres and 650 *ml*

Working

Ans.

26. Subtract 4 cm from 4.23 m

Ans.

27. A bag of nails weighs 1.75 kg and a bag of screws weighs 0.235 kg. Find the weight of one bag of nails and 3 bags of screws.

Ans.

28. There are approximately 1.6 km in 1 mile. How many kilometres are there in 2.5 miles.

Ans.

ROUNDING DECIMALS

Rounding off to the nearest whole number (to nearest one)

John weighs 57.8 kg, Maria weighs 42.3 kg and David weighs 68.5 kg. Round off these figures to the nearest whole number. (nearest kg)

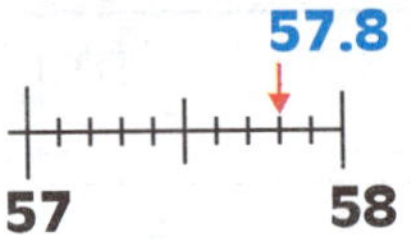

57.8 is nearer to 58 than to 57

John's weight is 58 kg when rounded off to nearest kg.

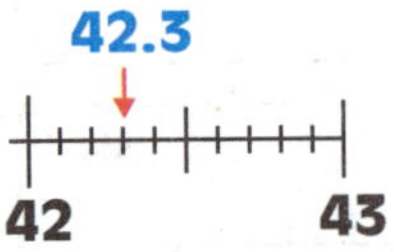

42.3 is closer to 42 than to 43

Maria's weight is 42 kg when rounded off to nearest kg.

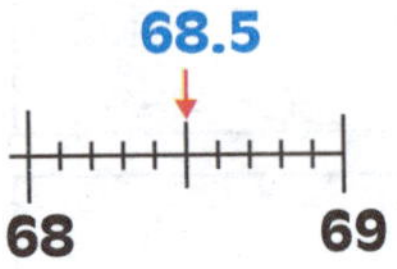

68.5 is half way between 68 and 69 and is rounded off to 69.

David's weight is 69 kg when rounded off to the nearest kg.

ROUNDING OFF TO THE NEAREST TENTH (TO ONE DECIMAL PLACE)

A tree is 6.84 m tall. Round off its height to 1 decimal place (1 dp).

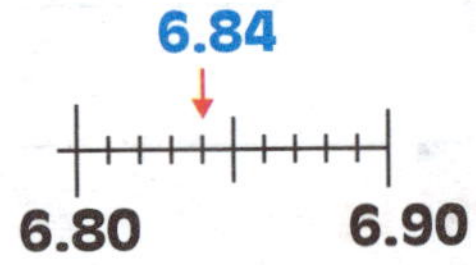

6.84 is closer to 6.80 than to 6.90

It is not always easy to round off a value with the help of number line, so we follow the rule given below :

Working Rule for rounding decimals:

1. Identify the place you want to round to.

   4 . 6(7)5 — ↑ Rounding place

2. Examine the digit to the right of the rounding place digit.

   4 . 6(7)5 — Digit to the right is 5.

3. If that digit is equal to 5 or greater than 5, add 1 to the rounding place digit. If that digit is less than 5, let the rounding place digit remain as such.

   4 . 6(7)5 → 4 . 6 8 (+ 1)

   In both cases, drop all the digits to the right of the rounding place digit.

Example 1. **Round off (a) 19.83 (b) 74.328 (c) 28.37 (d) 91.653 to the nearest tenth.**

Solution. To round off to the nearest tenth, we examine the digit to the right of the digit at the tenth's place *i.e.*, at the hundredth's place.

Examine

	Number	19.83	74.328	28.37	91.653
	Check	3 < 5	2 < 5	7 > 5	5 = 5
Hence	Rounded off to	19.8	74.3	28.4	91.7

Example 2. Round off (a) 23.594 (b) 17.3829 (c) 47.826 (d) 89.7056 to the nearest hundredth.

Solution. To round off to the nearest hundredth, we examine the digit to the right of the digit at the hundredth's place, *i.e.*, at the thousandths place.

Examine

	Number	23.594	17.3829	47.826	89.7056
	Check	4 < 5	2 < 5	6 > 5	5 = 5
Hence,	Rounded off to	23.59	17.38	47.83	89.71

Example 3. Express 23.8254 to (a) the nearest whole number (b) 1 decimal place (c) the nearest hundredths (d) to three decimal places.

Solution. **(a)** 23.8254 = 24, rounded off to the nearest whole number.

(b) 23.8254 = 23.8, rounded off to 1 decimal place.

(c) 23.8254 = 23.83, rounded off to the nearest hundredths.

(d) 23.8254 = 23.825, rounded off to 3 decimal places

Example 4. Evaluate (a) 4.83 + 19.465 correct to 1 decimal place

(b) 18 ÷ 0.8 to the nearest whole number

Solution. (a) 4.83 + 19.465

= 24.295 = **24.3** rounded off to 1 decimal place

$$\begin{array}{r} 4.830 \\ +19.465 \\ \hline 24.295 \end{array}$$

(b) $18 \div 0.8 = \frac{18}{0.8} = \frac{180}{8} = 22.5$ = **23** rounded off to the nearest whole number.

Round off each decimal to the nearest whole number :

1. 9.7 ☐ **2.** 43.5 ☐ **3.** 2.62 ☐

4. 12.78 ☐ **5.** 14.06 ☐ **6.** 100.95 ☐

7. 999.99 ☐ **8.** 110.11 ☐ **9.** 62.853 ☐

Round off each decimal to the nearest tenth (1 decimal place):

10. 3.28 ☐ **11.** 4.85 ☐ **12.** 9.48 ☐

13. 12.53		**14.** 111.16		**15.** 17.18	
16. 1.98		**17.** 29.39		**18.** 100.01	

Round off each decimal to the nearest hundredth (2 decimal places)

19. 0.318		**20.** 4.793		**21.** 2.847	
22. 16.555		**23.** 0.005		**24.** 99.985	
25. 100.001		**26.** 6.7498		**27.** 6.0025	

EXERCISE 35

I. Round off the decimal number to the nearest thousandth (3 decimal places)

1. 0.4683 **2.** 0.7809 **3.** 0.0025 **4.** 0.5823
5. 0.7948 **6.** 18.4972 **7.** 21.0007 **8.** 99.0002
9. 41.5627 **10.** 927.4875

II. Find the value of :

1. 3.0468 + 4.6251 + 9.6024 correct to 3 decimal places.
2. 7.8465 + 0.0073 – 1.0263 correct to 2 decimal places.
3. 5.798 × 0.8 correct to 3 decimal places.
4. 869.27 × 0.15 correct to **(a)** to the nearest whole number *(b)* to the nearest tenths.
5. **(a)** $\frac{7}{4}$ to the nearest tenths **(b)** $\frac{17}{8}$ to the nearest tenths.

III. Round these amounts of money and measurements up or down to the nearest whole number.

1. Rs 4.65 **2.** 8.75 kg **3.** 2.34 kg **4.** Rs. 12.28
5. 4.1 m **6.** 8.88 kg **7.** 7.06 cm **8.** 4.93 *l*
9. 16.08 *l* **10.** 5.89 cm.

10 BASICS OF GEOMETRY

The three building blocks of geometry are points, lines and planes.

Point: A point is an exact location in space. It has no length and no width. We represent a point by a **dot (.)** and label it with a capital letter.

Some representations of a point in everyday life are :

A distant star in the sky

The tip of your pencil

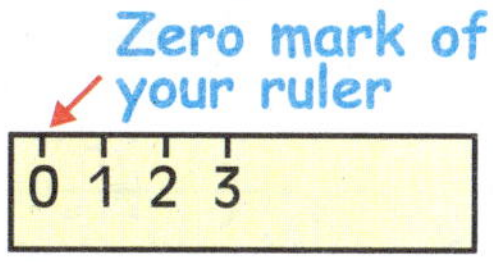

Ruler

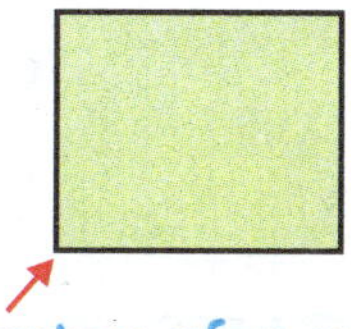
Vertex of a square

Line: A line is a straight path of points that extends on and on in both the directions without ever ending. The length of a line cannot be measured.

To show that a line extends endlessly in both the directions, you use arrowheads at both the ends.

A line is named by using any two points on a line.

line AB or line BA

Thus, the line given alongside is named as $\overleftrightarrow{AB}$ (line AB) or $\overleftrightarrow{BA}$ (line BA)

A **horizontal line** goes straight across.

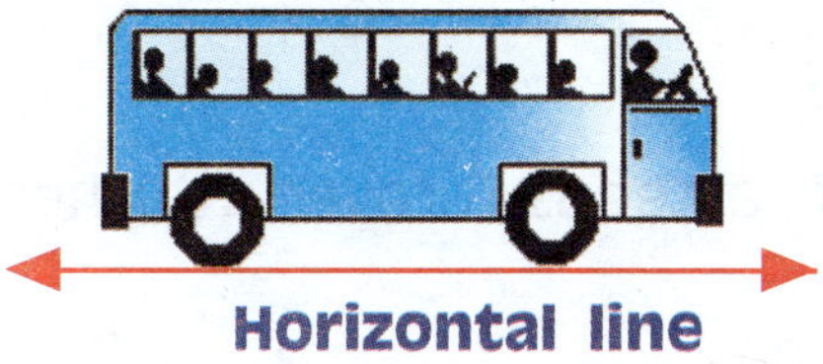

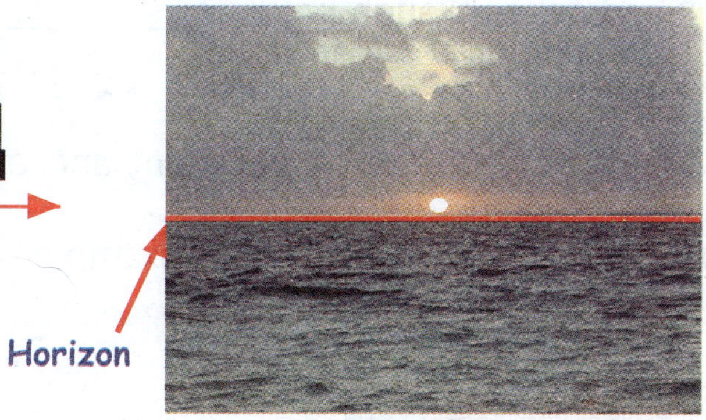

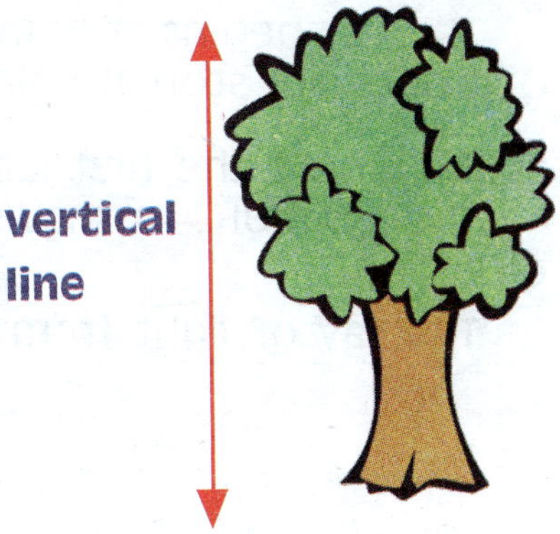

A **vertical line** goes straight up and down.

A line can also be curved. A curved line is called a **curve.**

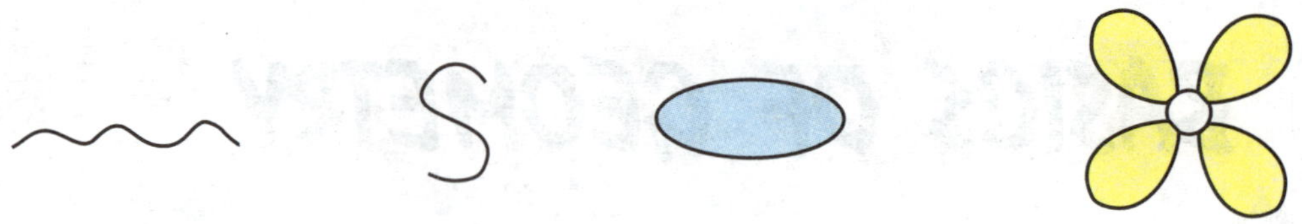

Line Segment

A part of a line is called a **line segment.** It has a **fixed length** that can be measured. The points at which the line segment begins and ends are called its **endpoints.** Every line segment has **two endpoints.** The given line segment has C and D as endpoints. It is named by using these endpoints. It is named by using these endpoints as $\overline{CD}$ or $\overline{DC}$.

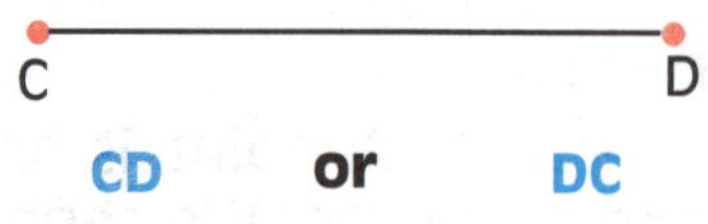

In the figure given below, which is the shortest path that can be followed by the rabbit to eat the carrot ?

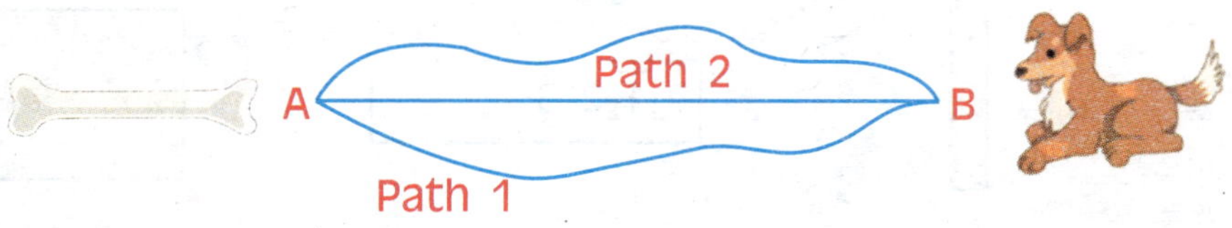

Path 2. As you can see Path 2 represents a line segment with carrot as one end point and rabbit as the other.

A line segment is also the shortest path between two points. Some representations of a line segment from your everyday life are:

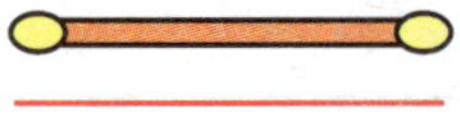

A matchstick

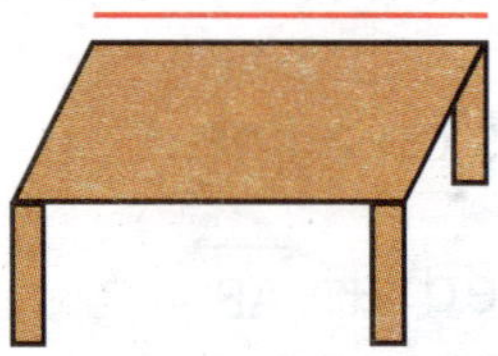

An edge of your table top

Lap top

Ray: A ray is a straight path that has one endpoint and goes on and on in one direction.

This ray begins at point P and goes through point Q. It does not stop at point Q. We name the given ray as $\overrightarrow{PQ}$ where the first letter is always the endpoint.

$\overrightarrow{PQ}$ or ray PQ

The symbol $\longrightarrow$ shows that a ray has a fixed endpoint and extends forever in other direction.

The ray of light from a torch is the most common example.

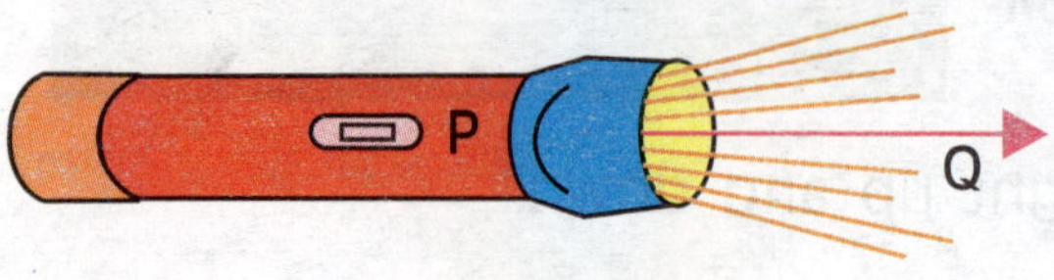

Plane: A plane is a flat surface. In mathematics, a plane goes on and on in all directions without end. We usually work with just a part of a plane. Points and lines lie on a plane. A plane can be named by using any three points on it. The given figure shows plane ABC. The order of the points does not matter.

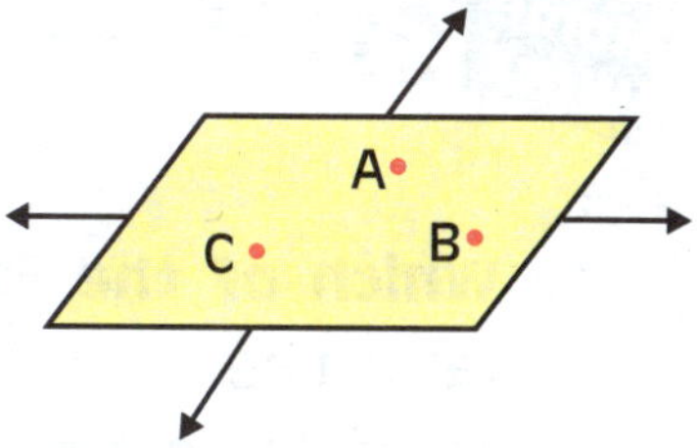

Plane ABC

Some representations of a plane surface from your everyday life are:

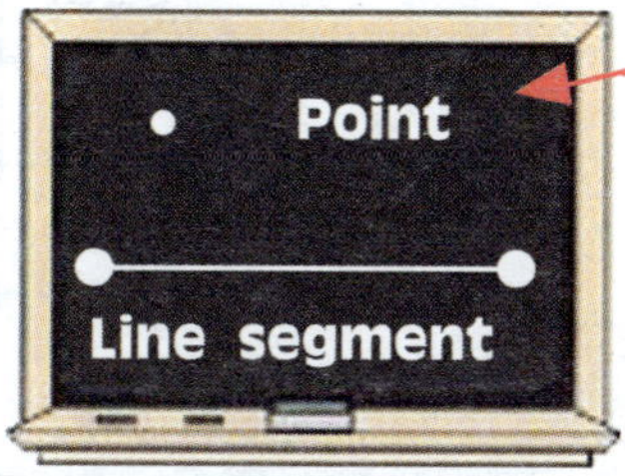

Surface of a blackboard

Surface of a book

A wall of your room

Screen of a flat TV

Upper surface of a stool

Your school playground

A. Name each figure and write an example representing the figure from your surroundings.

1. • P	2. A ——— B ——•	3. M ———— N
Name ______________		
Example ______________		

B. Draw and label the following figures in the given space.

1. Line XY	2. Plane XYZ	3. A curve

EXERCISE 36

1. **Which of the following has a fixed length ?**
   (a) A line **(b)** A ray **(c)** A line segment **(d)** A point
2. **Classify as rays, lines and line segments. Also, name them.**

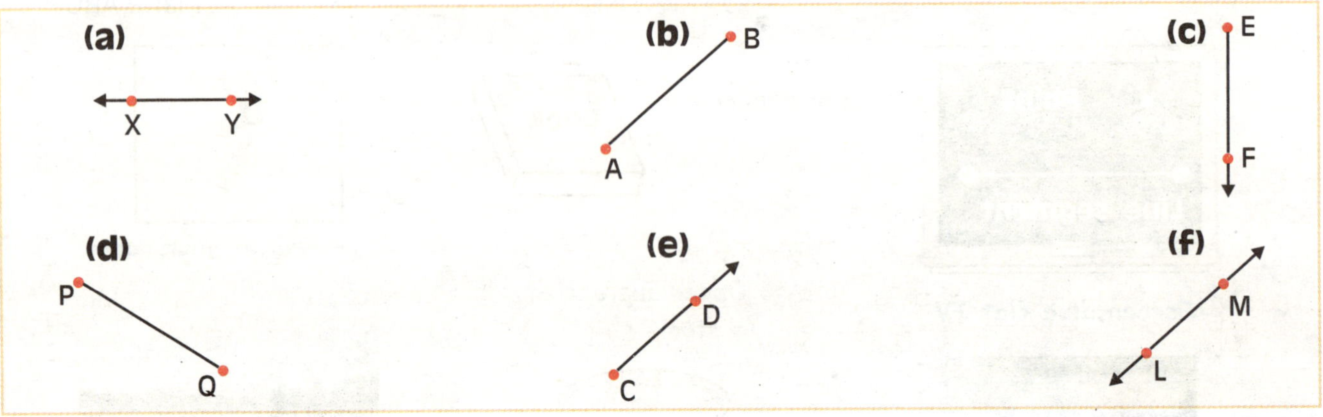

3. **State True (T) or False (F).**
   (a) A ray has a definite length
   (b) A line cannot be drawn on a piece of paper.
   (c) A ray has one endpoint.
   (d) A line segment has no endpoints.
   (e) A plane has fixed boundary.

4. **Count and name the line segments in each of the following figures.**

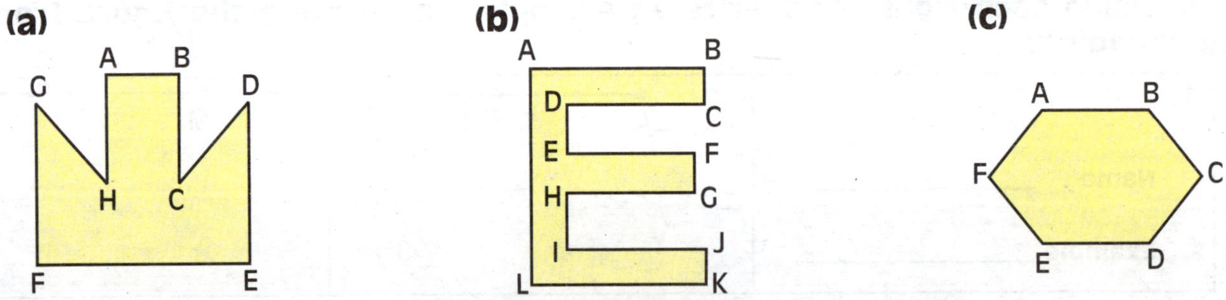

5. **Name the rays, and line segments in the following figures.**

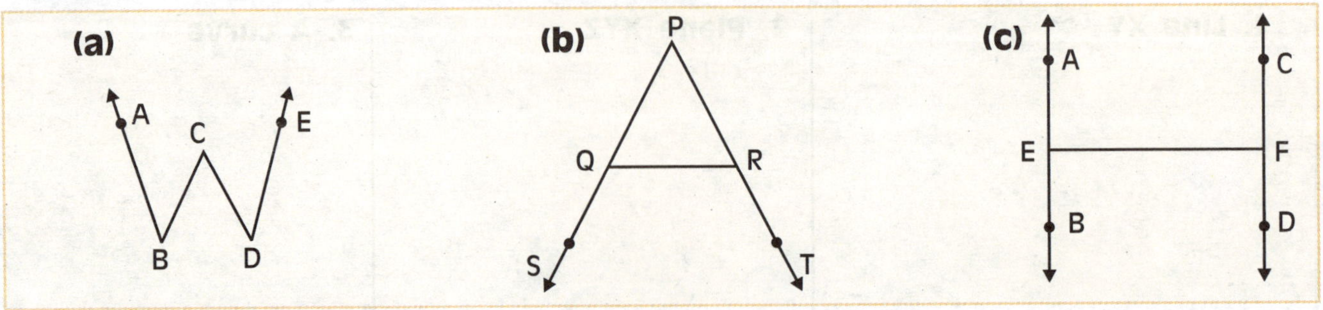

6. Identify 4 plane surfaces in your class room.

ANGLES

What is an angle ?

An angle is a figure formed by two rays meeting at a common endpoint. The common endpoint is called the **vertex** of the angle and the two rays are called the **arms** of the angle. Looking at these pictures, you can form some idea of an angle.

The symbol for the word angle is ∠.

You can name an angle in three ways:

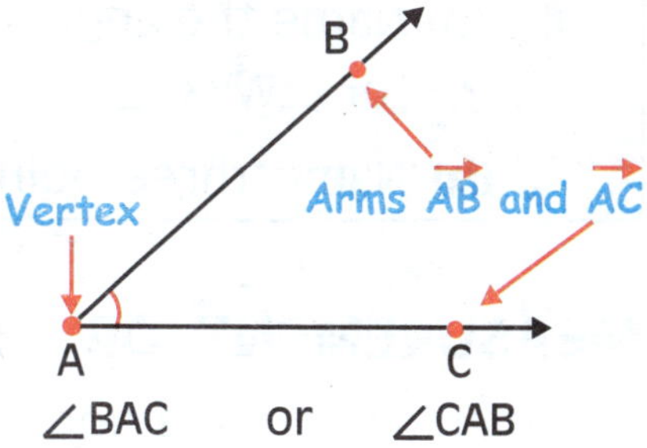

★ Using a three letter name in the order : a point on one ray, vertex and a point on the other ray.
∠BAC or **∠CAB.**

★ Using only one letter name *i.e.*, the vertex. ∠A.
(This can be used when there is only one angle with this vertex)

★ Writing a number within the rays of the angle and naming the angle using this number as ∠1.

INTERIOR AND EXTERIOR OF AN ANGLE

The inside of the angle, that is the region between the rays is called the **interior** of the angle. Point *P* is in the interior of ∠ABC.

Points A, B, C, D and E are on the arms (also called sides) of the angle. The points of the plane that do not lie on the arms or in the interior of the angle are in the **exterior** of the angle. The points Q and R are in the exterior of ∠ABC.

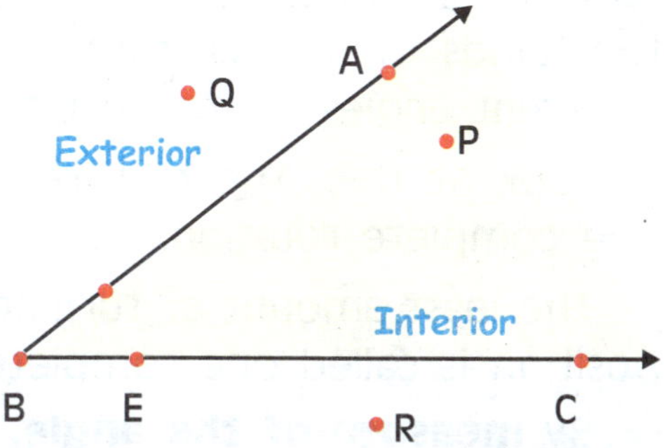

1. Underline the correct answer in each question given below :

(a) The vertex is (point P, point Q, point R).

(b) The angle may be named as (∠QRP, ∠QPR, ∠RQP).

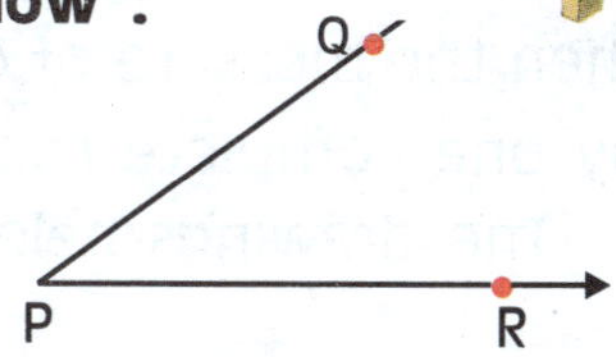

2. For the figure given on the right:

(a) Name the marked angle. __________

(b) Name the vertex of the angle. __________

(c) Name the arms of the angle. __________

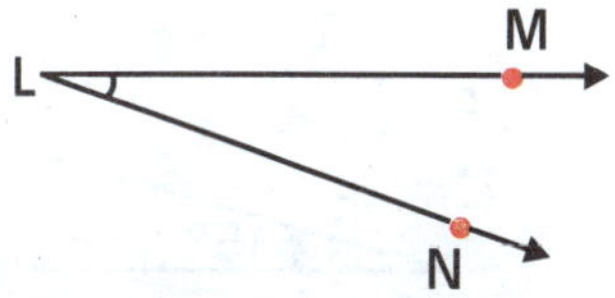

3. **For the given figure:**

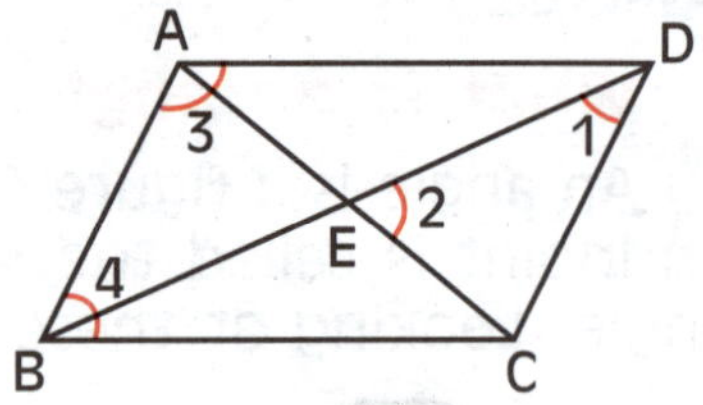

(a) Name the angle marked 1 in four different ways.

____

(b) Name the vertex and arms of angle marked 4.

Vertex ____________ Arms ____________

(c) Name the angles whose arms are $\overrightarrow{DB}$ and $\overrightarrow{DA}$

____

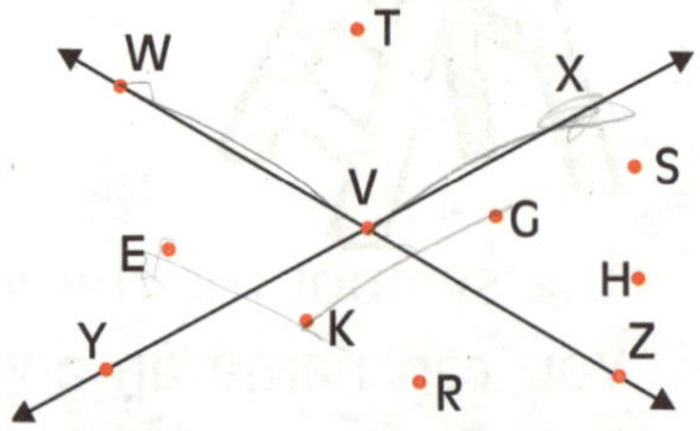

4. (a) Name the angle formed by the rays opposite to those of ∠WVX. ____________

(b) Name three points in the interior of ∠XVZ. ____________

MEASUREMENT OF ANGLES

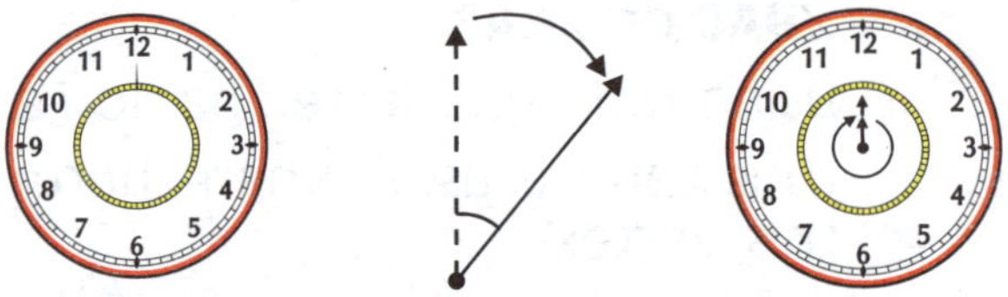

When a hand of a clock turns from one position to another, it changes its direction. You can show this change of direction or turning by drawing two arrows from one point.

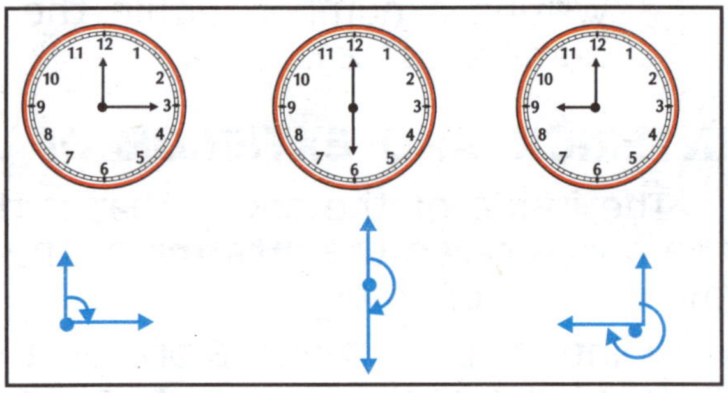

You can think of an angle as **an amount of turning** or **rotation.** The pictures here show how the hands of a clock turn through different angles as they rotate.

Look at this. The picture shows one complete rotation.

The least amount of turning that brings the hand back to its starting position is called one complete rotation.

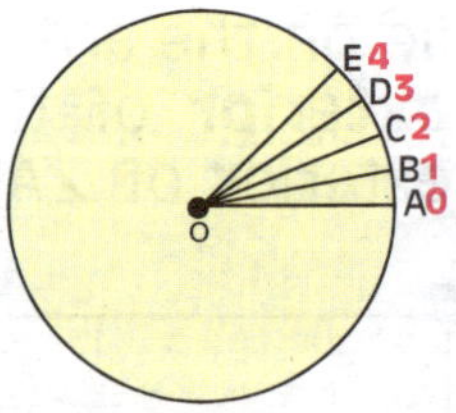

By **measure of the angle,** we mean the amount of rotation. Just as there are standard units of length, of area, of weight, etc., there are also standard units for the measurement of angles. The standard unit of measure for angles is the **degree.** The symbol for degree is °.

If one complete rotation is divided into 360 equal parts, such as AOB, BOC, COD, etc. then the measure of one part such as ∠AOB is called **one degree** or 1°. The angle formed by one complete rotation measures 360 degrees or 360°.

The drawings below show angles of 15°, 30° and 70°.

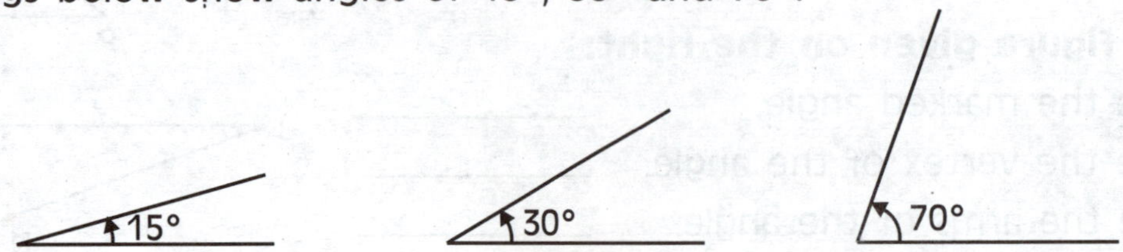

Example 1. If a circle is divided in 40 equal parts, what will be the measure of the angle of each part ?

Solution. Angle formed by one complete rotation = 360°

∴ Measure of angle of each of the 40 equal parts $= \frac{360°}{40} = \mathbf{9°.}$

Example 2. Calculate the angle between neighbouring spokes on each wheel.

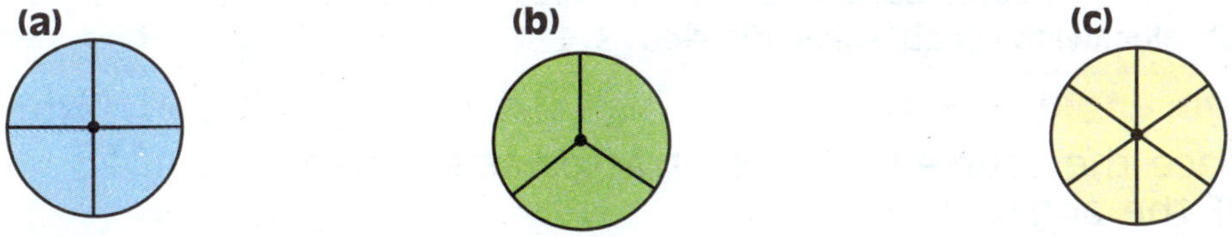

Solution.

(a) As the wheel (circle) is divided into 4 equal parts, the angle between the neighbouring spokes $= \frac{360°}{4} = \mathbf{90°.}$

(b) The wheel is divided into 3 equal parts, so the angle between the neighbouring spokes $= \frac{360°}{3} = \mathbf{120°.}$

(c) The wheel is divided into 6 equal parts so the angle between the neighbouring spokes $= \frac{360°}{6} = \mathbf{60°.}$

1. If a circle is divided into 12 equal parts, what will be measure of the angle of each part ?

2. Ten complete turns are needed to screw the nail right in. How many degree is it ?

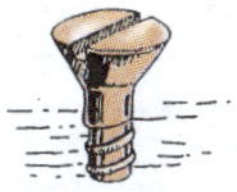

3. What fractions of a complete turn are 180°, 90°, 60°, 45°, 1°?

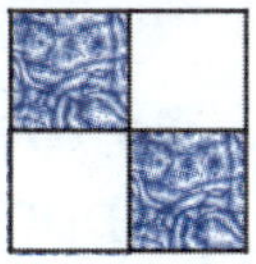

4. What is the measure of each angle at the centre of a floor tile ?

5. The electric cooker control turns clockwise. How many degrees does it turn through from :

(a) OFF to 1 ______________

(b) OFF to OFF ______________

(c) 3 to OFF ______________

(d) 1 to 4? ______________

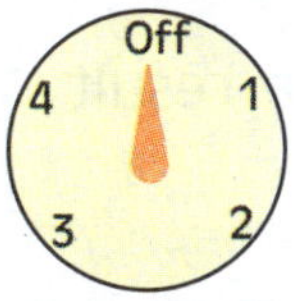

[**Hint:** The angle turned through in moving from off to 1, 1 to 2, 2 to 3, 3 to 4 and 4 to off $= \frac{360°}{5} = 72°$.]

MEASURING AN ANGLE

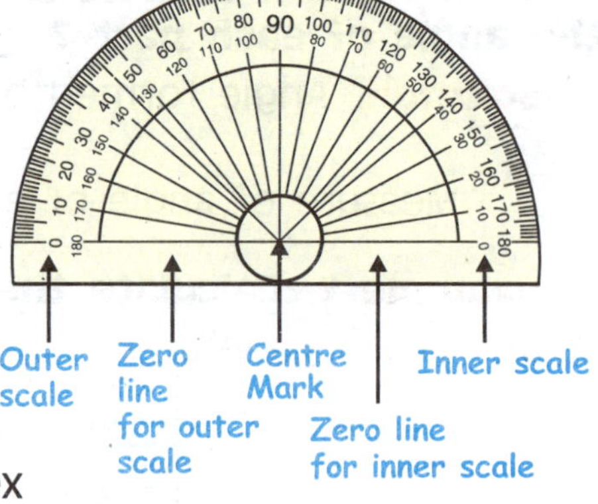

The instrument for measuring angles is called a **protractor.** The diagram on the right shows a protractor. It has the shape of a semi-circle and the angle at the centre is divided into 180°.

> **Note:** The inner scale of the protractor is marked from 0° to 180° anticlockwise. The outer scale of the protractor is marked from 0° to 180° clockwise. Each small division is 1°.

To measure an angle less than 180°.

Step 1. Place the centre of the protractor at the vertex of the angle.

Step 2. Make sure that 0° line of the line protractor is placed along one arm (AC) of the angle.

Step 3. Read the value on the protractor as indicated by the other arm (AB) of the angle.

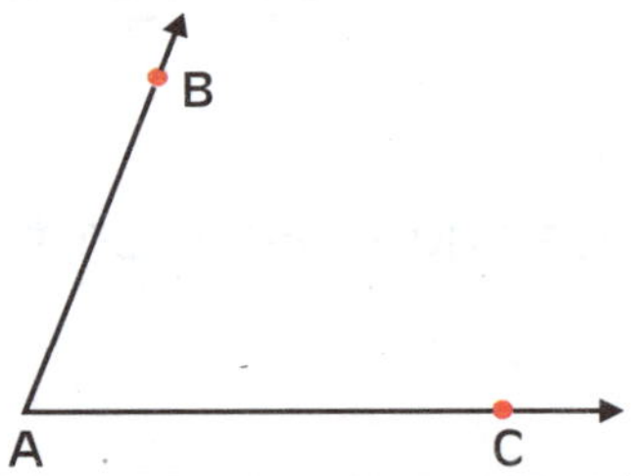

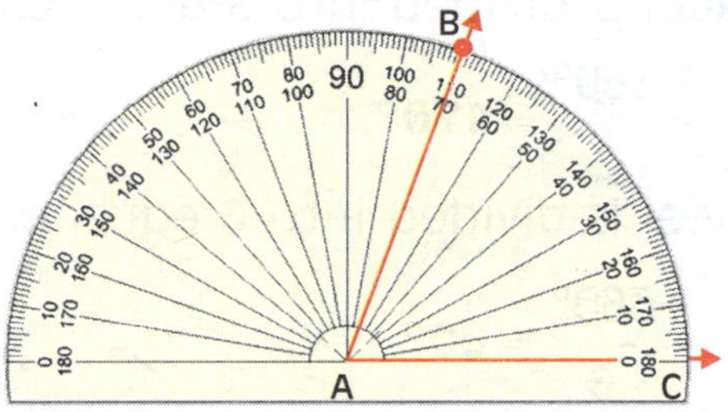

You find that $\angle BAC = 70°$

Always use the scale with 0° on one of the arms of the angle.

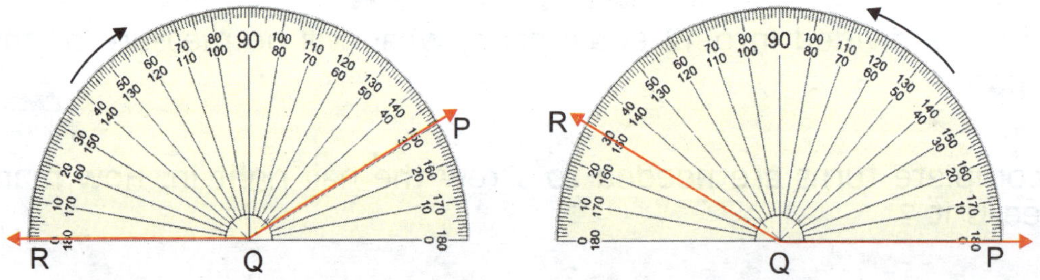

Using outside scale, $\angle RQP = 150°$, Using inside scale, $\angle PQR = 150°$

To measure a reflex angle, i.e., angle greater than 180°.

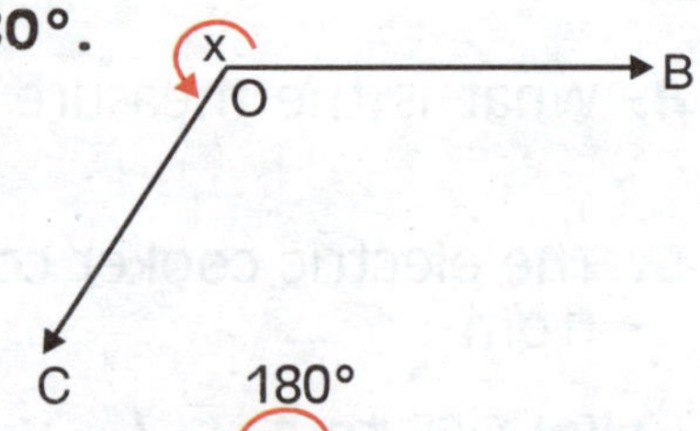

Suppose you are required to measure $\angle x$

Step 1. Extend either arm from the vertex to break the angle into two parts. We have extended ray $\overrightarrow{OB}$

Step 2. One part of the given angle is 180°. Measure the other part marked as **a.**

Step 3. Add 180° to a to get the result. Then

$\angle \mathbf{x} = \mathbf{180° + a}$

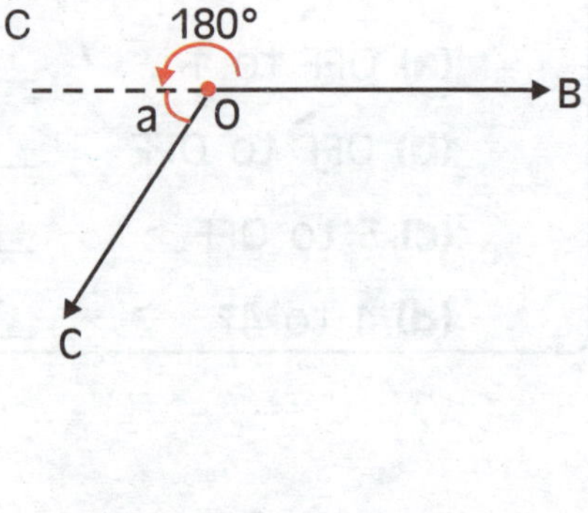

Skill Practice — **Class Work**

Use your protractor to measure these angles.

1. 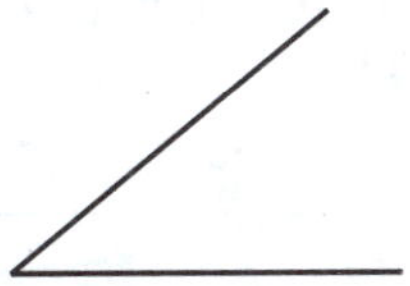

____

2. 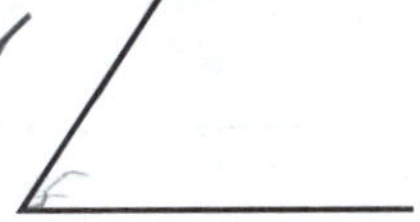

____

3. 

____

4. 

____

5. 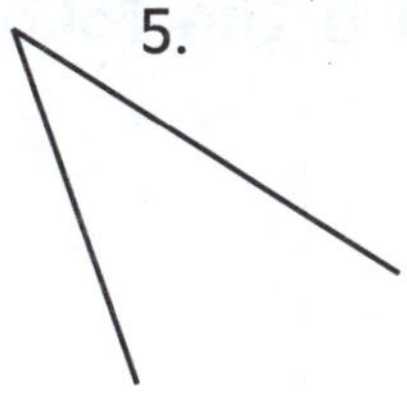

____

6. 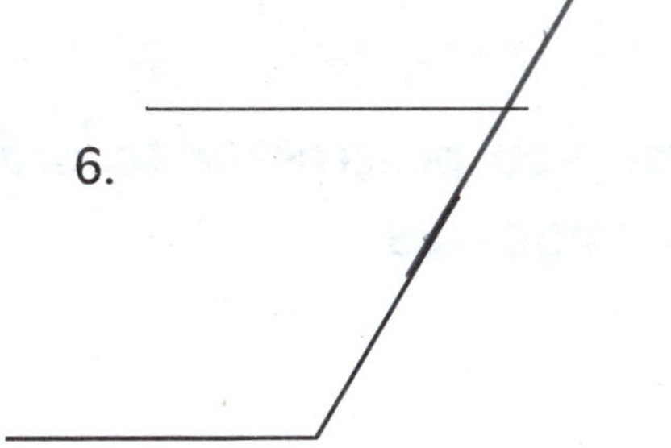

____

DRAWING ANGLES

To draw an angle that measures 70°.

Step 1. Draw any ray *OA*.

Step 2. Place the protractor such that its centre mark may fall on O, the endpoint of the ray. Align the ray with the 0° mark of one of the protractor scales. The end-point O of the ray will be the vertex of the angle.

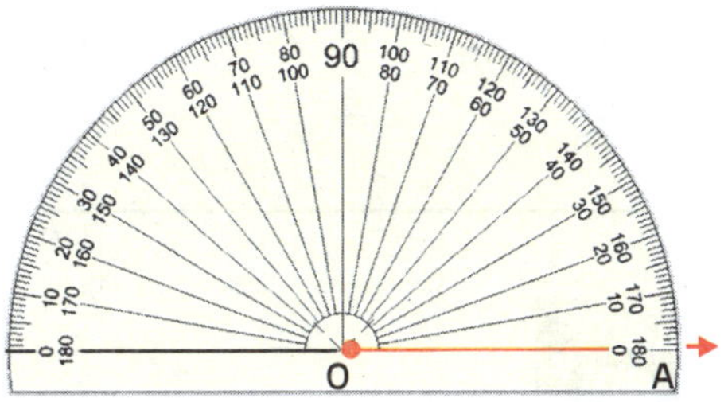

Step 3. Using the scale on which the ray OA aligns with 0°, mark the point at 70°. Label the point as *B*.

Step 4. Remove the protractor and join B to O. The ∠AOB is the required angle.

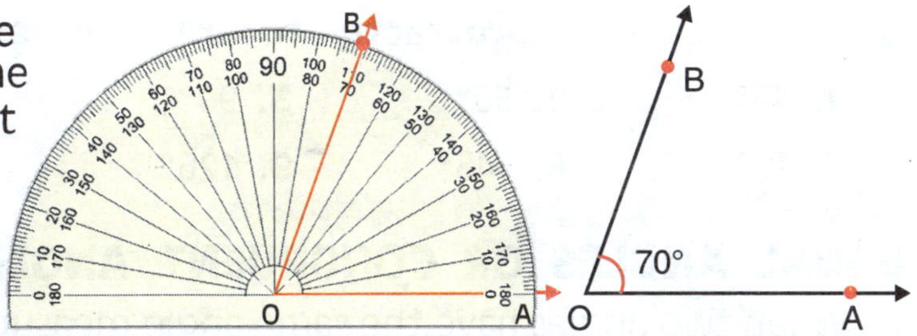

Note : As you can see, we have drawn this angle using the inner scale on the protractor. What will be the angle like if we use the outer scale on the protractor ?

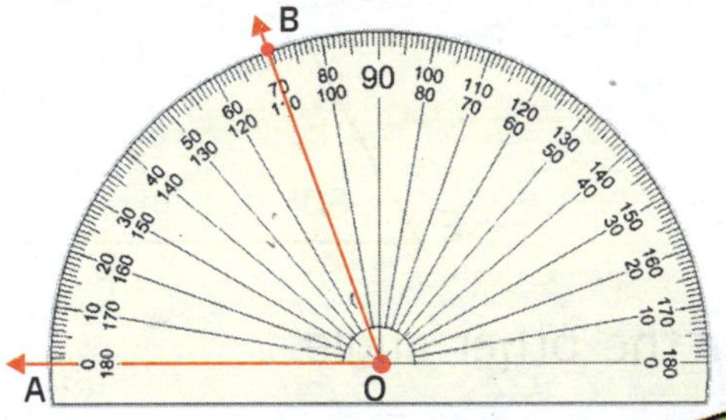

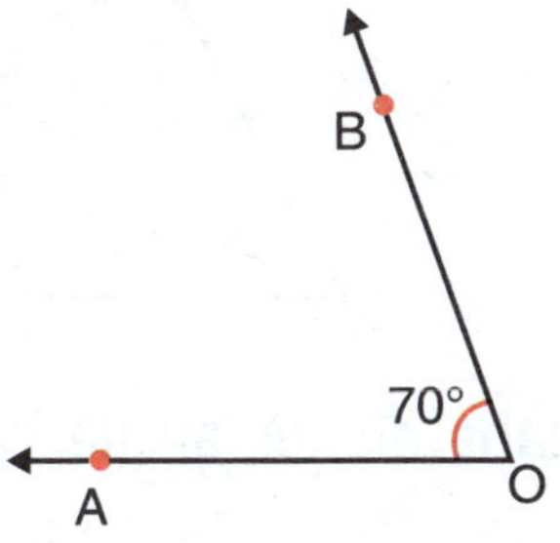

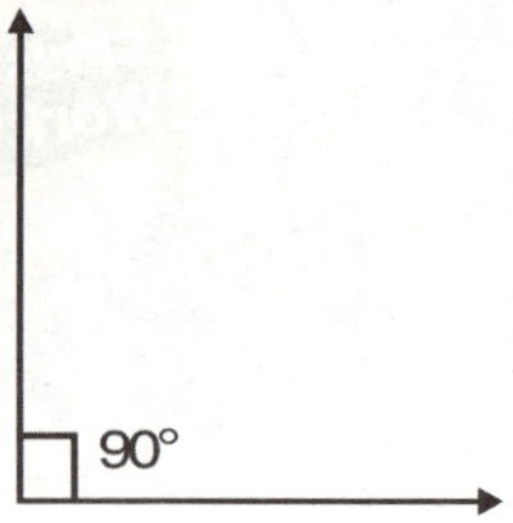

Maths Alert ! The sides (arms) of an angle are rays, so the lengths of the sides in drawing an angle does not affect the measure of the angle.

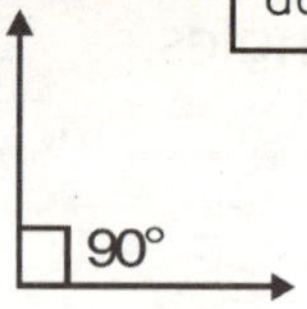

Irrespective of the length of the arms, each angle = 90°.

Skill Practice

Class Work

Use you protractor to draw angles with the following measures :

1. $\angle ABC = 85°$

2. $\angle PQR = 135°$

Home Work

EXERCISE 37

Use your protractor to draw angles with the following measures

1. 45° **2.** 56° **3.** 90° **4.** 110° **5.** 165° **6.** 25°
7. 10° **8.** 145° **9.** 126° **10.** 75°

EQUAL ANGLES OR CONGRUENT ANGLES

When two angles have the same angle measure, we say that they are equal or congruent and write $\angle ABC = \angle DEF$.

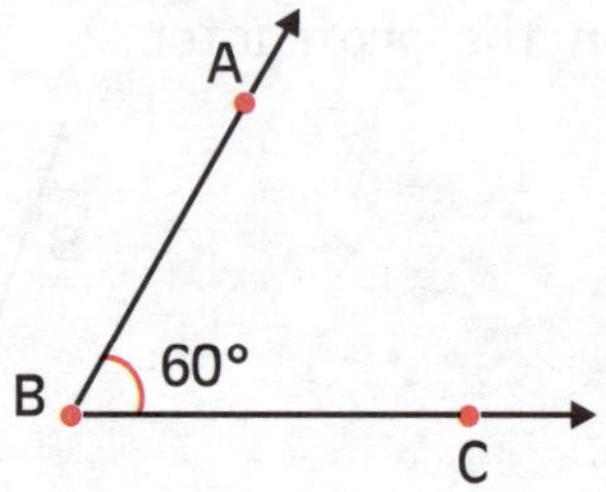

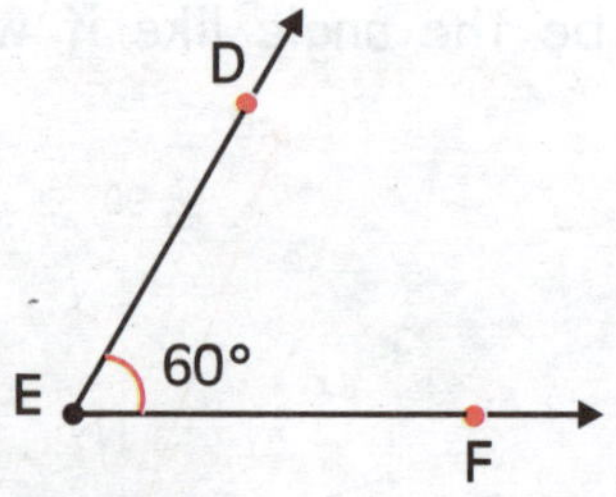

This means either angle can fit exactly over the other angle.

Skill Practice — **Class Work**

Using the diagram given below, determine the measures of each of the following angles: Some colour rays form an angle)

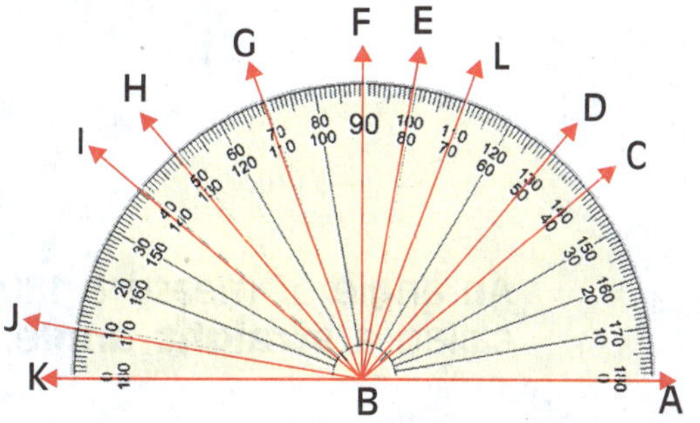

1. ∠KBC = ____________
2. ∠ABD = ____________
3. ∠ABE = ____________
4. ∠ABF = ____________
5. ∠KBH = ____________
6. ∠ABI = ____________
7. ∠KBJ = ____________
8. ∠KBF = ____________
9. ∠KBG = ____________
10. ∠KBL = ____________
11. How much is ∠ABK?
12. Can you identify pairs of equal angles ? If yes then name them.

KINDS OF ANGLES

Angles are named according to their measure between 90° and 180° and 180° and 360°.

The fan is closed : 0°

The fan is closed 0°

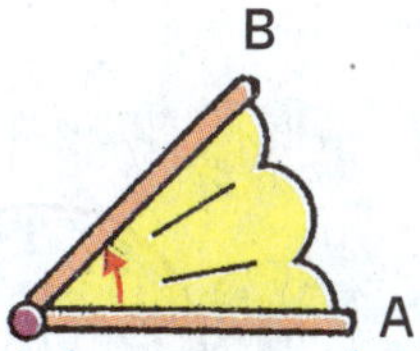

OA has turned through an acute angle between 0° and 90° to position OB

The side OA has turned through an angle between 0° and 90°. It is called an **acute angle.**

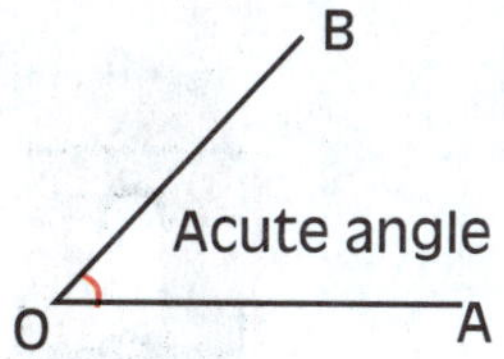

OA has turned through a right angle, i.e., 90° to position OB

OA has turned through 90°. It is called a **right angle.**

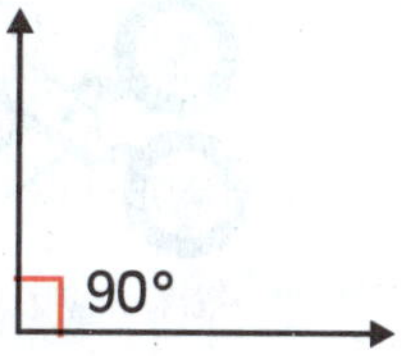

An obtuse angle between 90° and 180°

OA has turned through an angle between 90° and 180°. It is called an **obtuse angle**.

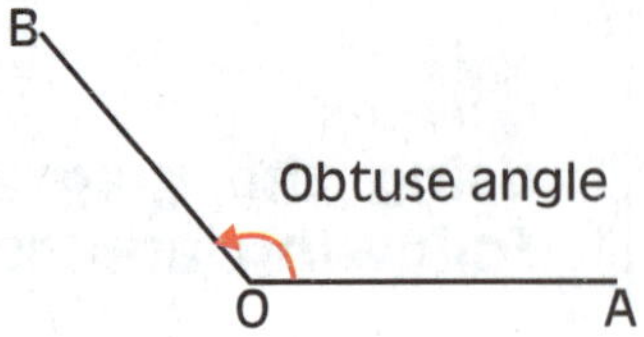

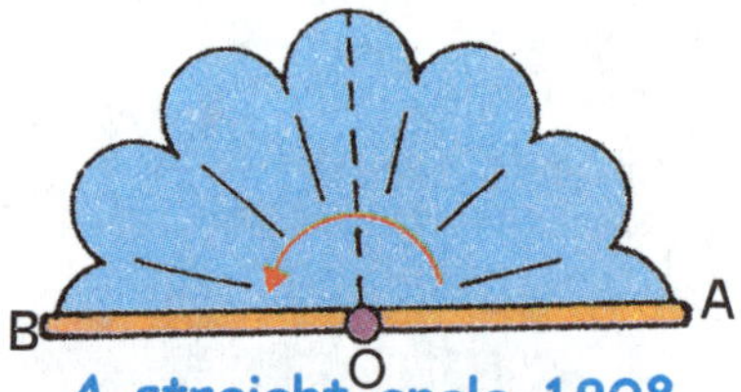

A straight angle 180°

An angle of measure 180° is called a **straight angle.**

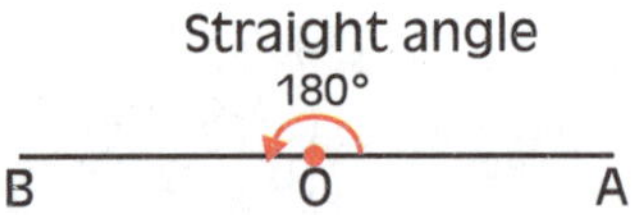

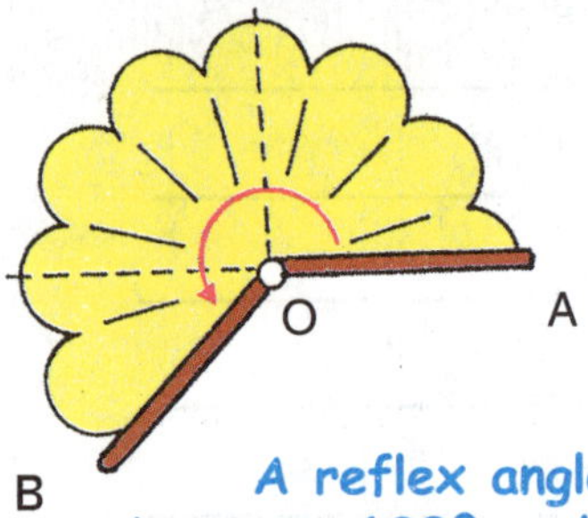

A reflex angle between 180° and 360°

OA has turned through an angle more than 180° but less than 360°. It is called a **reflex angle**.

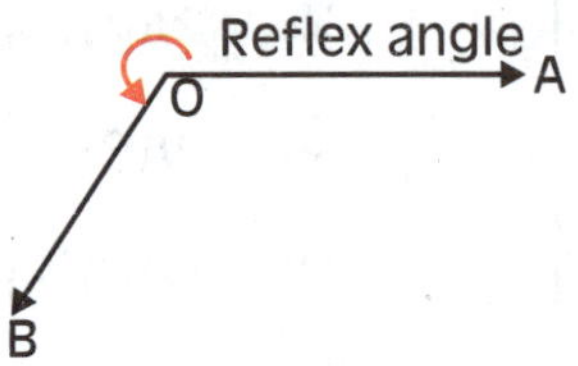

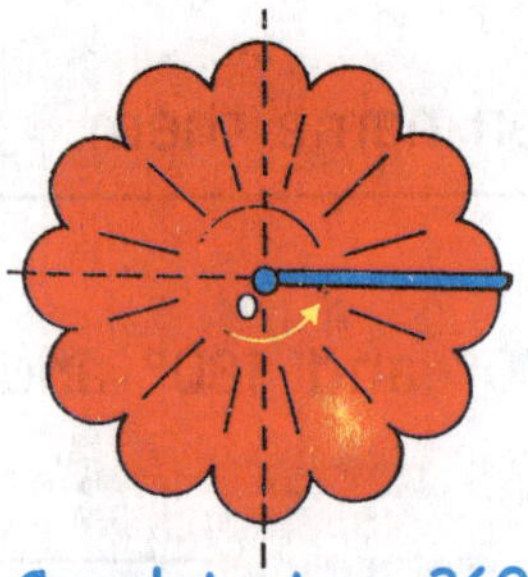

Complete turn 360°

OA has made one complete turn. This angle measures 360° and is called a **complete angle**.

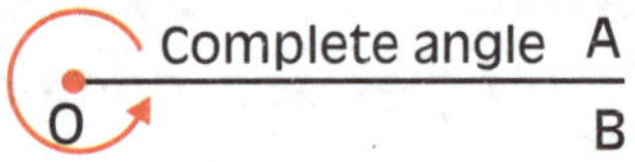

Identifying the different kinds of angles in your environment.

Right Angle

Actue Angle

What type of angles are shown below, acute, obtuse, or right angle ? Measure each angle. Give two names for each of these angles. Complete the table giving this information.

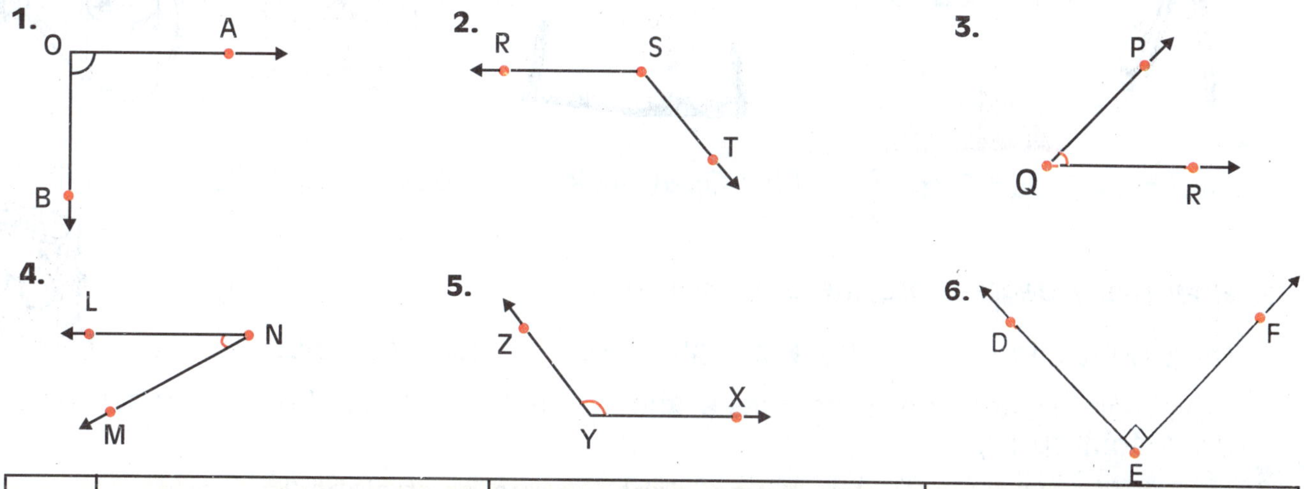

	Type of the angle	Measure of the angle	Name of the angle	
			(1)	(2)
1.				
2.				
3.				
4.				
5.				
6.				

EXERCISE 38

1. What kind of angle is shown in each figure?

(I) (II) (III) (IV) (V)

2. **(a)** What kind of angle is marked in the following pictures ?
   (b) Which of these angles are fixed, and which can change size ?

Compasses Bus Stop Flat Open Book Cake Watering Can

3. How many degrees are there in each of these ?

   (a) a right angle **(b)** a straight angle **(c)** a complete turn ?

4. How many right-angle turns does a soldier make one after the other when he does an "about turn".

5. Are the angles given below acute, right, obtuse or straight ?

   (a) 63° **(b)** 136° **(c)** 90° **(d)** 180° **(e)** 101° **(f)** 89°

6. **Name the number of right angles the hour hand on the clock face turns through in the direction from :**

   (a) 12 to 3 **(b)** 12 to 6 **(c)** 12 to 9

   (d) 12 to 12 **(e)** 3 to 6 **(f)** 3 to 9

   (g) 3 to 3 **(h)** 4 to 7 **(i)** 8 to 11

7. Write 3 examples of right angles from your classroom.

8. Can two acute angles be put together to form a straight angle ? Give reason for your answer.

9. Copy the table. Use a protractor to measure each angle in the figure and complete the table.

	Angle	Measure	Kind of angle
(a)	∠QPR		
(b)	∠RPS		
(c)	∠QPS		
(d)	∠RPT		
(e)	∠QPT		
(f)	∠RPU		
(g)	∠QPU		

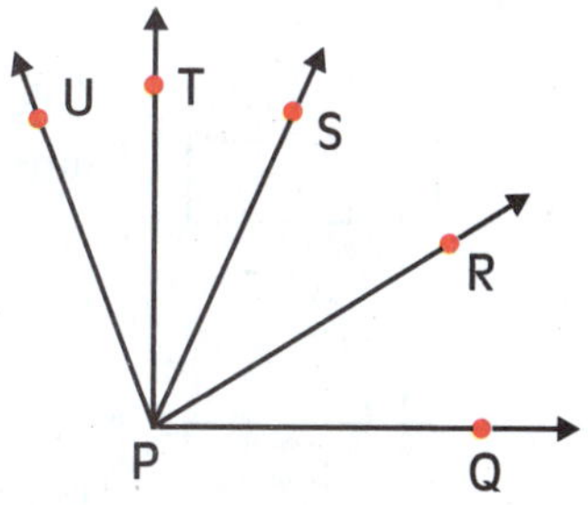

10. Observe the diagram drawn below and tell about each of the following angles whether it is acute, right, obtuse, or straight.

(a) ∠BAC....... **(b)** ∠ABC.......

(c) ∠ACB....... **(d)** ∠DCA.......

(e) ∠ADC....... **(f)** ∠ADB.......

11. Given below are all the letters of the English alphabet.

A B C D E F G H I J K
L M N O P Q R S T U V
W X Y Z

(a) List all the letters that have angles.

(b) List all the letters that have :

(i) acute angles **(ii)** obtuse angles **(iii)** right angles.

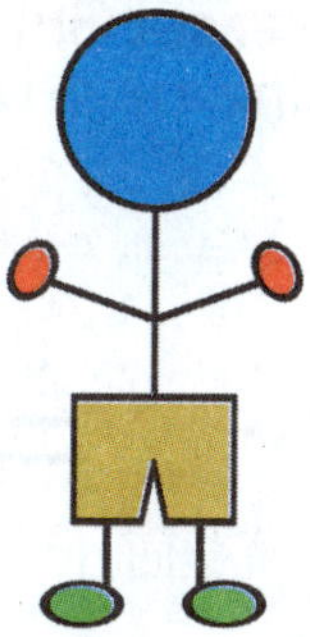

ACTIVITY

Take a square piece of paper and fold it as shown below: Identify the types of angles formed:

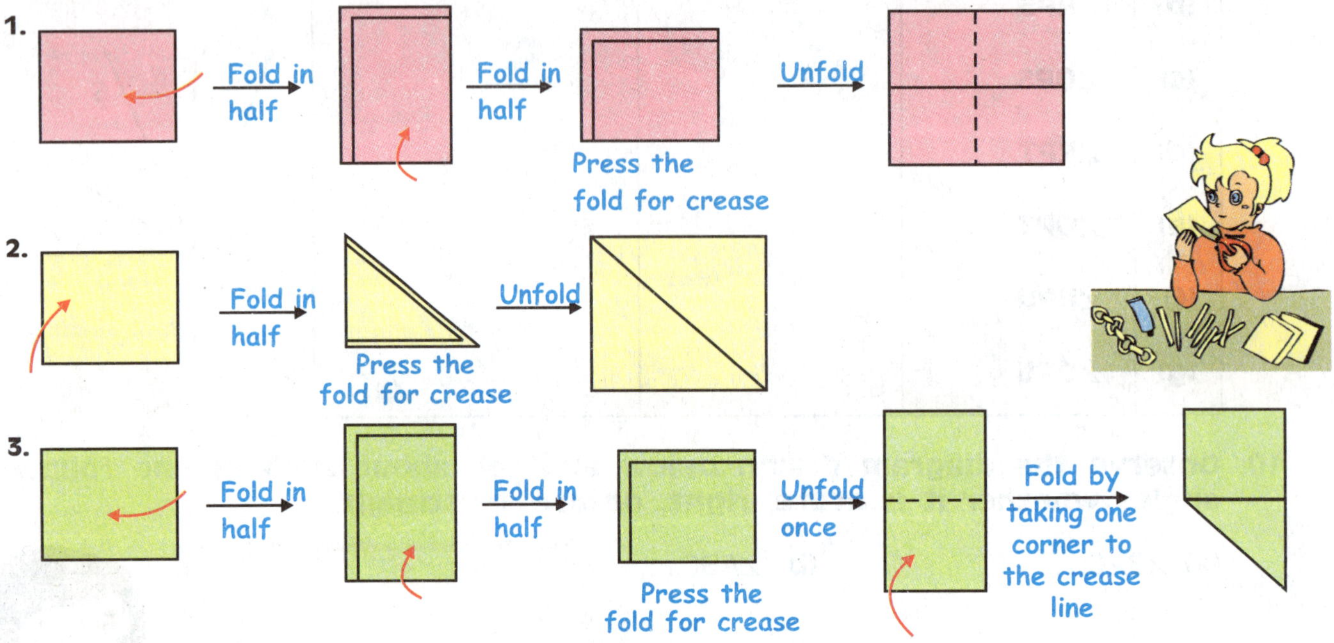

TYPES OF LINES

Parallel Lines are lines on the same plane that never meet, no matter how far they are extended.

They are always the same distance apart.

The symbol || is used to show **"is parallel to"**.

Here, Line AB is parallel to line XY and line PQ || line RS. The following are some of the representations of parallel lines in everyday life.

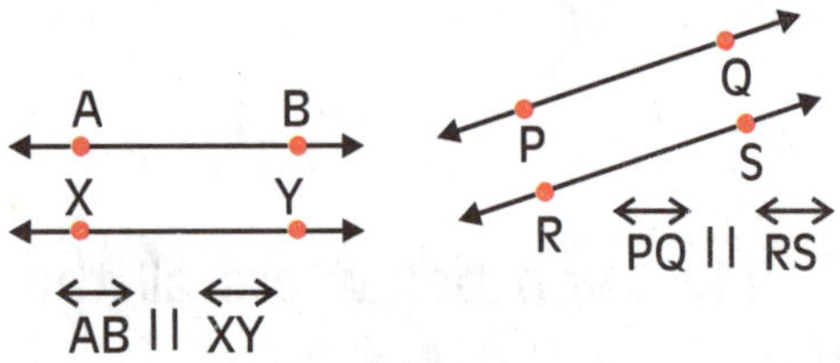

Two rails of a railway track

Stumps of a wicket

Opposite edges of a table

Intersecting Lines are lines that cross each other at a point.
Think of a red light where two roads meet and cross each other.

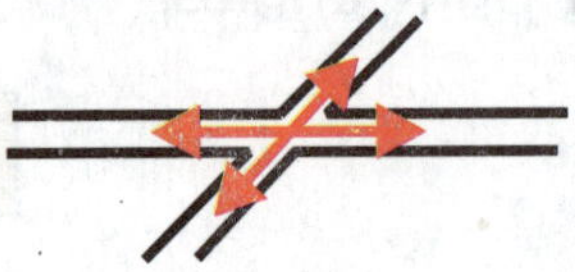

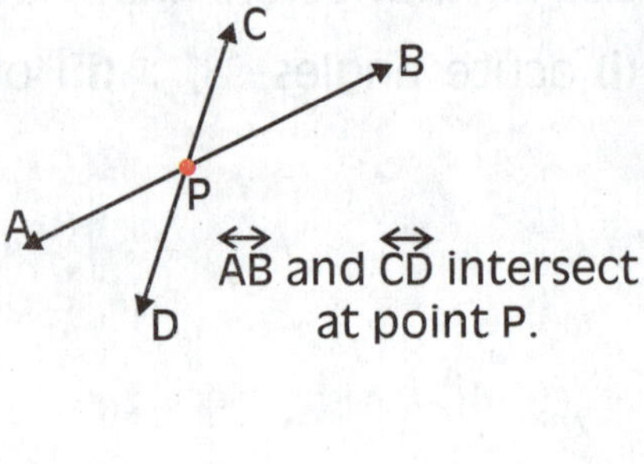

The letter X is also an example of intersecting line segments.

Perpendicular Lines. When two intersecting lines meet to form right angles they are called perpendicular lines. They are indicated by the symbol '⌉' (a square corner) in the diagrams.

Line *AB* is perpendicular to line CD and is written in short as $\overleftrightarrow{AB} \perp \overleftrightarrow{CD}$.

⊥ is the symbol for **"is perpendicular to"**.

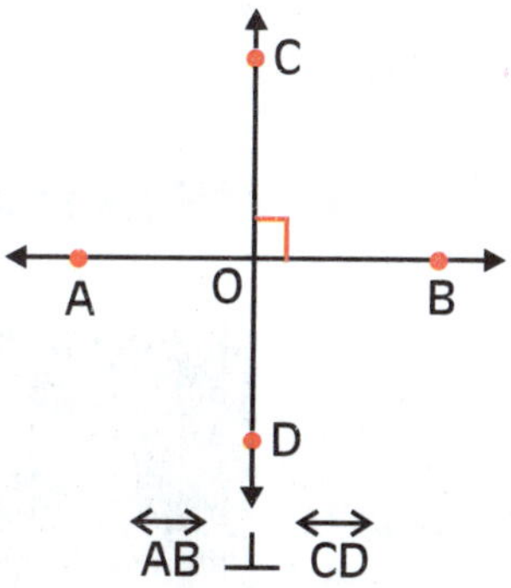

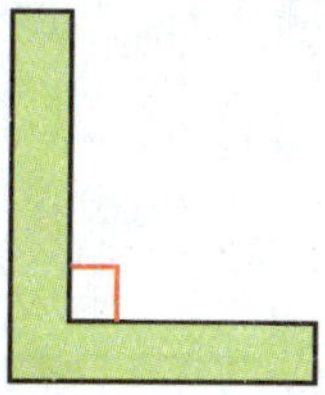

Using appropriate symbols, express the relationship between pairs of perpendiculars and parallel lines in the following diagrams.

1.

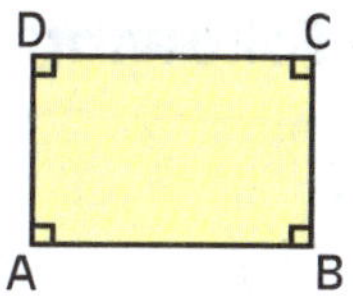

AB || DC, AB ⊥ AD

2.

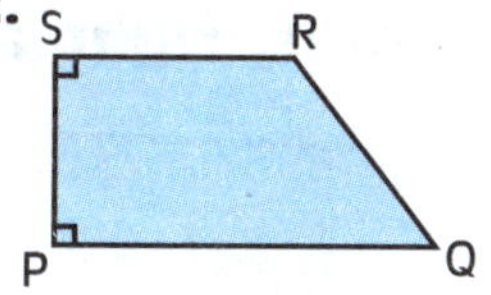

____

3.

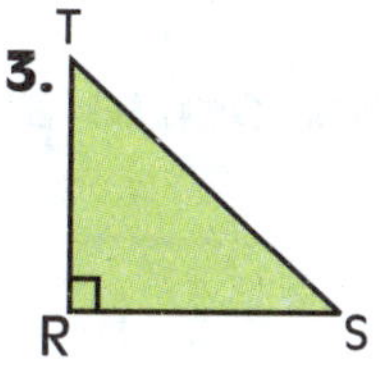

____

4.

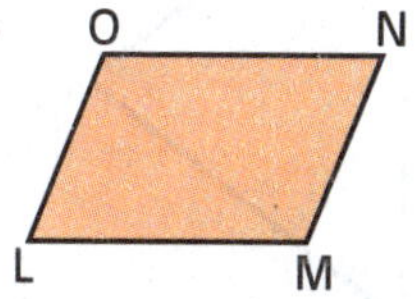

____

5.

____

6.

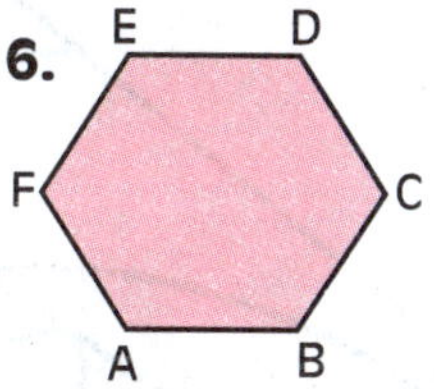

____

7. State True (T) or False (F):

(a) Parallel lines make a right angle with each other. ______________

(b) The distance between two parallel lines always remains the same. ______________

(c) All perpendicular lines are intersecting lines. ______________

(d) All intersecting lines are perpendicular lines. ______________

(e) A flag post is always perpendicular to the ground. ______________

(f) A flag is always parallel to the ground. ______________

ACTIVITY

1. Identify parallel lines, perpendicular lines and planes in the following picture.

2. Mark acute angles, obtuse angles, right angles in the following figures.

11 GEOMETRICAL SHAPES

TRIANGLES

Geometry is commonly found in architecture. Many bridges, buildings, space stations, towers etc. have triangles as a part of their construction, as they hold their shape very well and are kind of fixed.

You have already studied in class IV about closed figures and polygons.

A polygon is a simple closed figure made up of line segments. Some examples of polygons are:

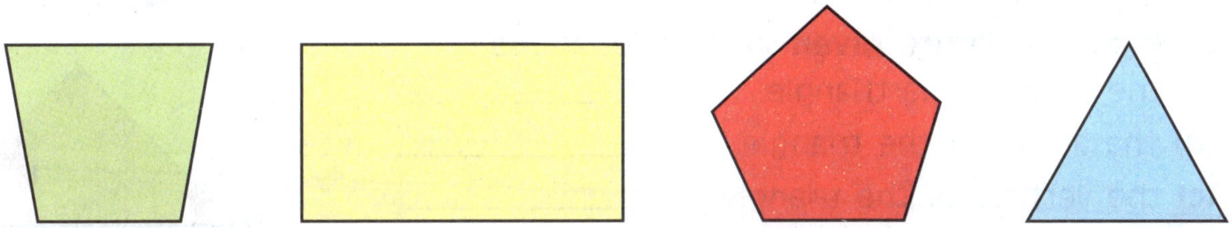

You can see a polygon with 3 sides in the figures give above. That is a triangle.

Triangles around us:

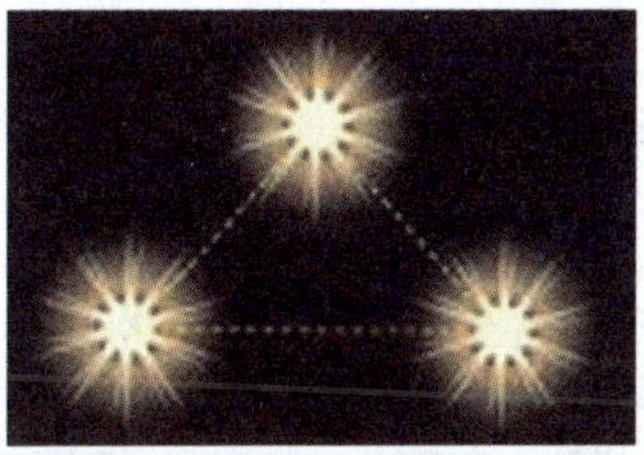

What is a triangle ?

A triangle is a simple closed plane figure made up of 3 line segments.

It can be formed by joining three non-collinear points.

It has :

★ three line segments as its sides, AB, BC and AC

★ three vertices – A, B and C

★ three angles – ∠A, ∠B and ∠C.

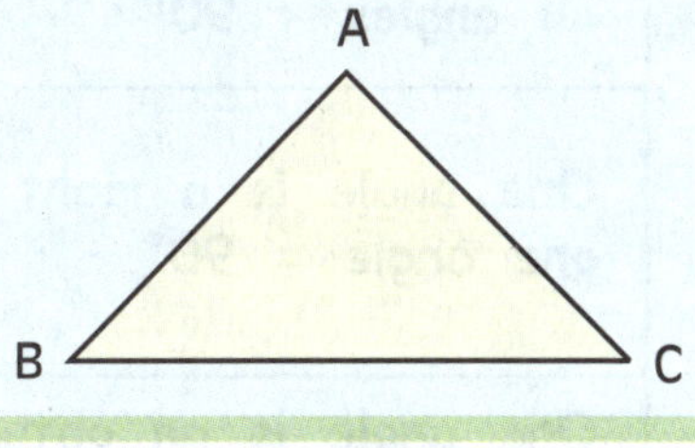

The word triangle is made up of two words **tri** and **angles**.

Tri means **three**, so Triangle means a figure having three angles. The symbol Δ is used for the word 'triangle'. So we name the above triangle as ΔABC.Other names for ΔABC are ΔBAC, ΔBCA, ΔCAB, ΔCBA, ΔACB. The order of vertices does not matter while naming a triangle.

I. Which of the following figures are triangles. Tick (✓)

1. 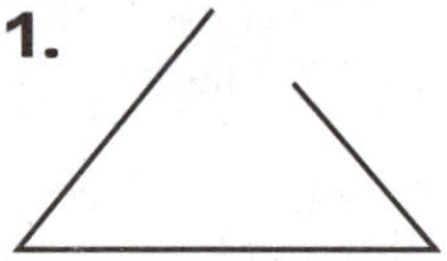

2. 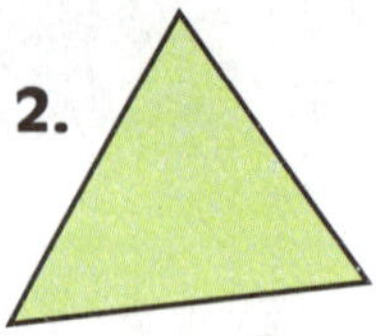

3. 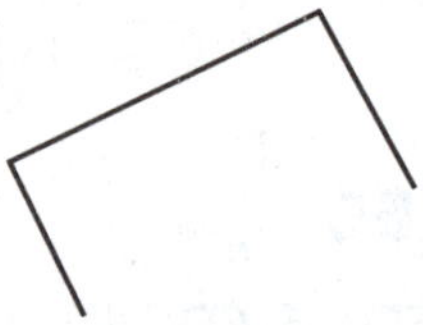

4. 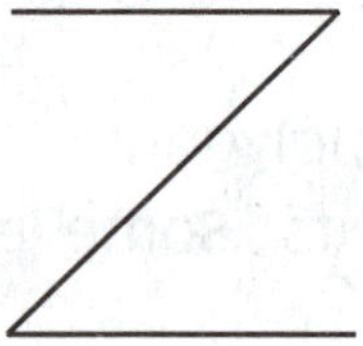

5. 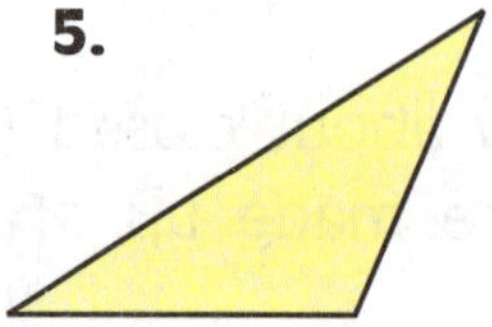

6. 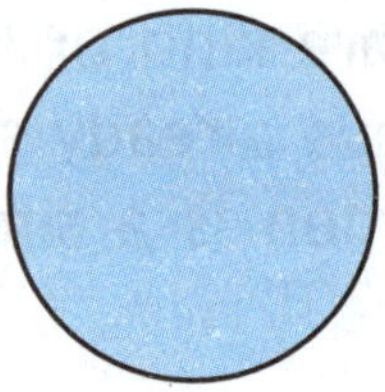

II. Look at the figure given alongside. Name

(a) the sides of the triangle. ________________

(b) the angles of the triangle. ________________

(c) the vertices of the triangle. ________________

(d) the name of the triangle. ________________

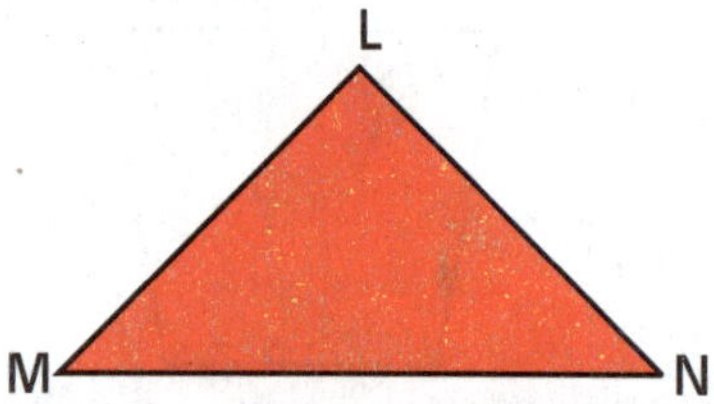

CLASSIFICATION OF TRIANGLES

Triangles are classified and named according to the measures of their angles and lengths of their sides.

★ **By size of its Angles**

For every triangle, there are at least 2 acute angles, therefore, the third angle of the triangle is used in classifying a triangle according to its angles.

Type of angle	Name	Example
All angles are acute, i.e., all angles < 90°	Acute-angled triangle or Acute triangle	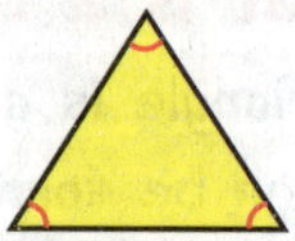
One angle is a right angle, i.e., one angle = 90°	Right angled triangle or Right triangle	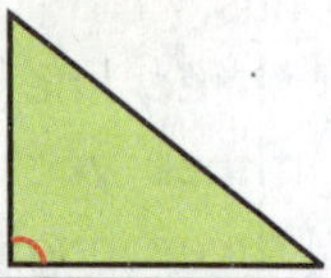
One angle is an obtuse angle, i.e., one angle > 90°	Obtuse angled triangle or Obtuse triangle.	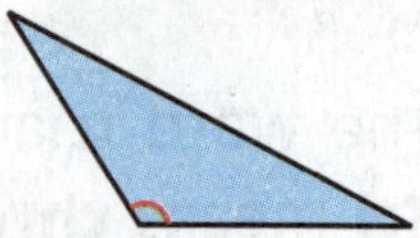

★ **By Lengths of its sides**

Just as triangles can be classified by the types of angles that they contain, they can also be classified according to the number of equal sides that thay have.

Length of Sides	Name	Example
All sides equal	Equilateral triangle	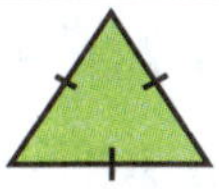
Two sides equal	Isosceles triangle	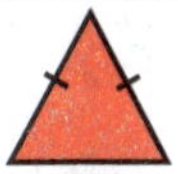
All sides of different lengths	Scalene triangle	

Note : Similar markings on the sides of a triangle show equal sides.
All equilateral triangles are isosceles triangles, but isosceles triangles are not equilateral triangles.

1. Measure the sides of the triangles and classify as equilateral, scalene or isosceles.

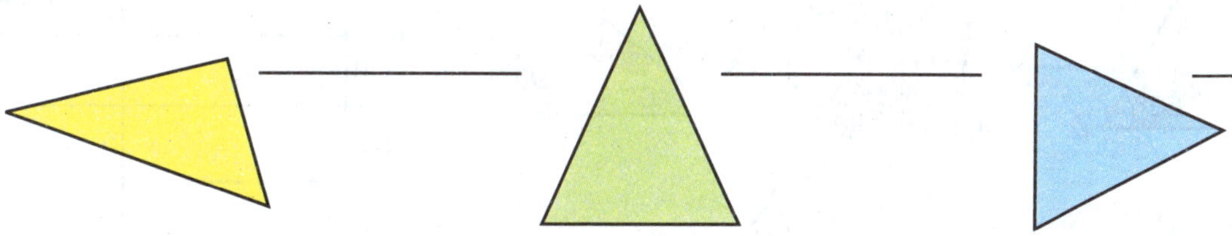

2. Measure the angles of the triangles and classify as acute, obtuse or right triangle.

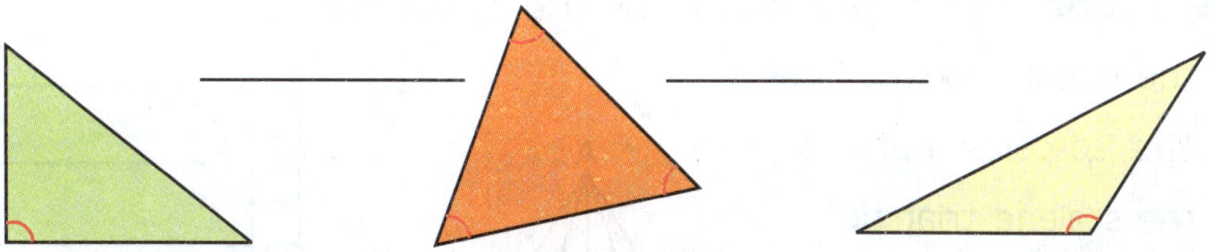

3. Name the vertices, sides and the triangle:

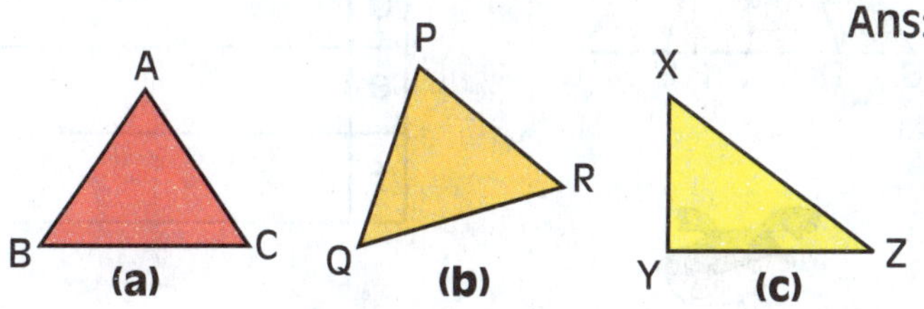

Ans:

	Name of the triangle	Vertices	Sides
1.			
2.			
3.			

4. Name the triangle according to sides:

(a) AB = 3 cm, BC = 4 cm, CA = 6 cm

(b) AB = 4 cm, BC = 5 cm, CA = 5 cm

(c) PQ = 5.2 cm, QR = 5.2 cm, PR = 5.2 cm

(d) PQ = 3.5 cm, QR = 4 cm, PR = 3.5 cm

Ans:

	Triangle	Kind of triangle
(a)		
(b)		
(c)		
(d)		

(e)

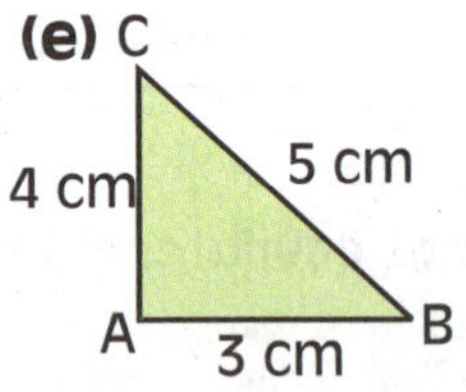

(f)

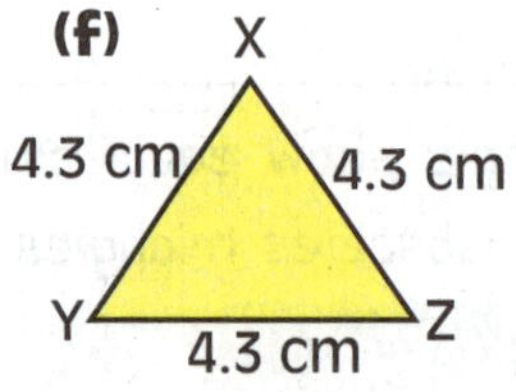

(g)

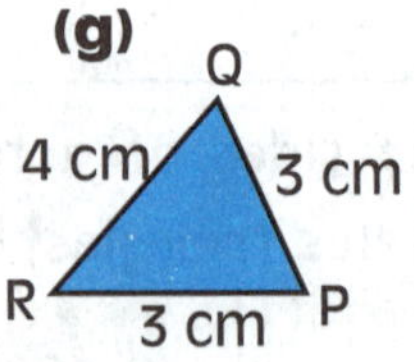

	Triangle	Kind of triangle
(e)		
(f)		
(g)		

5. Name the triangle according to angles:

(a)

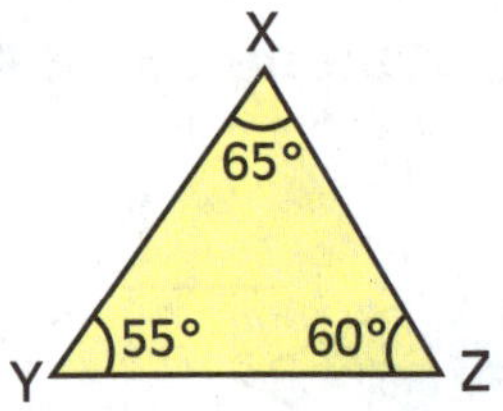

(b)

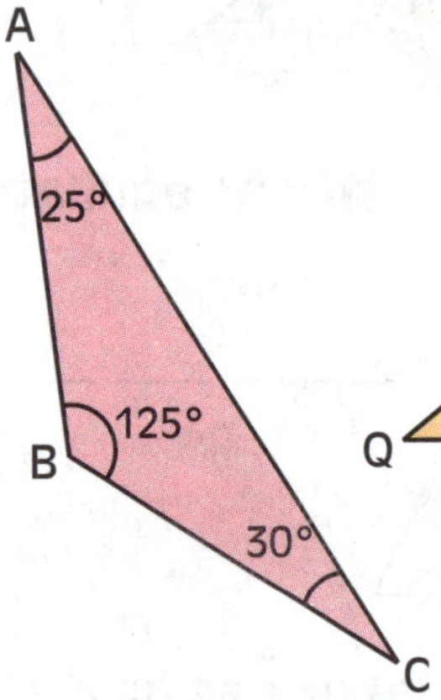

(c)

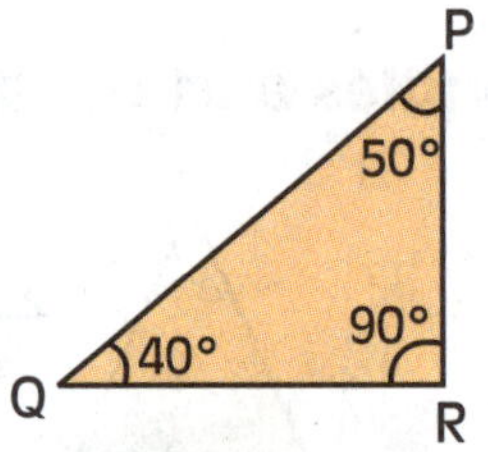

	Triangle	Kind of triangle
(a)		
(b)		
(c)		

6. Study the figure and answer the following questions:

(a) Name the equilateral triangles.

(b) Name the isosceles triangles.

(c) Name the scalene triangles.

(d) Name the acute triangles.

(e) Name the obtuse triangles.

(f) Name the right triangles.

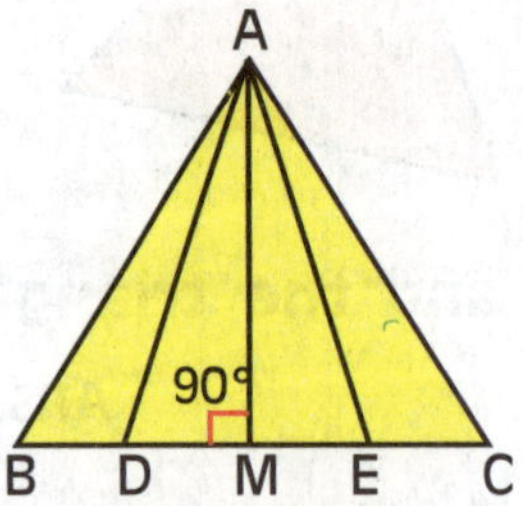

Ans:

a	
b	
c	
d	
e	
f	

7. Practical : Draw in your copybook.

1. 3 different acute triangles.

2. 3 different obtuse triangles.

3. 3 different right triangles.

ANGLE SUM PROPERTY OF A TRIANGLE

ACTIVITY 1

Given below are three triangles. Measure the three angles of each of them with a protractor and fill in the blank columns of the given table.

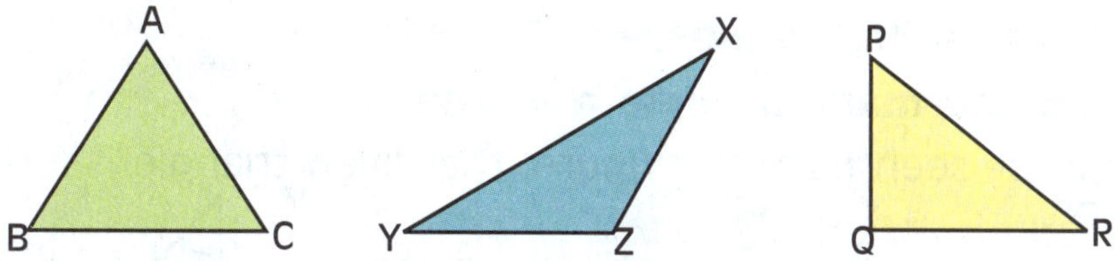

Triangle	First angle	Second angle	Third angle	Sum of the measures of the three angles
ΔABC	$\angle A =$	$\angle B =$	$\angle C =$	
ΔXYZ	$\angle X =$	$\angle Y =$	$\angle Z =$	
ΔPQR	$\angle P =$	$\angle Q =$	$\angle R =$	

What is the sum of the measures of the three angles in all the three triangles ? Is it 180°.

It follows from the above activity that

The sum of the measures of the three angles of a triangle is 180°.

ACTIVITY 2

The statement given above can also be shown true by the following activity.

Cut out a triangle from a piece of paper and follow the steps given below:

1. Label the three angles of the triangle as *a*, *b* and *c*.

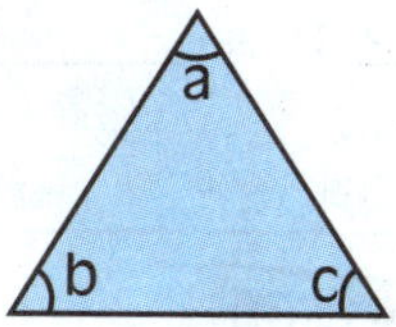

2. Cut the triangle into 3 pieces such that each piece has only 1 of the 3 marked angles.

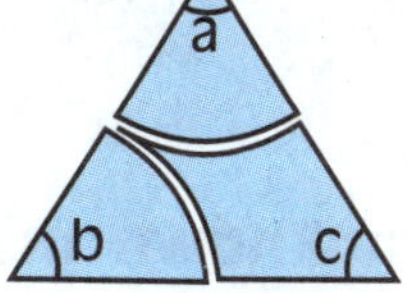

3. Rearrange the pieces as shown.

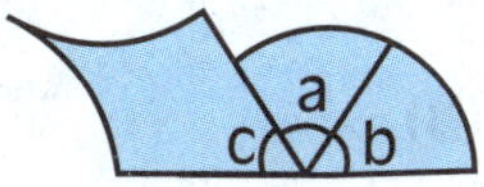

What do you see ?

The three angles of the triangle lie on a straight line. A straight line makes a straight angle = 180°

Therefore $\angle a + \angle b + \angle c = 180°$

It follows from the above activity also that

The sum of the measures of the three angles of a triangle is 180°.

Maths Alert : If the sum of the three angles is either less than or greater than 180°, they are not the angles of a triangle.

Example 1 Find the unknown angle x in the figure.

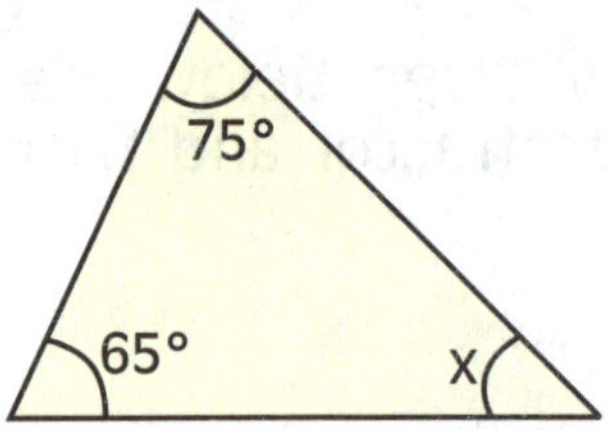

Solution. As the sum of the three angles of a triangle = 180°.

$\angle x = 180° - 65° - 75°$

$= 180° - 140° = \mathbf{40°}.$

Example 2. Find the marked angle a in the figure.

Solution. As can be seen from the figure, the given triangle is a right triangle. So one of its angles = 90°.

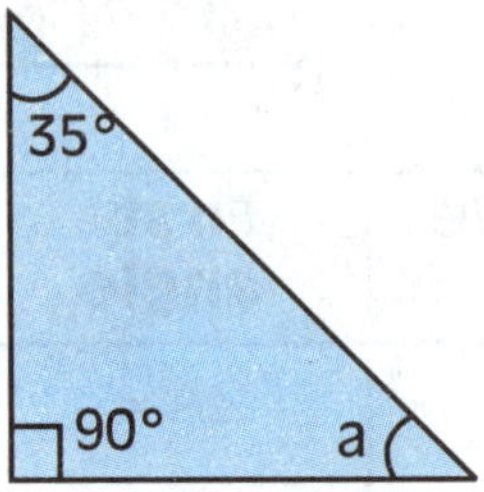

As the sum of three angles of a triangle = 180°, so

$90° + 35° + \angle a = 180°$

$\therefore \quad \angle a = 180° - 90° - 35°$

$= 180° - 125° = \mathbf{55°}.$

Skill Practice

Class Work

Find the unknown marked angle in each of the triangle given below.

1.

$\angle P =$ ______

2.

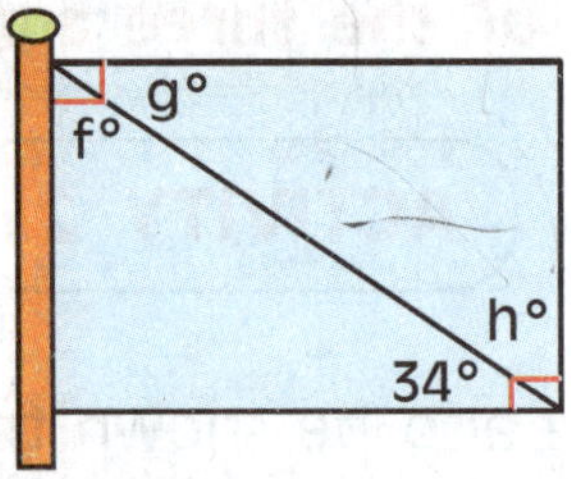

$\angle f =$ ____, $\angle g =$ ____, $\angle h =$ ____

3.

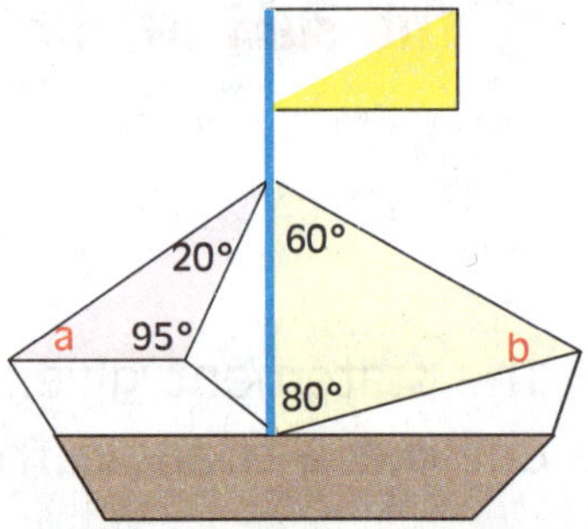

$\angle a =$ ______

$\angle b =$ ______

Home Work

EXERCISE 39

1. Find the unknown marked angle in each of the triangles given below.

(a)

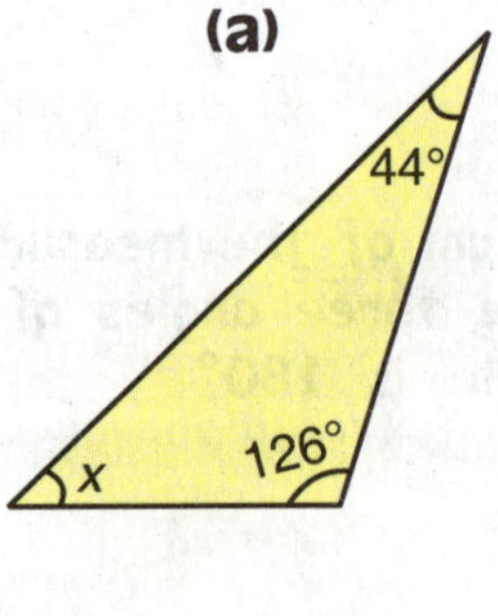

(b)

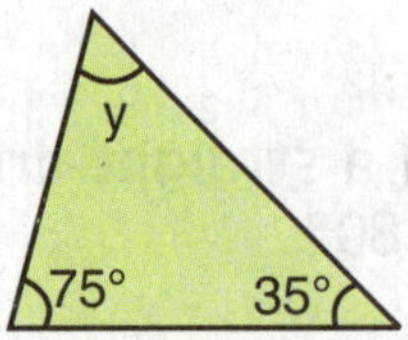

(c)

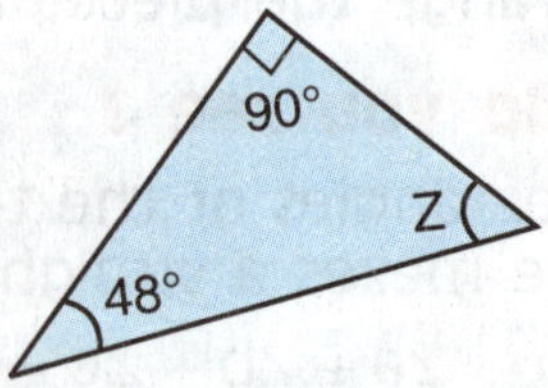

(d)

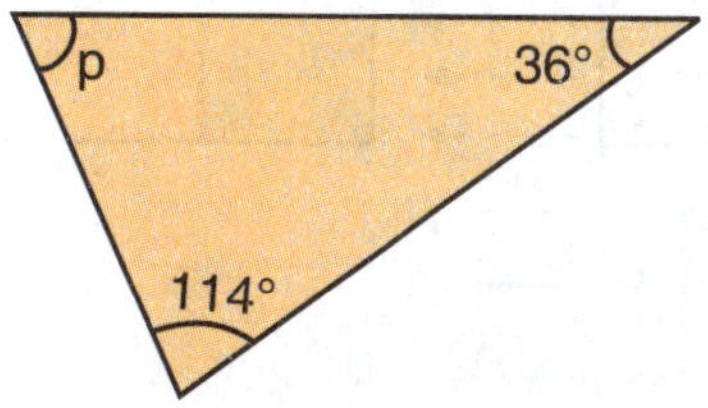

(e)

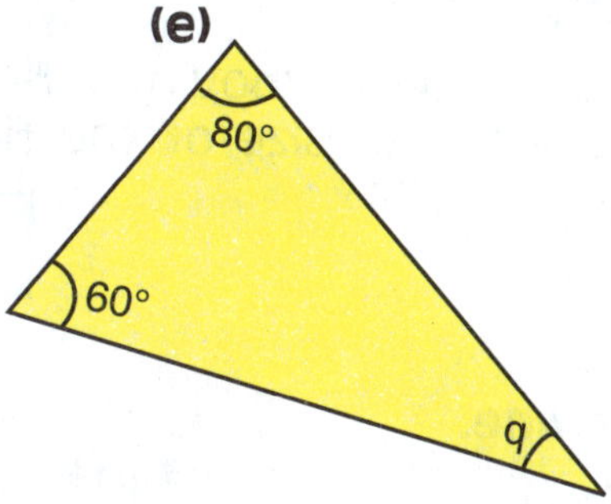

(f)

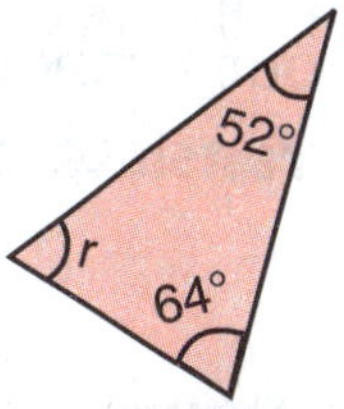

2. Find the unknown angle in each of the triangles below. Which of them are right triangles?

(a)

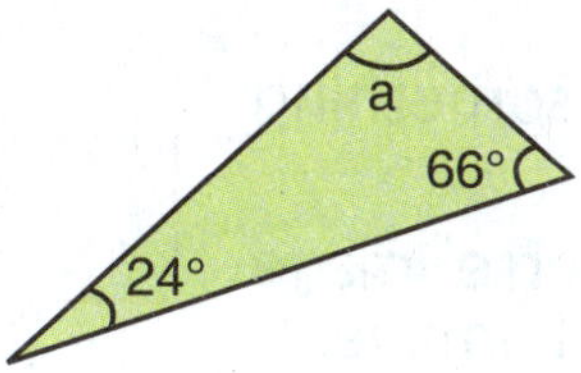

(b)

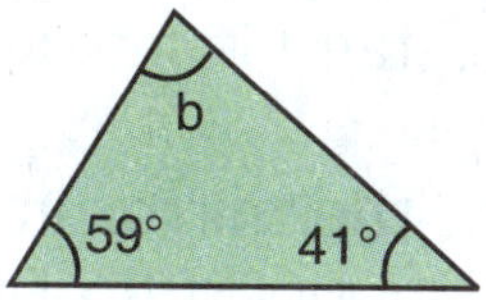

(c)

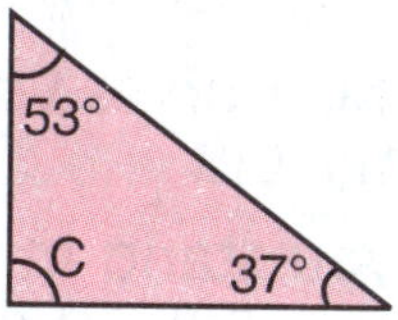

(d)

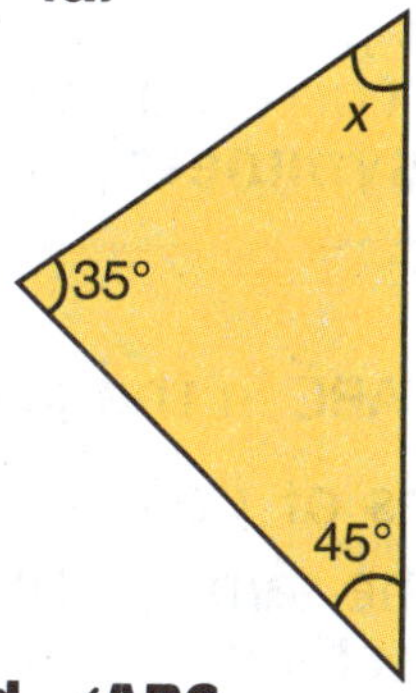

(e)

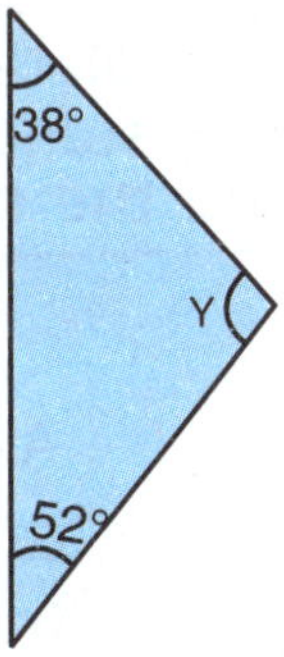

(f)

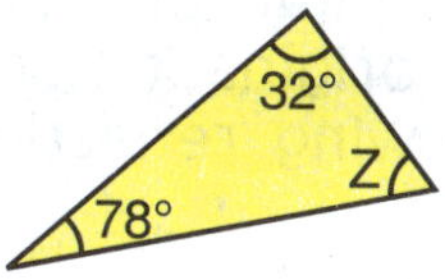

3. Find ∠ABC.

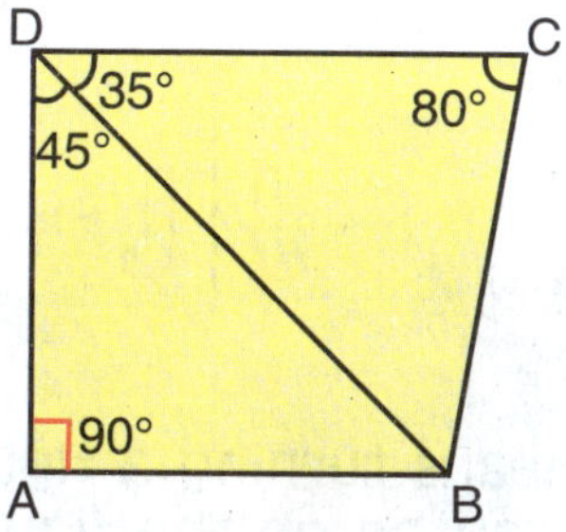

Hint : ∠ ABC = ∠ ABD + ∠ DBC

4. If all the angles of a triangle are of the same measure, find the measure of each angle.

5. In ΔABC, ∠A = 136° and ∠C = 21°, find ∠B.

6. In a right triangle, one acute angle measures 28°, find the other acute angle.

Can the following measures of three angles be the measures of the angles of a triangle?

7. 40°, 95°, 55° **8.** 61°, 30°, 90° **9.** 45°, 70°, 35° **10.** 108°, 42°, 30°

[**Hint:** The sum of the angles of a triangle should be equal to 180°. Find out the sum of the angles in each case. If the sum is more or less than 180°, then the given angles cannot be the angles of a triangle.]

REFLECTION AND ROTATION

There are broadly three different ways of moving a figure from one place to the other in a plane, without changing the shape or size of the figure.

1. Slide or translate a figure.

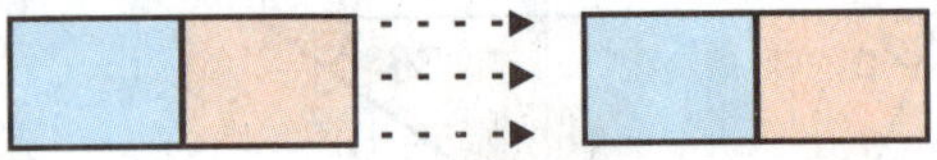

2. Reflect a figure across a line.

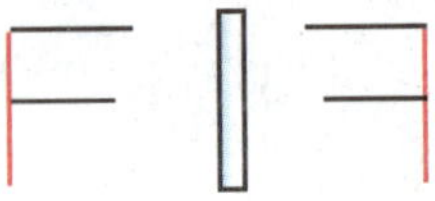

3. Rotate a figure around a point.

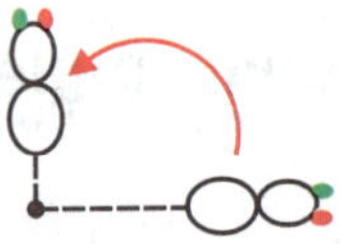

★ REFLECTION

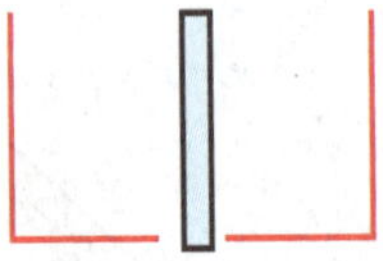

Reflection as you normally understand is seeing the image of something in a mirror.

If you stand in front of a mirror, the mirror image is the reverse image of yours. Right becomes left and left becomes right in a mirror image.

Similarly, reflection in geometry is flipping a figure across a line such that it appears to be the mirror image or reflection of the original object. Each point in the reflected figure (mirror image) is at the same distance from the line as the corresponding point in the original object. The line is called the **mirror line** or **line of symmetry.** Look at the following reflection.

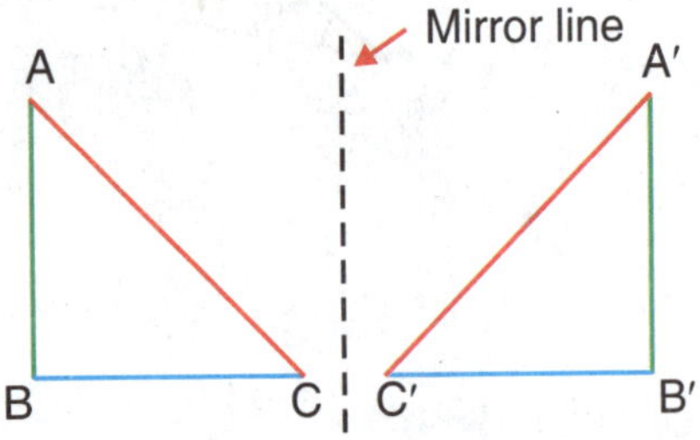

A′B′C′ is the reflection of ABC in the given line. As can be seen, the distances of points. A′, B′ and C′ from the mirror line are the same as the distances of points A, B and C respectively from the mirror line.

Some examples of reflection are:

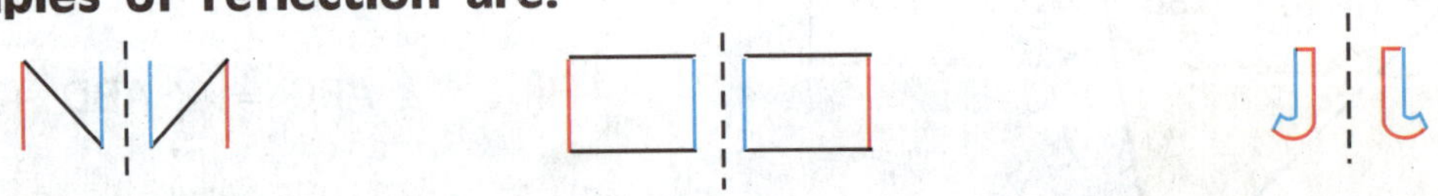

★ ROTATION

Rotation means turning. In geometry, rotating a figure means turning a figure about a given point, known as the **centre of rotation.** This point can be on the figure or any other point.

Some examples of rotation from real life situations are :

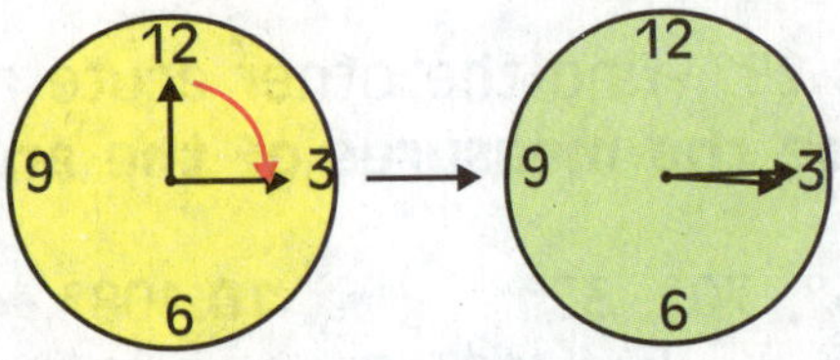

Rotation of the hands of a clock about the point where they meet

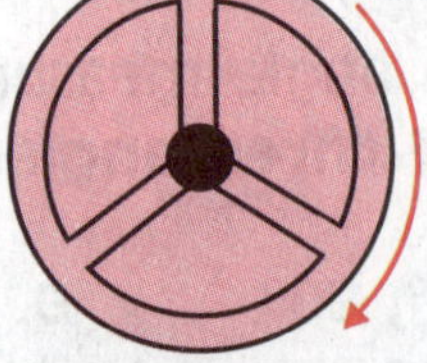

Steering wheel of a car about the centre

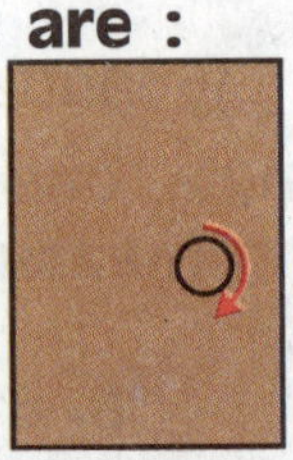

Knob of a door

Windmill

You can also rotate 2-D shapes or plane shapes about a point. Three things are needed to describe a rotation :

- the direction of rotation – clockwise or anticlockwise
- the angle of rotation
- the centre of rotation

The example given below shows the rotation of a square about a given point. The square has been further divided into coloured squares to make you understand clearly its movement.

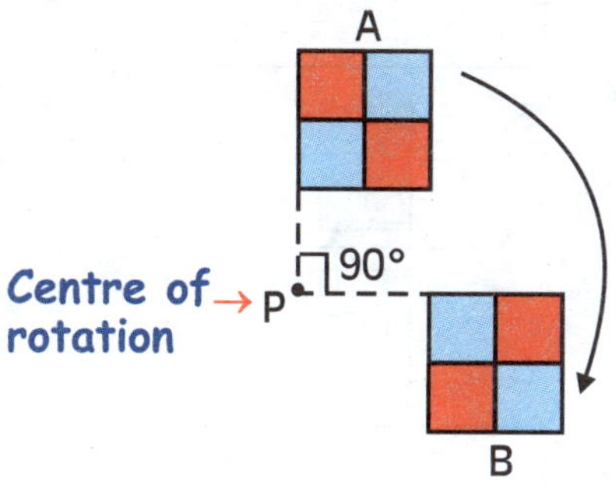

One quarter Turn

Shape *A* is rotated through an angle of 90° about the point P in the clockwise direction to get shape *B*.

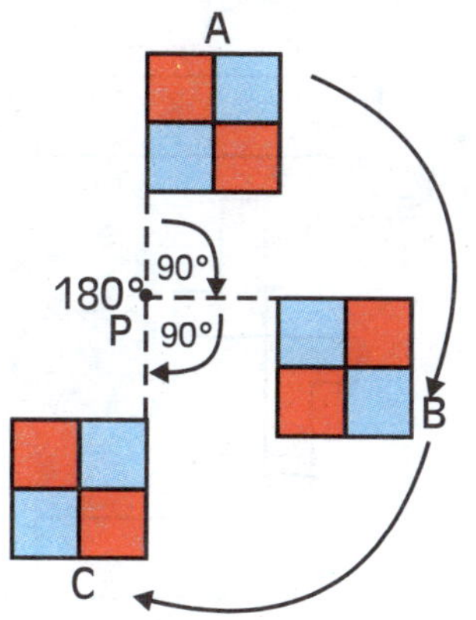

Half Turn

Shape *B* is further rotated through an angle of 90° about the same point *P* in the clockwise direction to get shape *C*. This is the same as if we would rotate shape *A* through an angle of 180° about point *P* in clockwise direction to get shape *C*.

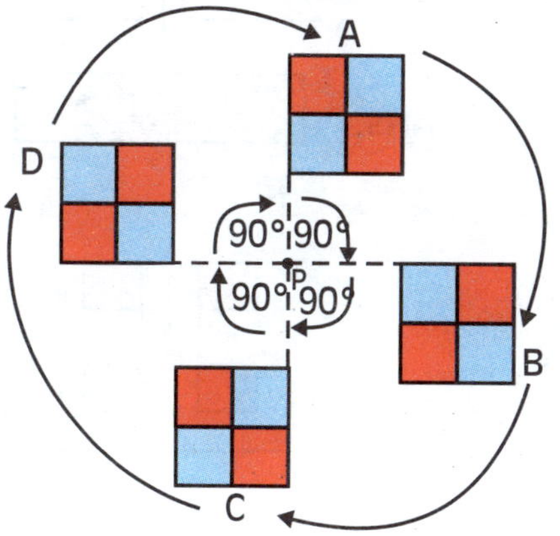

Full Turn

Shape *C* is again rotated by an angle of 90° about point *P* to get shape *D*.

What happens when shape D is further rotated through an angle of 90° about point P in clockwise direction?

We get back to shape *A*.

Some more examples are:

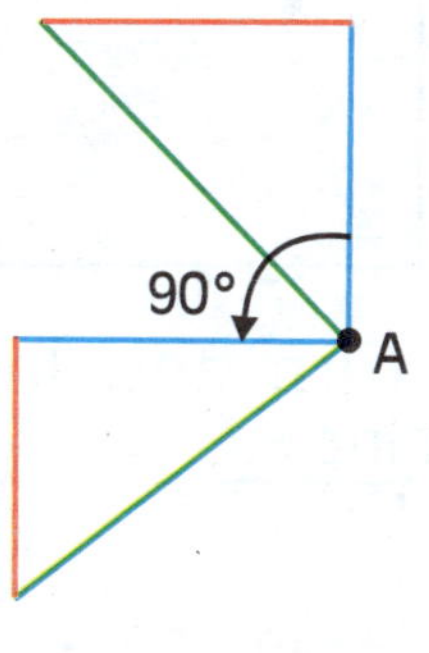

Given triangle is rotated about point A, through an angle of 90° in the anticlockwise direction.

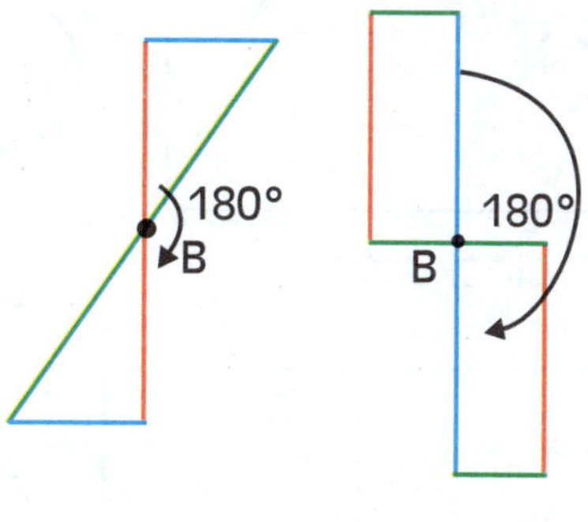

Given figures are rotated about point B, through an angle of 180° in the clockwise direction.

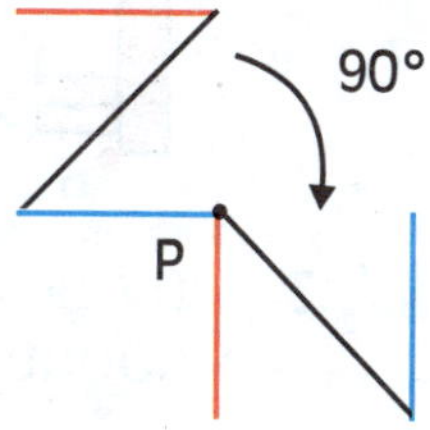

See what happens to letter Z when it turns through an angle of 90°, about the centre of rotation point P in a clockwise direction!

It becomes letter N.

Class Work

Skill Practice

I. Say whether the shape has been rotated or reflected:

1. 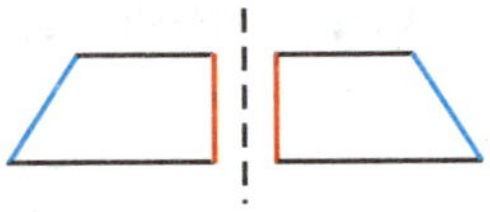

Ans. ____________

2. 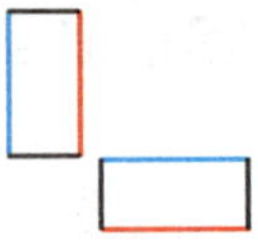

Ans. ____________

3. 

Ans. ____________

4. 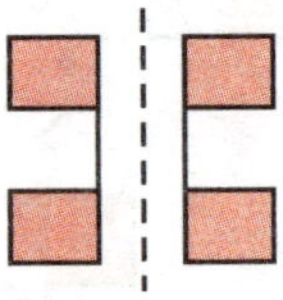

Ans. ____________

5. 

Ans. ____________

6. 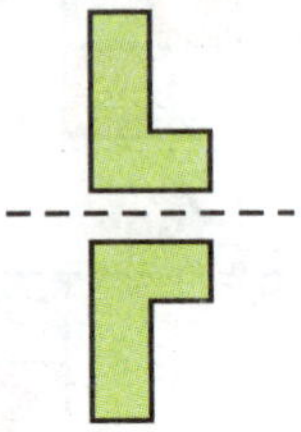

Ans. ____________

II. Reflect the following figures in the given mirror line

1. 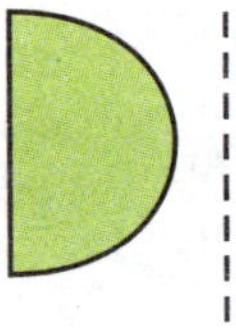

2. 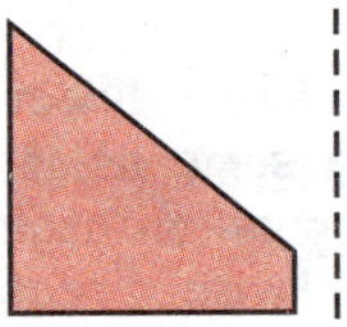

3. 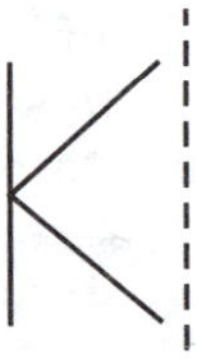

III. Rotate the following figures:

1. By 90° (quarter turn) clockwise about point *P*.

Ans.

2. By half turn (180°) anti clockwise about point *Q*.

Ans.

3. By 90° (quarter turn) clockwise about point *R*.

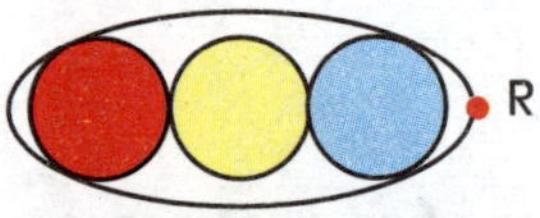

Ans.

4. By half turn anticlockwise about point *S*.

Ans.

Note: You can physically check the rotation of the above figures by cutting out the figures from hard paper and putting a pin through the point of rotation and then rotating about this point.

SYMMETRY IN SOLIDS

Symmetry is the exact likeness in a shape about a given line, point or plane.

You have already studied in class IV that plane shapes or figures which can be folded over in such a way that the two parts match exactly are called symmetrical figures and the fold line is called the line of symmetry.

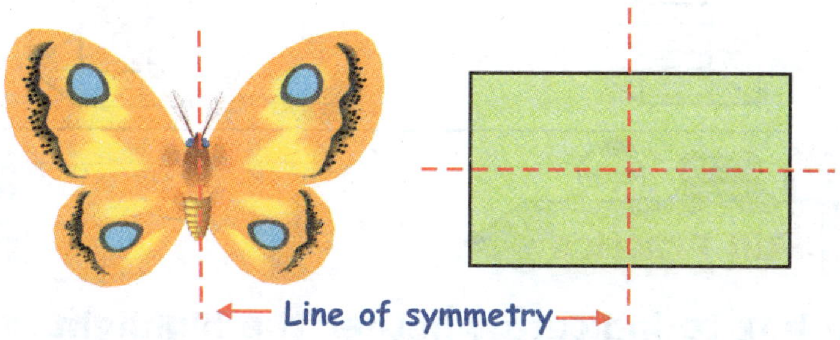

A symmetrical figure can have one line of symmetry or more than one lines of symmetry.

Solids also show symmetry. Some solids can be cut into halves so that both the halves are identical. The plane or that surface along which the solid is cut to give identical halves is called the **plane of symmetry**.

If we put a mirror (symbolising a plane) in front of each half, then each half is a reflection or mirror image of the other.

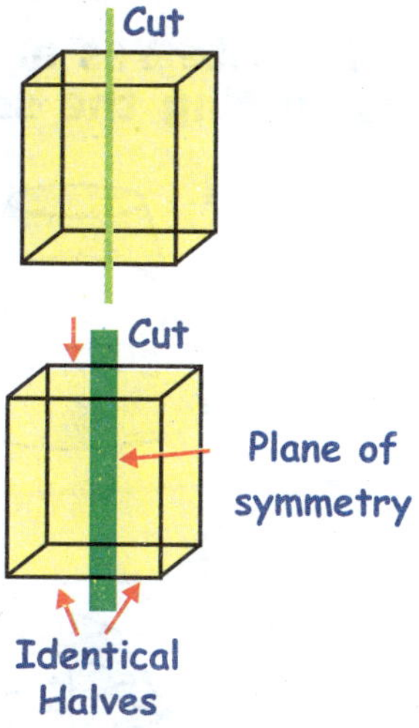

Just as flat (2-D) shapes may have more than one line of symmetry, similarly, solids can also have more than one plane of symmetry.

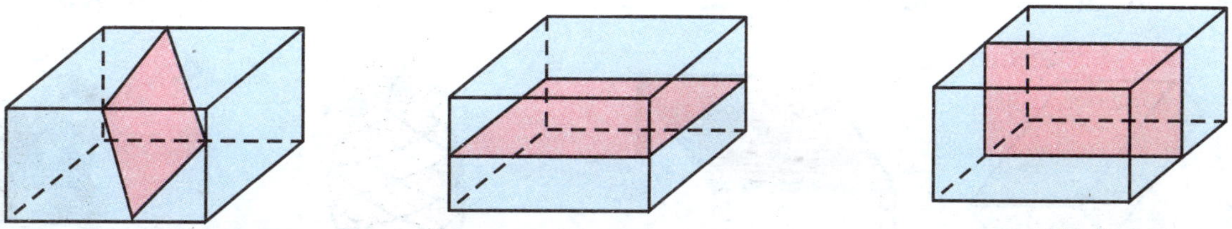

vertical cut   horizontal cut   vertical cut

Each cut is dividing the given solid into two identical halves. Here are some more examples:

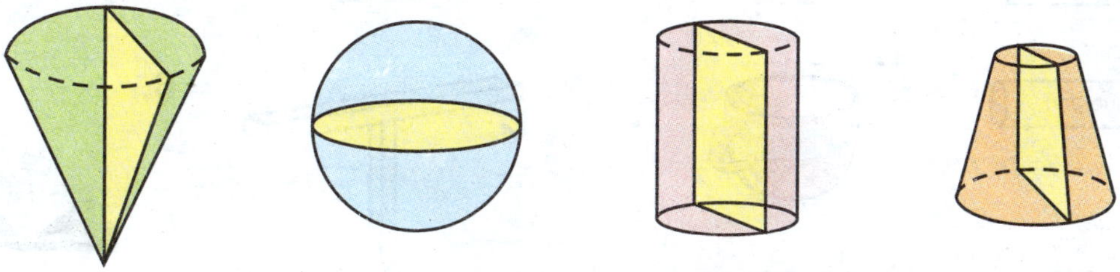

A cone and a sphere have infinite number of planes of symmetry.

A cube has many planes of symmetry. Here are some. Try to find out some with the help of your teacher by cutting a cubical box.

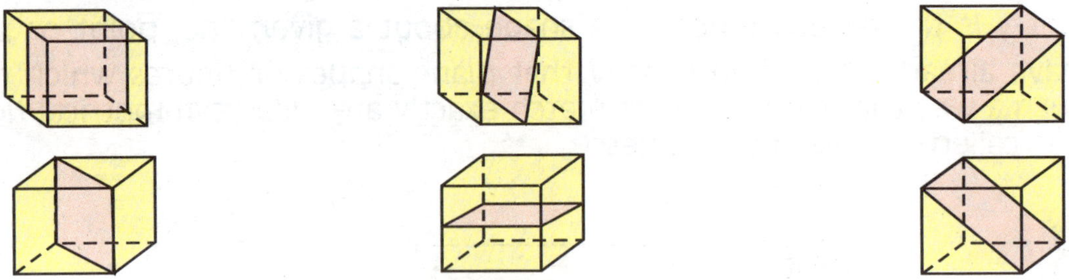

I. Write Yes or No in the given box to indicate whether the highlighted plane is cutting the solid into two identical halves or not.

1. 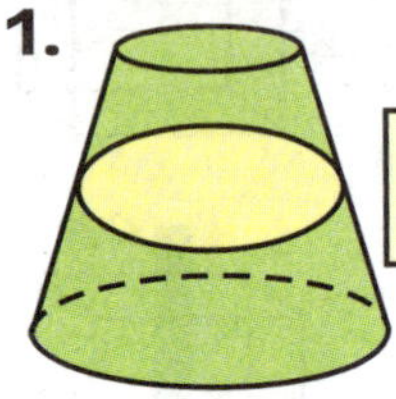 ☐

2. 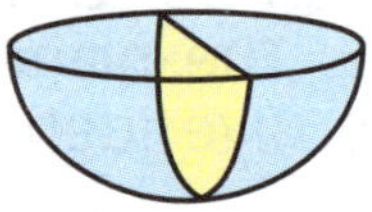 ☐

3. 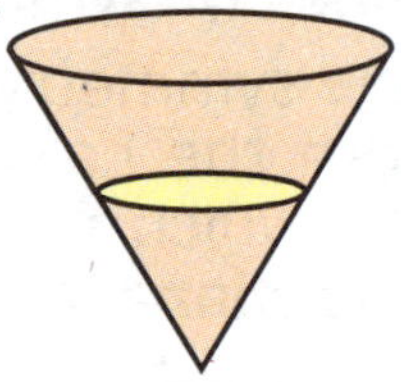 ☐

4. 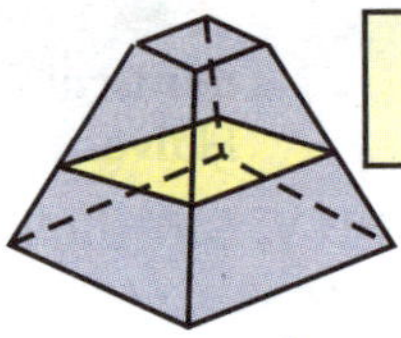 ☐

5. 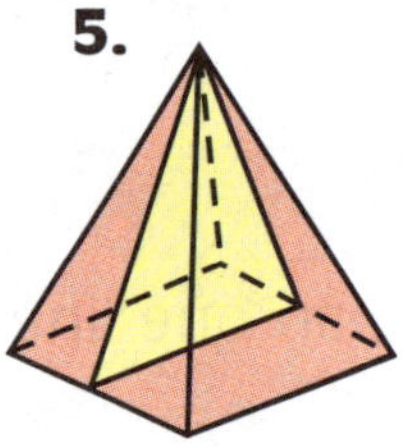 ☐

6.  ☐

II. Which of the following solids are symmetrical. Write Yes or No.

1. 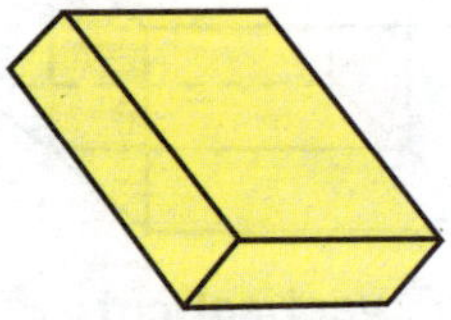 ☐

2. 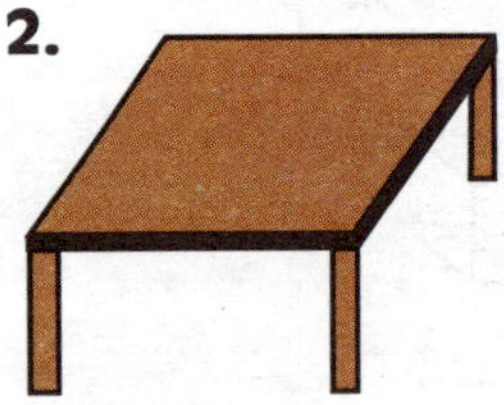 ☐

3.  ☐

4.  ☐

5. 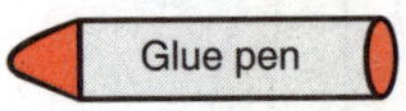

☐

6.  ☐

7. 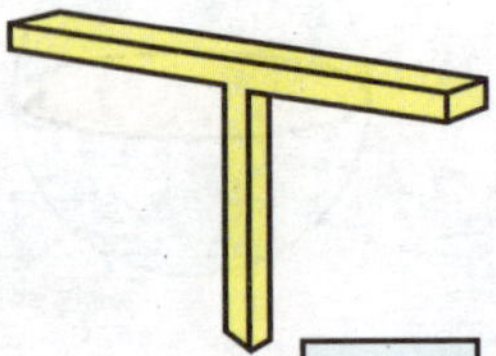 ☐

8.  ☐

NETS

If you could cut a 3D (solid) shape along some of its edges and open it out flat, you would get the **net** of the solid shape.

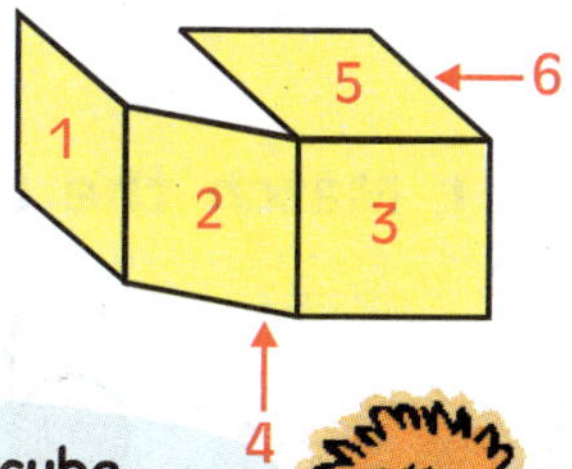

A net can be folded to make up the 3D shape.

If you open up a cube by cutting along its edges and then lay it flat you would get the net of the cube which looks like :

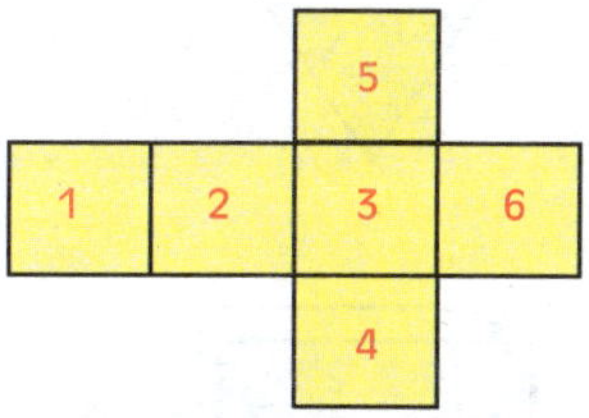

If you fold it up, it forms a cube.

Thus, we see that a net is a two dimensional figure which when folded forms a three dimensional figure.

There may be several possible nets for a 3D shape.

Here are the nets of some common 3D shapes you know.

Cube

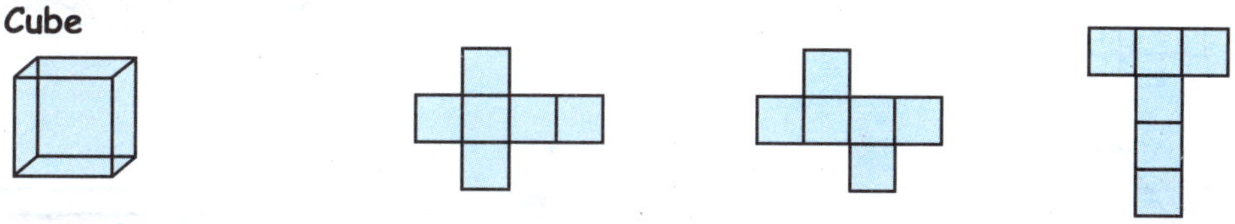

Cuboid

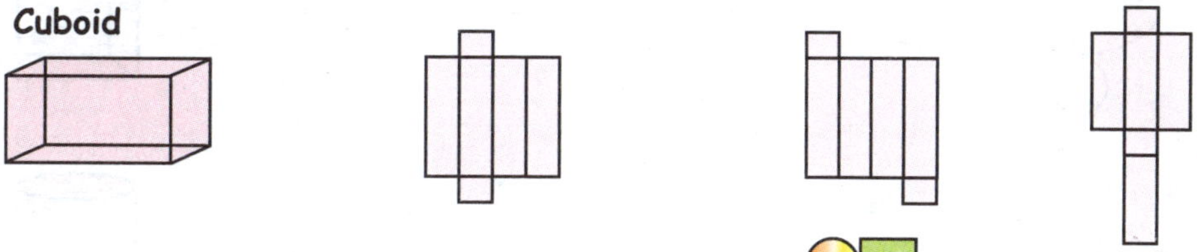

Cylinder

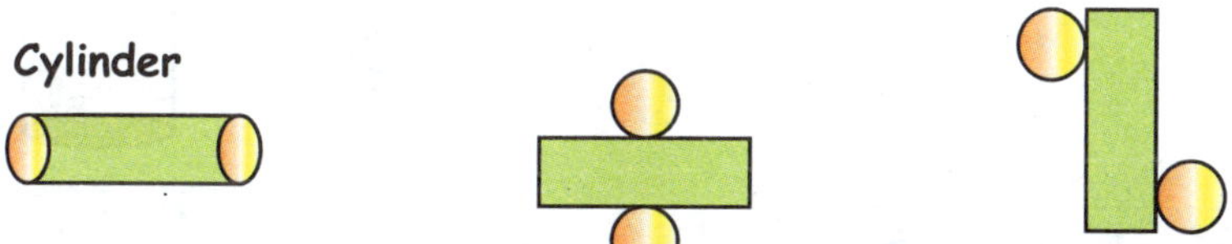

Cone

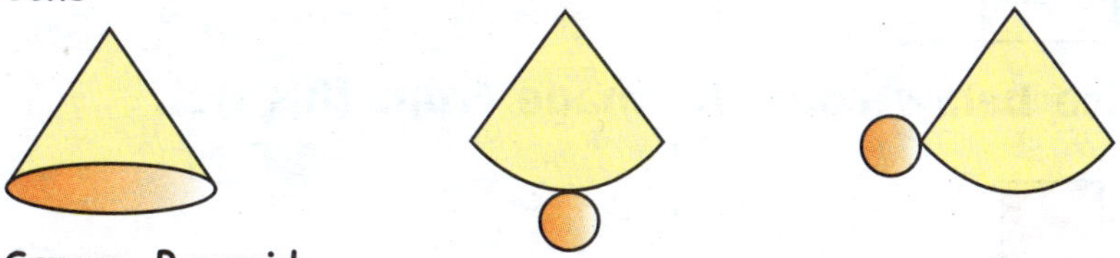

Square Pyramid

Triangular Prism

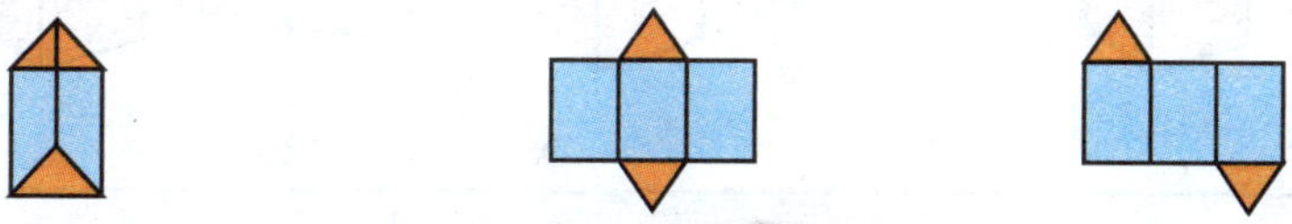

Skill Practice

Class Work

I. Match the nets of the respective solids:

1. 

A. 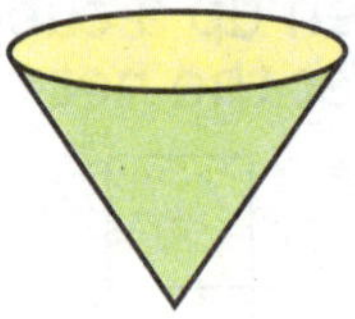

2. 

B. 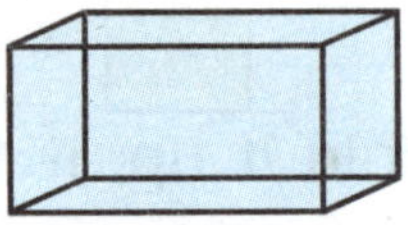

3. 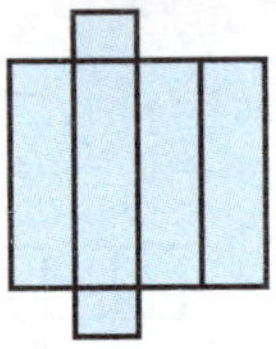

C. 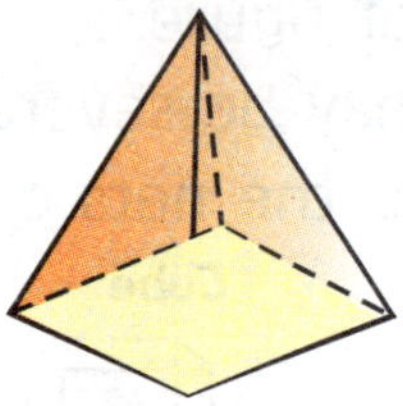

4. 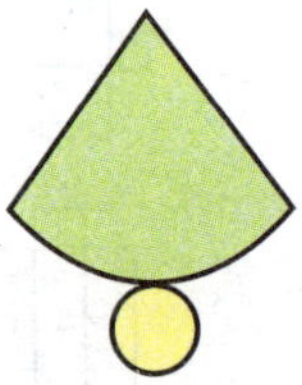

D. 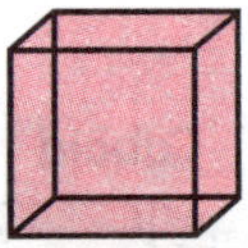

5. 

E. 

Ans:	1	2	3	4	5

II. Which of the dice below could be made from this net.

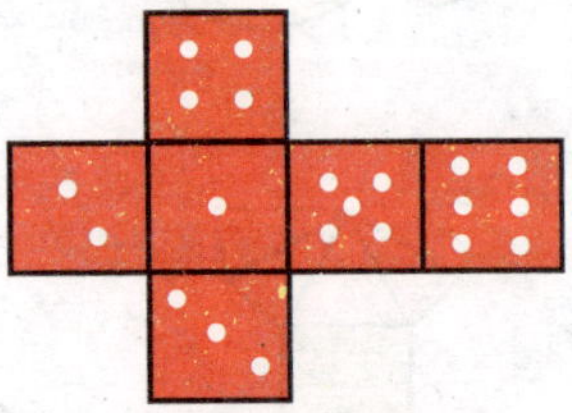

(a)  **(b)**  **(c)**  **(d)**  **Ans.** ☐

CIRCLES

You have already studied about circles in class IV. Let us revise what you have studied and extend our study further.

DEFINITIONS

Circle

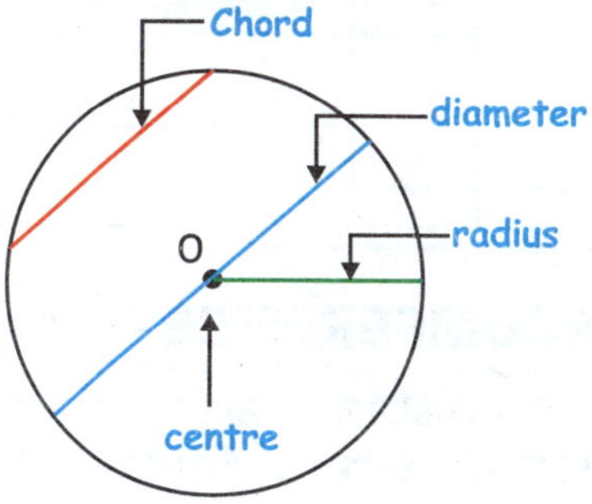

A **circle** is a simple closed figure, wherein, each point on the circle is at the same distance from a point inside the circle, called the **centre** of the circle.

We use the centre to name a circle. The given circle has centre marked as point O, so it is called circle O.

Parts of a Circle

Radius. A **radius** (Plural radii) is a line segment joining the centre to a point on the circle. All radii of a circle are equal.

Chord. A **chord** is a line segment joining any two points on a circle.

Diameter. A diameter is a line segment joining any two points on a circle and passing through the centre of the circle. Diameter is also a chord and the longest one. There are infinite diameters of a circle and all are equal.

A diameter divides a circle into two equal parts. Each part is called a **semi circle.**

Your protractor is an example of a semi-circle.

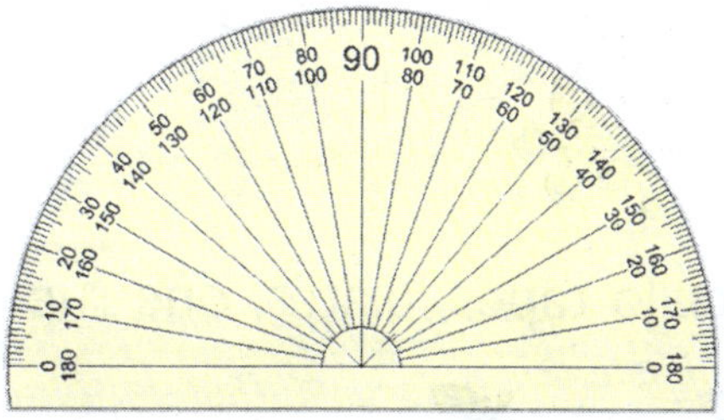

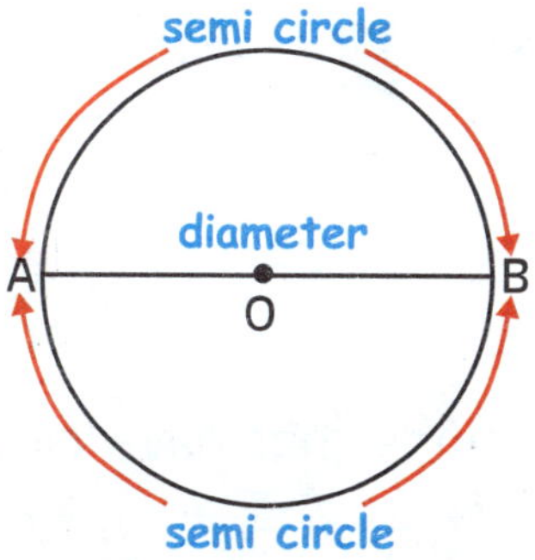

You also know that,

Diameter of a circle is twice the radius

Diameter = 2 × radius; Radius = diameter ÷ 2

I. Write the correct part of the circle drawn alongside against the given names.

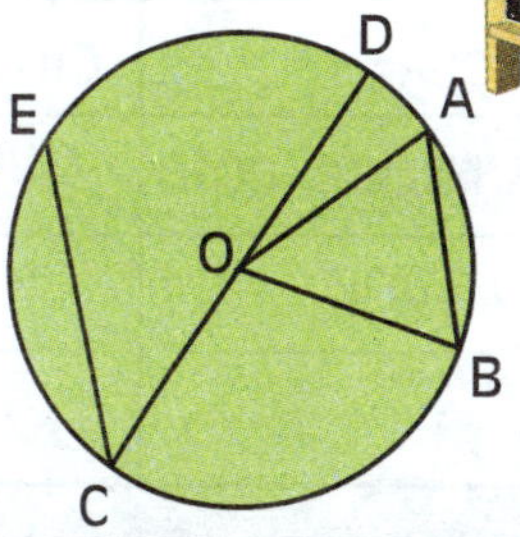

1. $\overline{CD}$ ______________
2. $\overline{AB}$ Chord
3. $\overline{OA}$ ______________
4. $\overline{CE}$ ______________

5. $\overline{OD}$ ______________ 6. $\overline{OC}$ ______________

7. point O ______________ 8. $\overline{OB}$ ______________

II. Fill in the blanks:

	Radius	**Diameter**		**Radius**	**Diameter**
1.	8 cm		**5.**	2.5 m	
2.	4.3 cm		**6.**		7 m
3.		18 cm	**7.**	11.25 cm	
4.		16.4 cm	**8.**		12.50 m

CIRCUMFERENCE

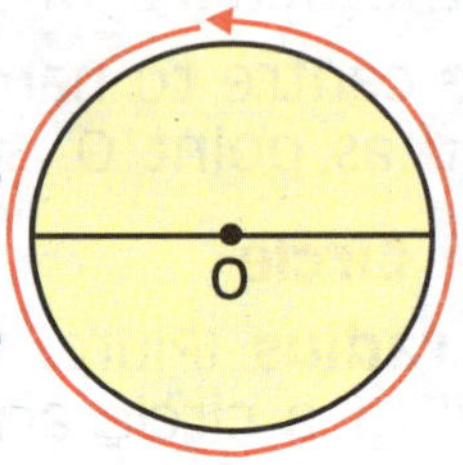

The distance around a circle or the perimeter of a circle is called its **circumference**.

As a circle is a curve, so you cannot measure the circumference with the help of a ruler or a measuring tape.

To measure the circumference of a circle, wrap a piece of string around the circumference of the circular object and then unwrap the string and measure the length of the string with a ruler.

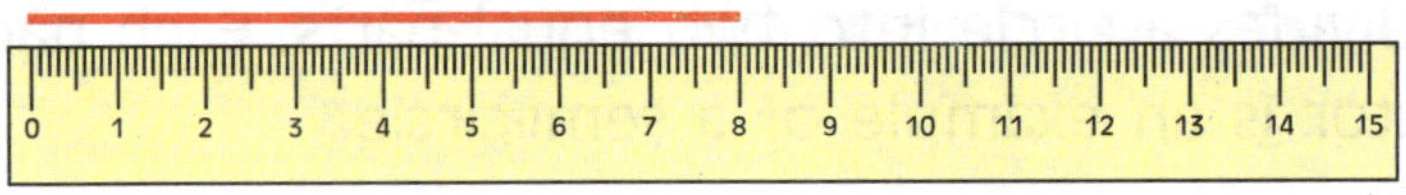

Collect things that have circular faces, such as lids, cello tapes, plates, one rupee coin, glue pots etc.

- ★ Measure the circumference of these objects by the method given above.
- ★ Measure the diameter of each of them by a ruler.

Make a table like this:

Object	Circumference (c)	Diameter (d)	$C \div d$
One rupee coin			
Cello tape			
Plate			

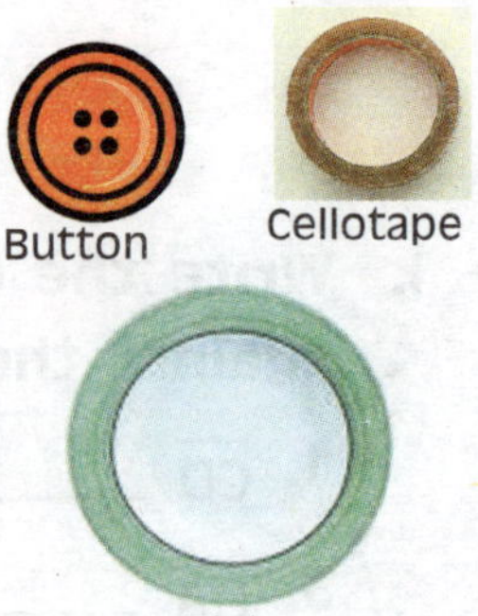

What do you notice ?

The circumference of all these objects is just over three times greater than the length of their respective diameters.

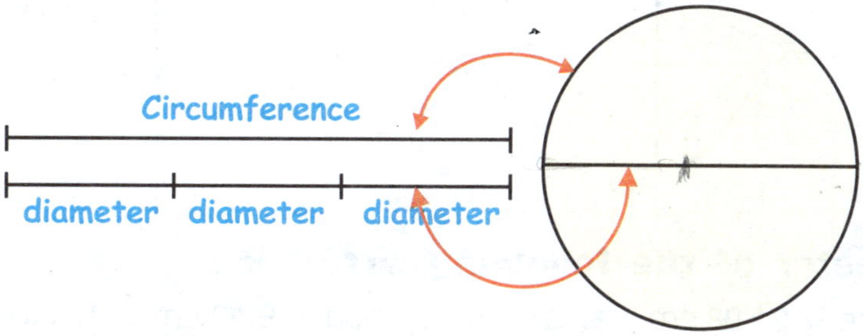

The circumference of a circle is equal to 3.14 times its diameter.

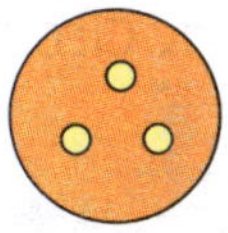

$$\text{Circumference} = 3.14 \times \text{diameter} \text{ or } \frac{22}{7} \times \text{diameter}$$

Example 1. Find the circumference of circle whose diameter is 8 cm.

Solution. Circumference = 3.14 × diameter

= 3.14 × 8 = **25.12 cm.**

Example 2. Find the circumference of a button whose radius is 2.5 cm.

Solution. You know that diameter = 2 × radius

∴ diameter of the button = 2 × 2.5 cm

= **5 cm.**

Circumference of the button = 3.14 × 5 cm

= **15.7 cm.**

Example 3. Find the diameter of a bangle whose circumference is 18.84 cm. Also find its radius.

Solution. Circumference = 3.14 × diameter

∴ diameter = circumference ÷ 3.14

$$\text{Diameter of the bangle} = \frac{18.84}{3.14}\text{ cm} = \frac{1884}{314}\text{ cm} = 6\text{ cm}.$$

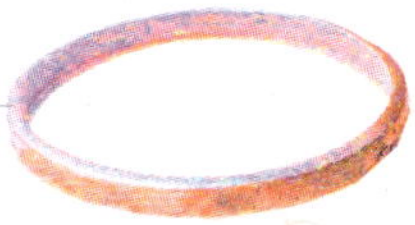

Radius = diameter ÷ 2

∴ radius of the bangle = 6 cm ÷ 2 = **3 cm.**

I. Find the circumference of the following circles if :

1. diameter = 4 cm

2. diameter = 10 cm

3. diameter = 20 cm

4. radius = 1 cm 4	**5.** radius = 4.5 cm 16	**6.** radius = 12 cm 48

II. Find the diameter of the following circles if :

1. Circumference = 21.98 cm	**2.** Circumference = 9.42 cm	**3.** Circumference = 125.6 cm

III. Find the radius of the following circles if :

1. Circumference = 31.4 cm	**2.** Circumference = 25.12 cm	**3.** Circumference = 62.8 cm

MENTAL MATHS–7

1. Name the type of angle formed in each case:

(a) (b) (c) (d)

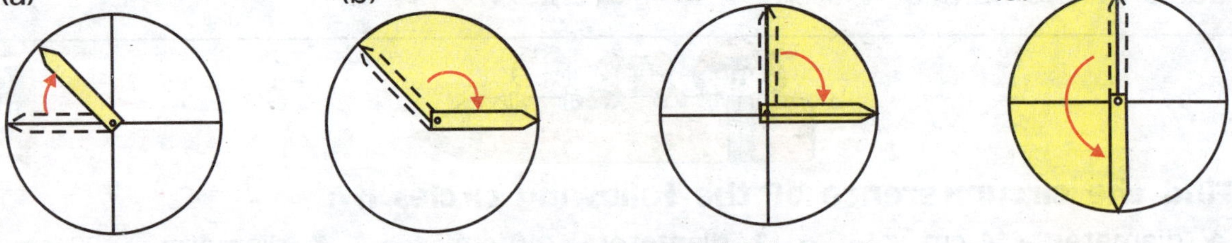

2. How many degrees are there in two complete turns ?

3. In the given diagram of a road crossing, name

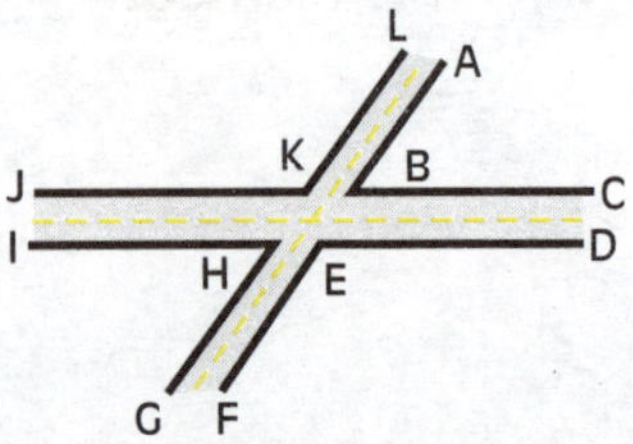

(a) two acute angles

(b) two obtuse angles

4. A plane figure with three sides is called a ________________ .

5. A triangle having sides 6.3 cm, 4.7 cm and 6.3 cm is an ________________ triangle.

6. The sum of the angles of a triangle is equal to 2 right angles. True of False ?

7. If the diameter of a circle is 100 mm, what is its radius in cm ?

8. The chair given in the adjoining diagram is made of line segments. Name

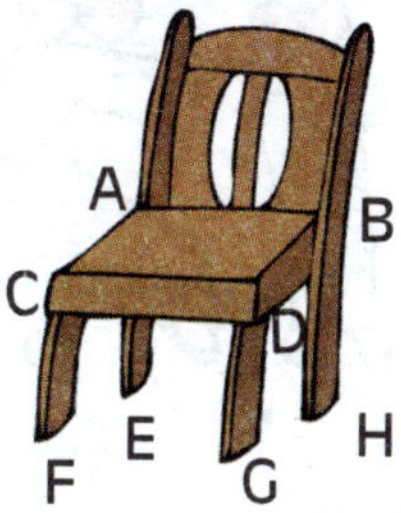

   (a) 2 pairs of parallel line segments.

   (b) 2 pairs of perpendicular line segments.

9. Circle the correct figure showing the letter after it is rotated clockwise through a half turn? (F)

(a)  **(b)**  **(c)**  **(d)** 

10. Circumference of a circle = 3.14 × ________.

KNOW YOUR NUMBERS

THE NUMBER FIVE

Five Olympic rings

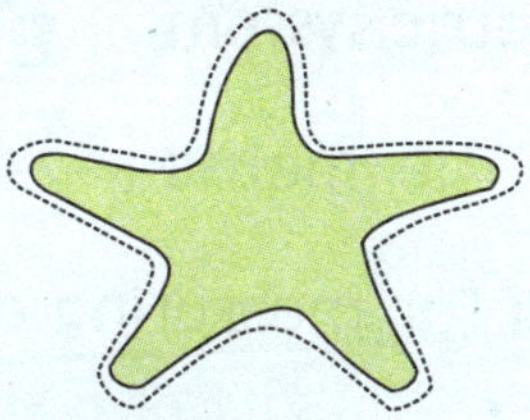

Starfish

5 is a prime number

We have five fingers on each hand. The Roman numeral V for five might have originated from the image of a hand.

Starfish usually have five arms. There are five rings in the Olympic symbol. Basketball is played with teams of five players each.

'Panch' in Sanskrit means "five". The number five has great importance in Indian history & Mythology. A set of 5 instruments used in South Indian temples is called *'Panchavadhya'*. The five principles of co-existence are called *'Panchasheel.*

The *five Ks* are worn by the Singhs following the Sikh religion. These are *kes,* long hair; *kangha*, a comb; *kirpan*, a small sword, *kachh*, short trousers; and *kara*, a steel bracelet.

A devout follower of Islam worships five times a day facing the holy city of Mecca. There are five vowels in the English alphabet. There are five senses. *Penta* means five. A *pentathlon* is an athletics contest with five events and a *pentagon* is a figure with five sides and five angles.

PENTAGON

12 MEASUREMENT

In your earlier classes you have seen how a change took place from the non-standard units of measurements such as hand span, cubit, pace etc., to the common standard units like kilometre, metre and centimetre of length; kilogram and gram of mass; litre and millilitre of capacity.

This set of standard units of measurement, that we use is known as the **metric system.** It uses numbers that are **multiples and sub-multiples of 10.**

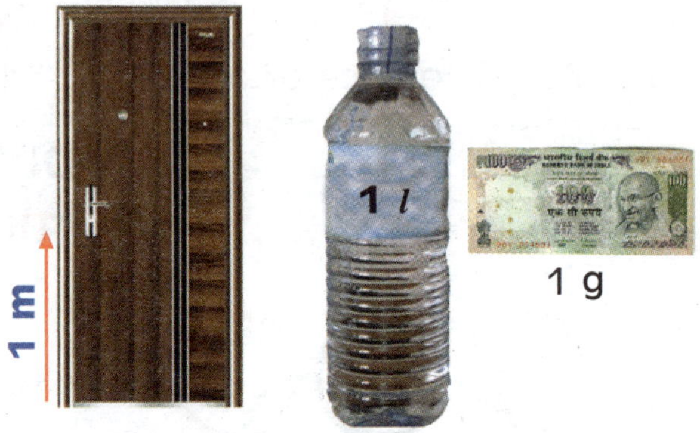

The standard unit of length in the metric system is a **metre (m).**

The standard unit of mass in the metric system is a **gram (g).**

The standard unit of capacity in the metric system is a **litre (*l*).**

These standard units are too small for measuring certain quantities and also too large for some.

Gram is too small a unit to measure the weight of a human being, whereas metre becomes too large to measure the length of a paper clip. Litre is too large to measure amount of liquid in a dropper, and too small to measure the capacity of a petrol tanker. So we introduce some more units of measures.

Some are higher than the standard units and some are lower. We name them by adding prefixes to the standard unit and the prefix shows how big or small the unit is compared to the standard unit (metre, gram, litre).

	Length	Mass	Capacity
Higher Unit	Kilometre	Kilogram	Kilolitre
	Hectometre	Hectogram	Hectolitre
	Decametre	Decagram	Decalitre
Standard Units	Metre	Gram	Litre
	Decimetre	Decigram	Decilitre
Lower units	Centimetre	Centigram	Centilitre
	Millimetre	Milligram	Millilitre

Meaning of Prefix

Prefix	Meaning	
Kilo	thousand	1000
Hecto	hundred	100
Deca	ten	10
Deci	one-tenth	$\frac{1}{10}$
Centi	one-hundredth	$\frac{1}{100}$
Milli	one-thousandth	$\frac{1}{1000}$

To name metric units other than the standard units (metre, gram, litre) a prefix precedes the standard unit.

Since, **kilo** means **1000,** therefore, kilometre (km) = 1000 metres.

As **milli** means $\frac{1}{1000}$, therefore, millilitre (ml) = $\frac{1}{1000}$ litre = 0.001 litre

Now we deal with all the measures i.e., length, mass and capacity and study how the various units relate with the standard unit and with each other.

LENGTH

Your height, the distance between Delhi and Bombay, the distance around a football ground etc. are all quantities which measure length.

The standard unit of length in the metric system is **metre (m).**

The chart given below shows the units of length in the metric system with metres as the reference unit.

1000	100	10	1	$\frac{1}{10}$	$\frac{1}{100}$	$\frac{1}{1000}$
kilometre (km)	hectometre (hm)	decametre (dam)	metre (m)	decimetre (dm)	centimetre (cm)	millimetre (mm)

This shows that the metric system is like the decimal system. From the above given chart, you can conclude the following:

10 millimetres = 1 centimetre	**1 dm = $\frac{1}{10}$ m = 0.1 m**
10 centimetres = 1 decimetre	**1 cm = $\frac{1}{100}$ m = 0.01 m**
10 decimetres = 1 metre	**1 mm = $\frac{1}{1000}$ m = 0.001 m**
100 centimetres = 1 metre	
1000 millimetre = 1 metre	
10 metres = 1 decametre	**1 km = 1000 m**
10 decametres = 1 hectometre	**1 hm = 100 m**
10 hectometres = 1 kilometre	**1 dam = 10 m**
100 metres = 1 hectometre	
1000 metres = 1 kilometre	

The **conversion of various units to metre and metre to various units** can be shown diagramatically as shown on the next page:

Note the direction of the arrows.

To change from higher unit to lower unit, you multiply.

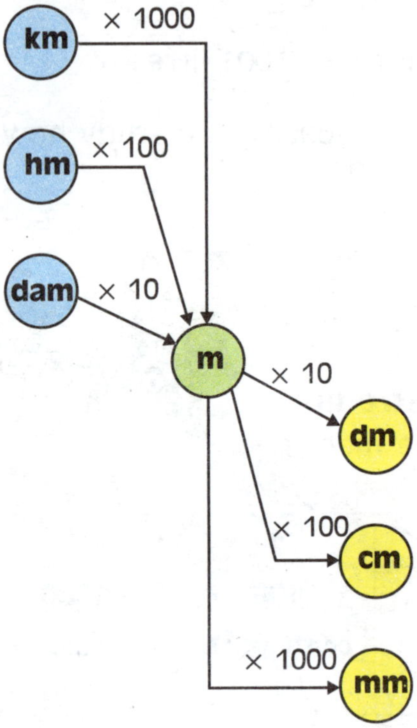

Examples :

4 km = 4 × 1000 m = 4000 m

6.5 hm = 6.5 × 100 m = 650 m

17.82 dam = 17.82 × 10 m = 178.2 m

11.5 m = 11.5 × 10 dm = 115 dm

6 m = 6 × 100 cm = 600 cm

8.07 m = 8.07 × 1000 mm = 8070 mm

To change from lower unit to higher unit, you divide.

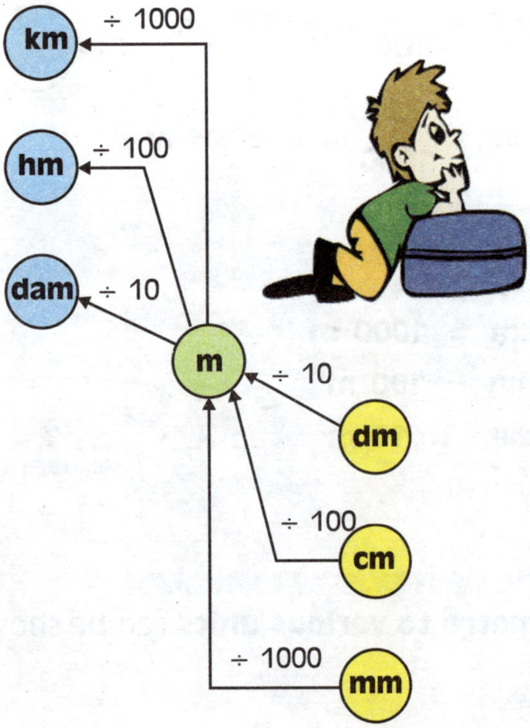

Examples :

92000 m = (92000 ÷ 1000) km = 92 km

870 m = (870 ÷ 100) hm = 8.7 hm

400 m = (400 ÷ 10) dam = 40 dam

5000 mm = (5000 ÷ 1000) m = 5 m

65 cm = (65 ÷ 100) m = 0.65 m

4.2 dm = (4.2 ÷ 10) m = 0.42 m

Class Work

Fill in the blanks:

1. 8 km	=	8 × 1000	m	=	8000	m
2. 1.26 hm	=		m	=		m
3. 16.5 dam	=		m	=		m
4. 32 m	=		cm	=		cm
5. 4.5 m	=		cm	=		cm
6. 9 m	=		mm	=		mm
7. 18.4 km	=		m	=		m
8. 7000 mm	=		m	=		m
9. 180 cm	=		m	=		m
10. 2 dm	=		m	=		m
11. 3400 m	=		hm	=		hm
12. 5000 m	=		Km	=		Km
13. 25 m	=		dam	=		dam
14. 6750 m	=		km	=		km

Conversion from one unit to the other units

The following diagram shows how to carry out these conversions.

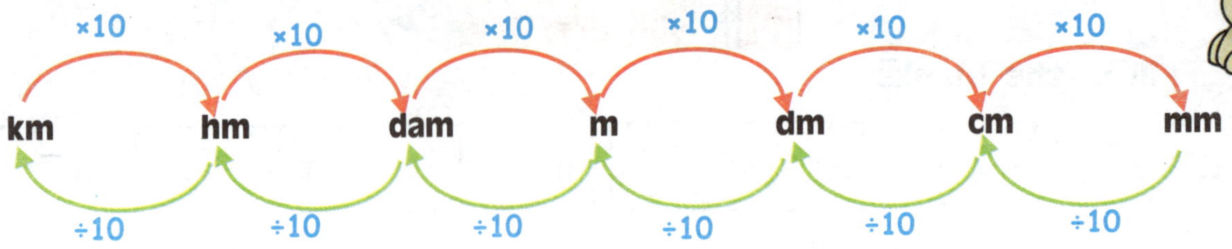

Moving right, you **multiply by 10** at each step.

Move one step right, multiply by 10. Move two steps right multiply by $10 \times 10 = 100$ and so on.

Moving left, you **divide by 10** at each step.

Move one step left, divide by 10, move two steps left, divide by 100 and so on.

Example 1. Convert (a) 46 hm to dam (b) 12 km to dam (c) 6.2 dm to mm (d) 4 cm to mm (e) 12 dam to cm.

Help Line : While doing problems relating to interconversion, you should first draw the reference chart like the one given below:

km hm dam m dm cm mm

(a) 46 hm = (46×10) dam = **460 dam.**

dam is one step to the right of hm, so multiply by 10.

(b) 12 km = (12×100) dam = **1200 dam.**

dam is two steps to the right of km, so multiply by $10 \times 10 = 100$.

(c) 6.2 dm = (6.2×100) mm = **620 mm.**

mm is two steps to the right of dm, so multiply by $10 \times 10 = 100$.

(d) 4 cm = (4×10) mm = **40 mm.**

mm is one step to the right of cm, so multiply by 10.

(e) 12 dam = (12×1000) cm = **12000 cm.**

cm is 3 steps to the right of dam, so multiply by $10 \times 10 \times 10 = 1000$.

Example 2. Convert (a) 800 dam to km (b) 60 mm to cm (c) 95 hm to km (d) 6250 dm to hm (e) 12000 cm to dam.

All these are conversions from lower units to higher units so divide.

(a) 800 dam = $(800 \div 100)$ km = **8 km**

km is two steps to the left of dam, so divide by 100.

(b) 60 mm = $(60 \div 10)$ cm = **6 cm**

cm is 1 step to the left of mm, so divide by 10.

(c) 95 hm = $(95 \div 10)$ km = **9.5 km**

km is 1 step to the left of hm, so divide by 10.

(d) 6250 dm = $(6250 \div 1000)$ hm = **6.25 hm**

hm is three steps to the left of dm, so divide by 1000.

(e) 12000 cm = $(12000 \div 1000)$ dam = **12 dam**

dam is three steps to the left of cm, so divide by 1000.

Fill in the blanks:

1. 1 kilometre = ______ hectometres
2. 1 decametre = ______ decimetres
3. 1 hectometre = ______ decimetres
4. 1 decimetre = ______ millimetres
5. 1 centimetre = ______ decametres
6. 1 millimetre = ______ decimetres
7. 1 decimetre = ______ hectometres
8. 1 hectometre = ______ kilometres

Using the diagram given below, fill in the blanks.

Kilometre km	hectometre hm	decametre dam	metre m	decimetre dm	centimetre cm	millimetre mm

9. 11.8 km = ______ dam
10. 15 hm = ______ dm
11. 6.25 dam = ______ cm
12. 1800 mm = ______ dm
13. 15.3 cm = ______ dam
14. 9250 dm = ______ hm
15. 25 hm = ______ m
16. 2.5 dm = ______ cm
17. 14 mm = ______ cm
18. 1.85 hm = ______ dam
19. 19 hm = ______ km
20. 41.9 dam = ______ km

WRITING METRIC MEASURES IN DECIMAL NOTATION

Since 100 centimetres = 1 metre, we can write 568 cm $= \frac{568}{100}$ m = 5.68m

5.68 m = 5 metres 68 centimetres
= **5 metres 6 decimetres 8 centimetres**

5.68 m = 5 m + 0.6 m + 0.08 m
= 5 m + 0.6 × 10 dm + 0.08 × 100 cm
= 5 m + 6 dm + 8 cm

Again, 2596 metres = 2000 metres + 500 metres + 90 metres + 6 metres
= 2 km + 5 hm + 9 dam + 6 m

$= 2 \text{ km} + \frac{5}{10} \text{ km} + \frac{9}{100} \text{ km} + \frac{6}{1000} \text{ km}$

= 2 km + 0.5 km + 0.09 km + 0.006 km

= 2.596 km.

2	.	5	9	6	km
↑		↑	↑	↑	
km		hm	dam	m	

The decimal point separates the kilometres from the smaller units.

Separates metre from smaller units

4	•	2	3	9
↑	↓	↑	↑	↑
m		dm	cm	mm

More Examples:

1. 0.708 km = 0 km 7 hm 0 dam 8 m
2. 8.015m = 8m 0 dm 1 cm 5 mm
3. 4.239 m = 4m 2 dm 3 cm 9 mm
4. 5 metres and 8 centimetres can be written as 5.080 m, which means 5 metres 0 decimetres 8 centimetres 0 millimetres. You can simply write it as 5.08 m as there are no millimetres. As there is no decimetre we put a zero at the decimetre's place.
5. 8 dam 3 m 5 dm 7 cm 2 mm can be written as 83.572 m as
   8 dam 3 m = 8 × 10 m + 3 m = 83 m.
   Therefore write 83, put the decimal point, and after the decimal point write the figures for dm, cm and mm respectively.

Express:

1. 412.92 m in **m** and **cm**. ______ m ______

2. 27.893 m in **m, dm, cm, mm** ______ m ______ dm ______ cm ______ mm

3. 6.893 km in **km, hm, dam, m** ______ km ______ hm ______ dam ______ m

Using decimal notation express:

4. 12 m 45 cm in m. ______ m

5. 2 m 6 dm 5 cm 3 mm in m. ______ m

6. 160 km 8 hm 9 dam 5 m in km. ______ km

Now we can use the above concept to devise a shortcut to make conversion between units easier.

Shortcut for conversion between units

Example 1. Convert 48.6 cm to m.

Solution.

$$1\,\text{cm} = \frac{1}{100}\,\text{m} \quad \therefore\ 48.6\,\text{cm} = \frac{48.6}{100}\,\text{m}$$

$$= \mathbf{0.486\ m.}$$

To divide by 100, shift the decimal point 2 places to the left.

Example 2. **Convert 170 dm to dam.**
Solution.

$$170\text{ dm} = \frac{170}{100}\text{dam} \quad = \mathbf{1.70\ dam.}$$

Shift the decimal point 2 places to the left.

Reference chart:

dam $\xleftarrow{\div 10}$ m $\xleftarrow{\div 10}$ dm

$\therefore \text{dm} = \frac{1}{100}\text{dam}.$

Example 3. **Convert 4.5 m to mm.**
Solution.

1 m = 1000 mm

$\therefore$ **4.5 m = (4.5 × 1000) mm**

= 4500 mm.

1. To multiply by 1000 shift the decimal point 3 places to the right.
2. As there are no digits after 5, add two zeros.

The same principles of conversions are used in case of units of measurement of mass and capacity. The only change is in the standard unit.

Home Work

EXERCISE 40

1. Express in metres and centimetres.

(a) 382.14 m **(b)** 15.08 **(c)** 895.40 m
(d) 500.00 m **(e)** 0.69 m **(f)** 0.03 m

2. Using decimal notation, express in metres.

(a) 5 m 13 cm **(b)** 721 m 58 cm **(c)** 29 m 17 cm
(d) 173 m 2 cm **(e)** 2 m 1 cm **(f)** 38 cm
(g) 4 cm **(h)** 50 cm **(i)** 938 cm

3. Express in m, dm, cm and mm.

(a) 549.321 m **(b)** 0.703 m **(c)** 605.203 m **(d)** 0.092 m
(e) 0.759 m **(f)** 354.018 m **(g)** 0.008 m **(h)** 935.001 m
(i) 300.000 m

4. Using decimal notation, express in metres

(a) 327 m 5 dm 6 cm 8 mm **(b)** 4 mm **(c)** 723 m 5 mm
(d) 4 dm 7 mm **(e)** 402 m 7 cm 9 mm **(f)** 3 dm 1 cm 4 mm
(g) 5 dm 3 mm

5. Express in km, hm, dam and m.

(a) 430.756 km **(b)** 0.807 km **(c)** 902.405 km **(d)** 0.078 km
(e) 0.652 km **(f)** 230.048 km **(g)** 0.008 km **(h)** 603.001 km
(i) 600.00

6. Using decimal notation, express in km

(a) 540 km 7 hm 3 dam 6 m **(b)** 5 m **(c)** 857 km 2 m
(d) 6 hm 3 m **(e)** 703 km 8 dam 5 m **(f)** 839 km 4 hm 1 m
(g) 6 hm 5 dam 9 m **(h)** 7 dam 6 m

MASS

The amount of matter in an object is its mass. The standard unit that is used to measure mass, in the metric system is **gram** (g).

The chart given below shows the units of mass in the metric system with gram as the reference unit.

1000	100	10	1	$\frac{1}{10} = 0.1$	$\frac{1}{100} = 0.01$	$\frac{1}{1000} = 0.001$
kilogram (kg)	hectogram (hg)	decagram (dag)	gram (g)	decigram (dg)	centigram (cg)	milligram (mg)

Important Relationships

1 kg = 1000 g	**1 mg = $\frac{1}{1000}$ g = 0.001 g**
1 hg = 100 g	**1 cg = $\frac{1}{100}$ g = 0.01g**
1 dag = 10 g	**1 dg = $\frac{1}{10}$ g = 0.1g**

Conversions

The given diagrams show conversions from the various higher and lower units to gram and vice versa. They also depict the conversions between the various units.

To convert higher units to lower units, multiply

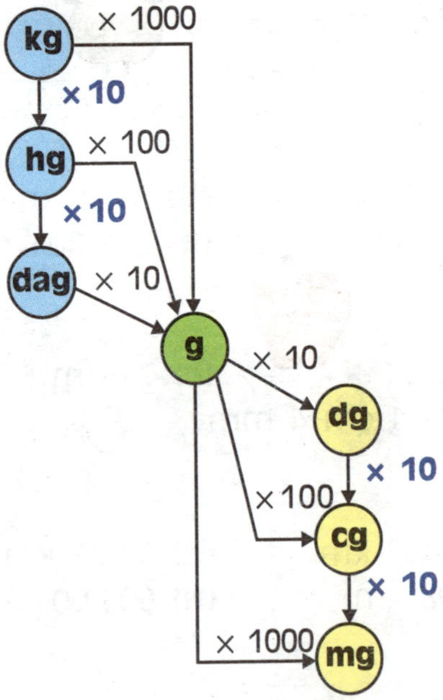

To convert lower units to higher units divide

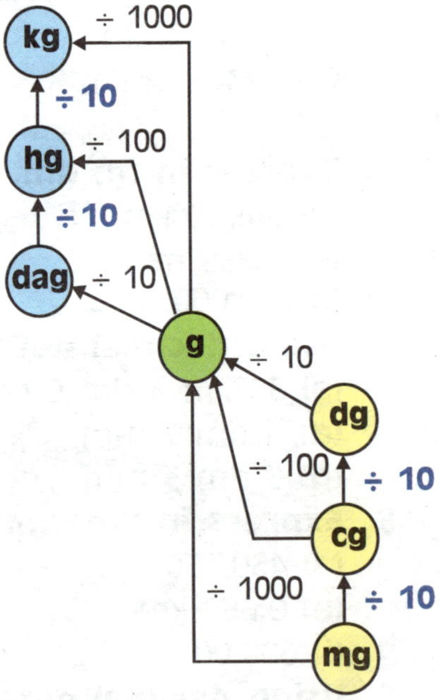

Examples:

Conversion of various units to grams and grams to various units

18 kg = 18 × 1000 g = 18000 g

1.53 dag = 1.53 × 10 g = 15.3 g

12 g = 12 × 10 dg = 120 dg

4.375 g = 4.375 × 1000 mg = 4375 mg

450 cg = 450 ÷ 100 g = 4.5 g

1900 mg = 1900 ÷ 1000 g = 1.9 g

120 g = 120 ÷ 100 hg = 1.2 hg

418.7 g = 418.7 ÷ 1000 kg = 0.4187 kg

Conversion between units.

Note : To solve probelms of interconversions, it will help you to first draw a reference chart like **kg hg dag g dg cg mg.**

25 kg = 25 × 100 dag = 2500 dag

14.6 dg = 14.6 × 10 cg = 146 cg

5 hg = 5 × 1000 dg = 5000 dg

28 mg = 28 ÷ 10 cg = 2.8 cg

500 dg = 500 ÷ 100 dag = 5 dag

1580 cg = 1580 cg ÷ 1000 dag = 1.580 dag

kg → dag: 2 steps, × 100

dg → cg: 1 step, × 10

hg → dg: 3 steps, × 1000

cg ← mg: 1 step, ÷ 10

dag ← dg: 2 steps, ÷ 100

dag ← cg: 3 steps, ÷ 1000

1. 1 kilogram = ________ grams
2. 1 gram = ________ milligrams
3. 1 decagram = ________ grams
4. 1 hectogram = ________ decigrams
5. 1 decagram = ________ decigrams
6. 1 decigram = ________ centigrams
7. 1 gram = ________ kilogram
8. 1 milligram = ________ centigram
9. 1 hectogram = ________ kilogram
10. 1 decigram = ________ decagram
11. **Complete the following table:**

kg	hg	dag	g	dg	cg	mg
13.7						
			4500			
				217.8		
		70.02				
						80,000
					145.72	
	15					

Note : You do not have to fill the coloured parts.

12. The weight of a baby elephant is 96 kg. Its weight in grams is ________ .
13. For Rachana's birthday 3250 g cake was ordered. The weight of the cake in kilograms is ____________ .

14. To get rid of his throat infection, Rohit needs to take 3 tablets each of 250 mg everyday. The total dosage for 1 day in grams is ____________ grams.

15. **Express in kg, hg, dag and g.**

(a) 451.378 kg = ______ kg ______ hg ______ dag ______ g

(b) 0.602 kg = ______ kg ______ hg ______ dag ______ g

(c) 0.056 kg = ______ kg ______ hg ______ dag ______ g

(d) 0.002 kg = ______ kg ______ hg ______ dag ______ g

16. **Using decimal notation, express in kg.**

(a) 418 kg 2 hg 5 dag 9 g = ______ kg

(b) 4 hg 8 dag 7 g = ______ kg

(c) 23 kg 8 g = ______ kg

(d) 4 dag 2 g = ______ kg.

The two higher units of mass which are frequently used are :

1 quintal = 100 kg

1 tonne = 1000 kg

CAPACITY

The standard unit of capacity in the metric system is litre (l). The chart given below shows the units of capacity in the metric system with litre as the reference unit.

1000	**100**	**10**	**1**	$\frac{1}{10}$	$\frac{1}{100}$	$\frac{1}{1000}$
kilolitre *kl*	Hectolitre *hl*	decalitre *dal*	litre *l*	decilitre *dl*	centilitre *cl*	millilitre *ml*

Important relationships

$1\ kl = 1000\ l$	$1\ ml = \frac{1}{1000}\ l = 0.001\ l$
$1\ hl = 100\ l$	$1\ cl = \frac{1}{100}\ l = 0.01\ l$
$1\ dal = 10\ l$	$1\ dl = \frac{1}{10}\ l = 0.1\ l$

CONVERSIONS

The conversion from higher and lower units to litres and conversion of litres to various units, as well as the interconversions between the various units are shown in the following diagrams.

To convert higher units to lower units, multiply

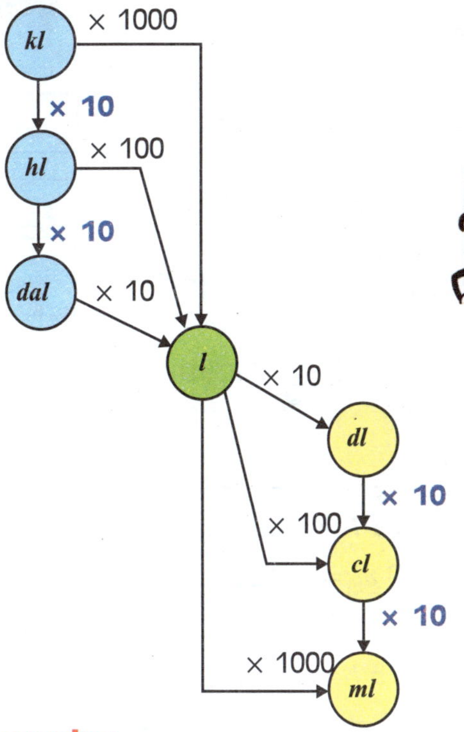

To convert lower units to higher units, divide

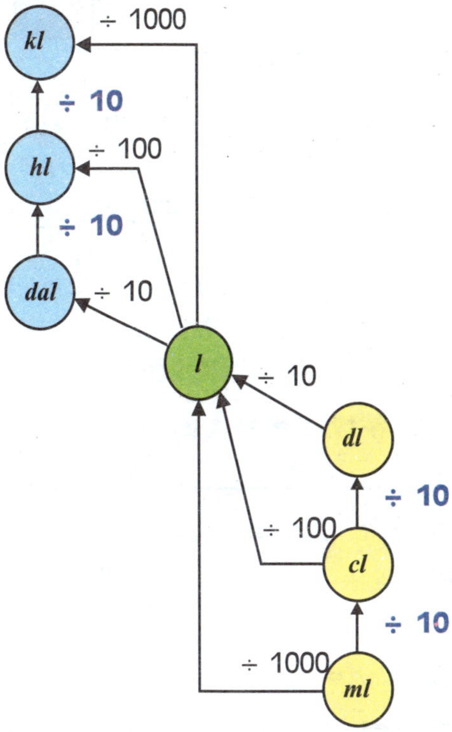

Examples

1.8 *kl* = 1.8 × 1000 *l* = 1800 *l*

42 *hl* = 42 × 100 *l* = 4200 *l*

0.5 *dal* = 0.5 × 10 *l* = 5 *l*

4.8 *l* = 4.8 × 10 *dl* = 48 *dl*

6 *l* = 6 × 100 *cl* = 600 *cl*

12.7 *l* = 12.7 × 1000 *ml* = 12700 *ml*

4.6 *l* = 4.6 ÷ 1000 *kl* = 0.0046 *kl*

500 *l* = 500 ÷ 100 *hl* = 5 *hl*

60 *l* = 60 ÷ 10 *dal* = 6 *dal*

1250 *ml* = 1250 ÷ 1000 *l* = 1.25 *l*

164.3 *cl* = 164.3 ÷ 100 *l* = 1.643 *l*

20 *dl* = 20 ÷ 10 *l* = 2*l*

Keep in mind the reference chart while making conversions.

kl	*hl*	*dal*	*l*	*dl*	*cl*	*ml*

Conversions between units

14 *hl* = 14 × 10 *dal* = 140 *dal*. — *hl* → *dal* (1 step, × 10)

15.2 *dl* = 15.2 × 100 *ml* = 1520 *ml*. — *dl* → *ml* (1 step, × 100)

500 *ml* = 500 ÷ 10 *cl* = 50 *cl*. — *cl* ← *ml* (1 step, ÷ 10)

1800 *dl* = 1800 ÷ 1000 *hl* = 1.8 *hl*. — *hl* ← *dl* (3 steps, ÷ 1000)

400 *dal* = 400 ÷ 100 *kl* = 4 *kl*. — *kl* ← *dal* (2 steps, ÷ 100)

4.875 *hl* = 4.875 × 1000 *dl* = 4875 *dl*. — *hl* → *dl* (3 steps, × 1000)

State True or False (T/F).

1. 1 kilolitre = 1000 millilitre ☐
2. 1 litre = 100 centilitres ☐
3. 1 decilitre = $\frac{1}{100}$ litre ☐
4. 1 hectolitre = $\frac{1}{10}$ kilolitre ☐
5. 1 decalitre = 100 decilitres ☐
6. 1000 millilitres = 1 litre ☐
7. 1 centilitre = 100 litres ☐
8. 1 litre = $\frac{1}{1000}$ kilolitres ☐

Fill in the blanks:

9. 8 *l* = ______________ *ml*
10. 150 *dl* = ______________ *dal*
11. 16 *kl* = ______________ *dal*
12. 1875 *l* = ______________ *kl*
13. 485 *cl* = ______________ *l*
14. 120 *dl* = ______________ *l*
15. 6.254 *kl* = ______________ *l*
16. 15 *hl* = ______________ *dl*
17. 700 *dal* = ______________ *hl*
18. 2475 *l* = ______________ *hl*
19. 2.5 *l* = ______________ *dl*
20. 19.4 *cl* = ______________ *dl*
21. A small car has a petrol tank whose capacity is 30 litres. Its capacity in millimetres is ______________ .

22. A tanker carries 12000 litres of milk, which in kilolitres is ______________.
23. Ankita used 500 ***ml*** of chocolate syrup in the cake she prepared for the dinner party. The chocolate syrup in litres is ______________ .

24. **Express in *l*, *dl*, *cl*, *ml***
    - **(a)** 529.183 *l* = ________ *l* ________ *dl* ________ *cl* ________ *ml*
    - **(b)** 0.308 *l* = ________ *l* ________ *dl* ________ *cl* ________ *ml*
    - **(c)** 0.062 *l* = ________ *l* ________ *dl* ________ *cl* ________ *ml*
25. **Using decimal notation, express in litres.**
    - **(a)** 12 *l* 4 dl 5 cl 2 ml = ________ *l*
    - **(b)** 40 *l* 6 cl 9 ml = ________ *l*
    - **(c)** 1 dl 9 cl = ________ *l*
    - **(d)** 215 *l* 5 ml = ________ *l*

THE FOUR BASIC OPERATIONS ON METRIC MEASURES

Addition:

Example 1. Three books of thickness 18 mm, 13 mm and 7 mm are piled on top of each other. What is the total thickness of the pile ? Express in (a) cm and mm (b) cm using decimal notation.

Solution. The total thickness of the pile

= 18 mm + 13 mm + 7 mm

= 38 mm

= 3 cm 8 mm = 3.8 cm.

$$\begin{array}{r} 18 \\ 13 \\ +7 \\ \hline 38 \end{array}$$

For addition and subtraction of metric measures write them in proper columns (*i.e.*, same units one below the other) and then add or subtract like whole numbers.

Example 2.

(a)

	2 cm	6 mm
+	9 cm	8 mm
	12 cm (1)	4 mm

14 mm = 10 mm + 4 mm
= 1 cm + 4 mm

= 12 cm 4mm

= **12.4 cm.**

(b)

	11 dm	6 cm
+	13 dm	7 cm
	25 dm (1)	3 cm

13 cm = 10 cm + 3 cm
= 1 dm + 3 cm

= 25 dm 3cm

= **25.3dm**

(c)

	42 kg	305 g
+	8 kg	725 g
	51 kg (1)	030 g

1030 g = 1000 g + 30 g
= 1 kg + 30 g

= 51 kg 30 g

= **51.03 kg.**

(d)

	9 *l*	718 ml
+	8 *l*	632 ml
	18 *l* (1)	350 ml

1350 ml
= 1000 ml + 350 ml
= 1 *l* + 350 ml

= 18 *l* 350 ml

= **18.35 *l*.**

Example 3.

(a)

	~~10~~ 9 m	~~45~~ 145 cm
–	7 m	62 cm
	2 m	83 cm

45 cm < 62 cm
Borrow 1 m = 100 cm
100 cm + 45 cm = 145 cm

= **2.83 m**

(b)

	~~27~~ 26 cm	~~5~~ 15 mm
–	16 cm	9 mm
	10 cm	6 mm

5 mm < 9 mm
Borrow 1 cm = 10 mm
10 mm + 5 mm = 15 mm

= **10.6 cm.**

(c)

	~~38~~ 37 kg	~~275~~ 1275 g
–	14 kg	500 g
	23 kg	775 g

275 g < 500 g
Borrow 1 kg = 1000 g
1000 g + 275 g = 1275 g

= **23.775 kg.**

(b)

	~~7~~ 6 kl	~~200~~ 1200 *l*
–	2 kl	450 *l*
	4 kl	750 *l*

200 *l* < 450 *l*
Borrow 1 kl = 1000 *l*
1000 *l* + 200 *l* = 1200 *l*

= **4.750 kl.**

MULTIPLICATION

Arrange numbers in columns unit wise and then multiply as you would multiply whole numbers.

Example 4. A toy car is 9 cm 5 mm long. If 9 such toy cars are placed front to back, how far will they stretch in cm ?

Solution. Length of 1 toy car = 9 cm 5 mm

Length of 9 toy cars

	cm	mm
	4	
	9	5
×		9
	85	(4)5

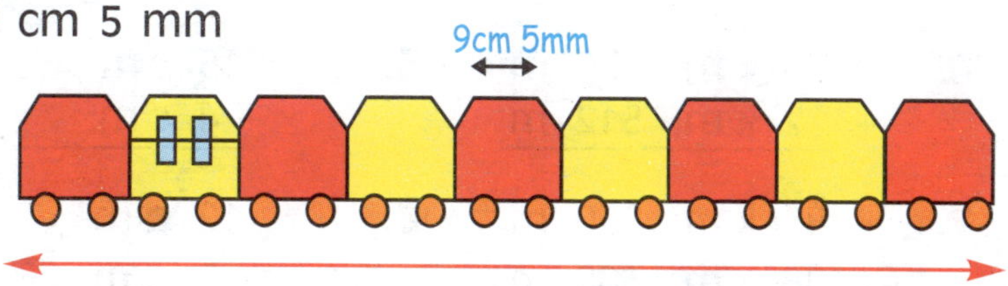

Length of 9 toy cars = 85 cm 5 mm = **85.5 cm**

Example 5. A box of apple weighs 7 kg 325 g. What is the weight of 18 such boxes in kg ?

Solution. Weight of one box = 7 kg 325 g = 7.325 kg

∴ Weight of 18 boxes = (7.325 × 18) kg

= 131.850 kg

= 131 kg 850 g.

```
   7325
    ×18
  58600
   7325
 131850
```

Put decimal point after 3 places.

DIVISION

In division also arrange the numbers in columns unit wise and then divide like whole numbers.

Example 6. A cooler jug has 3 *l* 156 *ml* of orange juice. Distribute it equally among 12 children in ml ?

Solution. Juice in the cooler jug = 3*l* 156 ml = 3.156 *l*

When distributed among 12 children

each child gets = (3.156 ÷ 12) *l*

= 0.263 *l*

= **263 *ml*.**

```
12)3.156
   0.263
```

Use multiplication table of 12 to divide

Example 7. There are 42 flats in a society. Each flat has an overhead tank, the capacity of all tanks being the same. The total capacity of water in these 42 tanks is 64 kl 176 *l*. What is the capacity of each tank in kl ?

Solution. Total Capacity of all the tanks = 64 *kl* 176 *l*

= 64.176 *kl*

∴ Capacity of each tank = (64.176 ÷ 42) *kl*

= 1. 528 *kl*

= 1 *kl* 528 *l*.

```
     1528
42)64176
  -42
   221
  -210
    117
   - 84
     336
    -336
       0
```

Move 3 places to the left and put the decimal point

EXERCISE 41

ADD OR SUBTRACT

1. 5 m 8 dm + 11 m 9 dm
2. 16 cm 3 mm + 15 cm 8 mm
3. 9 dm 6 cm + 23 dm 7 cm
4. 45 km 750 m + 92 km 512 m
5. 15 cm 8 mm − 4 cm 9 mm
6. 14 m 65 cm −2 m 80 cm
7. 183 km 250 m − 106 km 500 m
8. 12 dm 2 cm − 5 dm 8 cm
9. 85 kg 196 g + 15 kg 924 g

10.

	5 g	750 mg
+	6 g	523 mg

11.

	234 kg	750 g
−	96 kg	893 g

12.

	16 g	105 mg
−	12 g	375 mg

13.

	118 *l*	225 m*l*
+	56 *l*	780 m*l*

14.

	12 kl	525 *l*
+	15 kl	490 *l*

15.

	35 kl	5 *l*
−	12 kl	126 *l*

16.

	240 *l*	500 m*l*
−	130 *l*	600 m*l*

17. Find in metres the height of a pile of 25 books if each book is 3 cm 5 mm thick.

18. Three pieces of metal bars of lengths 12 m 56 cm, 5 m 43 cm and 16 m 24 cm are joined end to end. What is the total length of the resulting bar in m ?

19. Ritu drew a line segment of length 15 cm 4 mm. Then she erased a portion of it. The remaining line segment measured 7 cm 6 mm. What is the length in mm of the erased line segment ?

20. Vivek jogs 8 times around a square field of perimeter 140 m. How many kilometres and metres does he jog ? He had planned to jog 1.5 km. How much in metres is still left to jog ?

21. Manish's weight is 83 kg 450 g and Ashish is 6 kg 900 g heavier than him. Find Ashish's weight in kg.

22. Find the net weight in grams of the contents of a breakfast cereal package if the gross weight is 1 kg 200 g and the packaging weighs 15 g.

23. 15 bags of wheat were delivered at a grocery shop, the total weight being 144 kg 780 g. If all the bags contain equal amount of wheat, find the weight of each bag ?

24. A shopkeeper bought 6 sacks of peanuts, each sack weighing 25 kg. What is the total amount of peanuts he has in as (a) Kilograms (b) grams ? He repacks the peanuts into 400 equal packets. How much peanuts in grams are there in each packet ?

25. For a birthday party a 10 *l* oil can was purchased. After the party 6 *l* 450 ml of oil was still left in the can. How much oil in litres was used in the party ?

26. How many 150 *ml* glasses can I fill with 5 bottles of soft drinks each holding 1.2 litres ?

27. A bottle holds 1 *l* 140 *ml* of shampoo, Radhika uses 95 ml of shampoo each week.

(a) How many weeks will the bottle of shampoo last ?

(b) If she accidently spills 475 *ml* of shampoo, how long will the bottle last ?

ESTIMATION IN MEASURES

You cannot be always exact in measures. In our day to day life we estimate the measures. To be able to estimate measures correctly, you should have a fair idea of how to relate common place things with the commonly used units of measures.

Length

For long lengths or distances, we use kilometres (km) as unit

Length of India = 3214 km

Breadth of India = 2933 km

For short lengths, we use centimetres (cm)

Length of your small ruler = 15 cm

For medium lengths, we use metres (m)

Your height 1.32 m

For very short lengths, we use millimetres (mm)

The tip of your pencil = 2 mm

Mass

For very large masses, we use tonne (t) and quintal (q)

Truck = 4 tonne Sack of wheat = 1 quintal

For medium masses, we use kilograms (kg)

Your weight = 30 kg

For small masses, we use grams (g)

Your maths book = 150 g

For very small masses, we use milligrams (mg)

A grain of sand = 1 mg

Capacity

For large capacities, we use kilolitre (*kl*)

Capacity of a swimming pool = 600 *kl*

For medium capacities, we use litres (*l*)

A normal bucket = 20 *l*

For small capacities, we use millilitres (*ml*)

A bottle of shampoo = 200 *ml*

Skill Practice

Class Work

Circle the suitable measure.

1. length of a teaspoon

m **cm** **km**

2. width of a table

mm **km** **m**

3. a stack of newspapers

g **kg** **mg**

4. a hairbrush

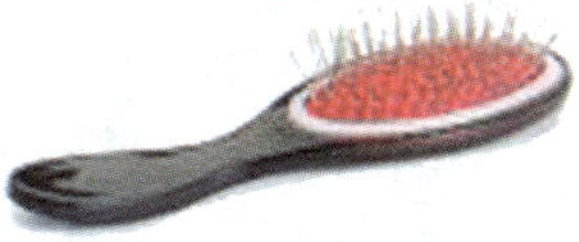

g **mg** **kg**

5. perfume in a bottle

ml *l* *kl*

6. distance of train trip

m *km* *cm*

7. Width of an eraser

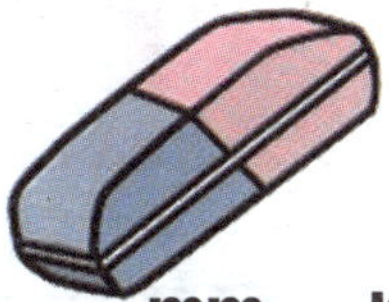

cm **mm** **Km**

8. A tablet of Vitamin C

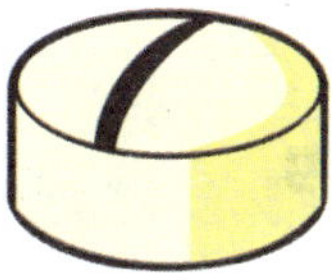

mg **g** **kg**

9. a loaf of bread

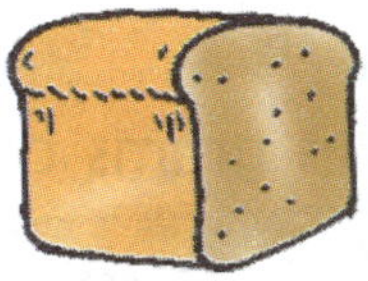

kg **g** **mg**

10. Water in a fish tank

ml *l* *kl*

11. The Length of Yuvraj's sixer in 20-20 world cup

m **km** **cm**

II. Write the correct unit, commonly used to measure the following things from the given box.

ml	*km*	*m*	*g*	*kg*	*mg*	*t*	*l*	*kl*	*mm*

1. The thickness of this page ☐
2. The diameter of earth.  ☐
3. The mass of an egg.  ☐
4. The capacity of a tea cup ☐
5. The amount of water drawn from a well.  ☐
6. The weight of a pinch of salt. ☐
7. The weight of a bus. ☐
8. The capacity of a refrigerator. ☐
9. The mass of a ceiling fan. ☐
10. The height of Mt. Everest.  ☐
11. The length of a grain of wheat ☐
12. The amount of water in a fish pond. ☐

APPLYING FRACTIONS OF QUANTITIES

Example 1. **Shalini bought 5 kg 544 g of flour. She used $\frac{1}{6}$ of the flour in making buns. How much flour in grams did she use for making buns? If each bun uses $\frac{1}{12}$ of the flour, what is the weight in grams of each bun ?**

Solution. Amount of flour used for making buns

$$= \frac{1}{6} \text{ of } 5 \text{ kg } 544 \text{ g}$$

$$= \frac{1}{6} \text{ of } 5.544 \text{ kg} = 0.924 \text{ kg}$$

$$\mathbf{= 924 \text{ g.}}$$

$$6\overline{)5.544} = 0.924$$

Flour used for each bun $= \frac{1}{12}$ of 924 g

= 77 g

$$12\overline{)924} = 77; \quad 924 - 84 = 84,\ 84 - 84 = 0$$

Weight in grams of each bun **= 77 g.**

Example 2. **Kiran had 3 *l* 180 *ml* of pineapple juice. She gave $\frac{2}{5}$ of it away to her neighbour and poured the rest equally into 6 bottles. How many millilitres of pineapple juice did each bottle contain ?**

Solution.

Amount of juice kiran has = 3*l* 180 *ml* = 3.180 *l*

Amount of juice given to neighbour $= \frac{2}{5}$ of 3.180 *l*

$$= \frac{2 \times 3.180\,l}{5} = 2 \times 0.636\,l$$

$$5\overline{)3.180} = 0.636$$

= 1.272 *l*

Remaining juice which is filled in 6 bottles = 3.180*l* − 1.272 *l* = 1.908 *l* = 1908 m*l*

$$3.180 - 1.272 = 1.908 \qquad 6\overline{)1908} = 318$$

∴ Juice in 1 bottle = (1908 ÷ 6) *ml*

= 318 *ml*

Example 3. **Madhuri drew a line segment of length 20 cm 5 mm. She accidently erased $\frac{2}{5}$ of it. What is the length of the remaining line segment in cm?**

Solution. Length of the line segment drawn = 20 cm 5 mm = 20.5 cm

Length of the erased line segment

$= \frac{2}{5}$ of 20.5 cm

= (2 × 4.1) cm

= 8.2 cm

∴ Length of remaining line segment

= 20.5 cm − 8.2 cm

= 12.3 cm = **12 cm 3 mm.**

$$5\overline{)20.5}$$
$$4.1$$

$$\begin{array}{r} 20.5 \\ -\ 8.2 \\ \hline 12.3 \end{array}$$

Conversion of larger fractional units into complete smaller units

Example 1. Express $\frac{1}{4}$ kg in grams and decagrams.

Solution. 1 kg = 1000 g and 1 kg = 100 dag

$\frac{1}{4}$ kg $= \frac{1}{4} \times 1000$ g

= **250 g.**

$\frac{1}{4}$ kg $= \frac{1}{4} \times 100$ dag

= **25 dag.**

Example 2. The length of a register is $\frac{2}{5}$ m. Express the length in cm and mm.

Solution. 1 m = 100 cm and 1 m = 1000 mm

$\frac{2}{5}$ m $= \frac{2}{\cancel{5}_1} \times \cancel{100}^{20}$ cm

= 2 × 20 cm = **40 cm.**

$\frac{2}{5}$ m $= \frac{2}{\cancel{5}_1} \times \cancel{1000}^{200}$ mm

= 2 × 200 mm = **400 mm.**

Example 3. A jug contains $\frac{1}{2}$ litre of water. How many millilitres and decilitres of water are there is the jug ?

Solution. 1 *l* = 1000 *ml*   1 *l* = 10 *dl*

$\frac{1}{2}$ *l* $= \frac{1}{2} \times 1000$ *ml*

= **500 *ml*.**

$\frac{1}{2}$ *l* $= \frac{1}{2} \times 10$ *dl*

= **5 *dl*.**

EXERCISE 42

Express:

1. $\frac{1}{4}$ g in mg. **2.** $\frac{3}{5}$ *l* in cl. **3.** $\frac{4}{5}$ kg in g. **4.** $\frac{3}{4}$ km in m.

5. $\frac{1}{5}$ *kl* in dal. **6.** $\frac{1}{2}$ g in mg. **7.** $\frac{2}{5}$ *kl* in *hl*. **8.** $\frac{4}{5}$ m in cm.

9. $\frac{3}{4}$ *dl* in *ml*. **10.** $\frac{1}{2}$ hm in m. **11.** $\frac{1}{20}$ kg in g. **12.** $\frac{1}{10}$ *l* in *dl*.

13. Manju bought $\frac{3}{4}$ m of string. She used $\frac{1}{3}$ of it to tie a parcel. What is the length of the string in cm that she used to tie the parcel ?

14. Aryan bought $\frac{4}{5}$ l of milk. On his way home he spilt $\frac{3}{4}$ of it. How many millilitres of milk did he spill ?

15. A basket contains 3 kg 705 g of mangoes. $\frac{2}{3}$ of the mangoes are eaten by Mr. Khanna. Rinki, his daughter gets $\frac{2}{5}$ of the remaining mangoes. What is her share in grams?

16. Mrs. Tewari has 1 l 160 ml of honey. She gave $\frac{5}{8}$ of the honey to her friend. She poured $\frac{4}{5}$ of the remaining honey over hot paranthas. How much honey in millilitres did she pour over the paranthas ?

17. Vedant walks $\frac{7}{10}$ km from his home to the bus stop. He boards a bus at the bus stop and travels $\frac{3}{4}$ km to his school. What is the total distance in (a) metres (b) kilometres Vedant travels?

18. A shopkeeper sold $\frac{3}{7}$ of 8 kg 750 g of his vegetables in the morning and sold $\frac{7}{10}$ of the remaining vegetables in the evening ? How much more of the vegetables in grams did he sell in the morning than in the evening ?

19. Anshu had $\frac{4}{5}$ kg of wafers. He packed all the wafers equally into 5 small packets. How many grams of wafers were there in each packet ?

20. Sita bought a blue ribbon of length $\frac{7}{8}$ m. She cut $\frac{3}{5}$ of it. What was the length in millimetres, of the ribbon that she cut ?

MENTAL MATHS–8

1. How many centimetres are there in 17 metres ?
2. Decilitres is two steps to the ________ of decalitres, so to convert dal to dl, multiply by ________ .
3. Express in metres: 6 dm 5 cm 8 mm.
4. How many milligrams is one-tenth of a 1 gram ?
5. 21.89 g = 2 dag 1 g 8 dg 9 cg. True or False !
6. 9000 litres = ________ hectolitres = ________ kilolitres

7. 1 cm 9 mm + 1 mm = ________ cm

8. What unit of length is $\frac{1}{100}$ m?

9. The normal serving of juice for one person at breakfast would be

Tick ☑ the correct answer.

(a) 20 ml (b) 200 ml (c) 750 ml (d) 1 litre.

10. Fill in the correct unit.

(a) 49.6 m ÷ 100 = 0.496 ________ (b) 12.5 hg × 1000 = 12500 ________

(c) 6.5 cm ÷ 10 = 0.65 ________ (d) 9 dal × 1000 = 9000 ________

ACTIVITY

Choose the correct unit for measuring the lengths of these objects and then measure the length of each of them

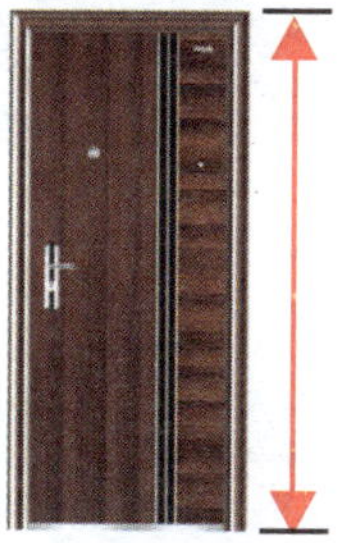

1. the height of your classroom door

____

2. the height of your cricket bat

____

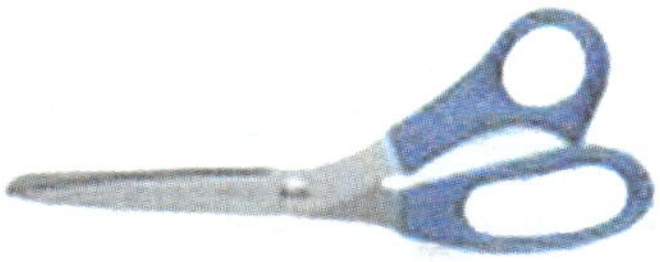

3. the length of a pair of scissors

____

4. the diameter of a 5 rupee coin

____

5. the width of your maths textbook

____

6. the length of your friend's foot

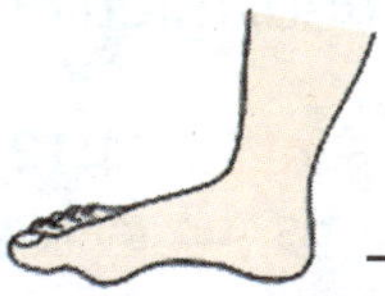

____

7. the width of your classroom

____

At the end of the activity you will get a fair idea of estimating lengths.

13 TIME

RECALL

Just as we measure quantities like lengths, mass, capacity etc, similarly we measure time also. The measuring tools are clocks and calendars. There are two main types of clocks or watches that we use.

Analog clock: It has a minute hand, hour hand and also a second hand sometimes.

Digital clock: The left side shows the hours from 1 to 12 and the right side the minutes from 00 to 59.

Here are some units for measuring time.

Second (s)		**Week**	1 week = 7 days
minute (min)	1 min = 60 s	**Month**	1 month = 28, 29, 30 or 31 days.
hour (h)	1 h = 60 min	**year**	1 year = 12 months
day (d)	1 day = 24 h		= 365 or 366 days (leap year)

Importants Terms.

A.M. (ante meridian : before midday): Is the time between midnight and noon.

Breakfast at 9:00 a.m. sharp

P.M. (Post meridian : after midday): Is the time between noon and midnight.

Dinner time at 8:00 p.m.

MATHS ALERT ! 12 O'clock, noon time is written as 12 noon and 12 O'clock, midnight is written as 12 midnight. It is wrong to write 12 A.M. or 12 P.M.

The railways, airlines, bus timetables, generally use the 24 hour clock. Instead of starting again at 1:00 P.M after 12 noon, the clock continues as 1300 hours, 1400 hours and so on.

6:45 A.M is 0645 hours.

6:45 P.M is 1845 hours.

HOURS, MINUTES AND SECONDS

Now that you are familiar with clocks and watches and you also know how to measure time in hours and minutes, we introduce you to the commonly used, unit of time **'second'.** A second is a very short period of time, say, time taken to blink your eye, to snap your fingers etc. It is a smaller unit of time than a minute.

> 1 minute has 60 seconds.
> 1 min = 60 s

Most of the clocks have three hands. The long thin one which moves very fast measures seconds.

Look at the following clocks and tell what time they show.

9:05 and 25 sec.

____

____

____

Remember:

60 seconds = 1 minute,

1 second = $\frac{1}{60}$ minute

60 minutes = 1 hour,

1 minute = $\frac{1}{60}$ hour

A. Changing from minutes to seconds and second to minutes.

Minutes to Seconds: 1 min = 60 s

Examples :

1 min = 1 × 60s = **60s.**
2 min = 2 × 60s = **120s.**
12 min = 12 × 60s = **720s.**
3 min 25 s = 3 min + 25s
= 3 × 60s + 25s
= 180s + 25s = **205s.**

Seconds to Minutes : s = $\frac{1}{60}$ min

Examples :

180 s = (180 ÷ 60) min = **3 min.**
600 s = (600 ÷ 60) min = **10 min.**
105 s = 60s + 45s
= 1 min + 45s
= **1 min 45s.**
or
105 s = (105 ÷ 60) min
= 1 min 45s.

$$\begin{array}{r|l} & 1 \leftarrow \text{min} \\ 60 & 105 \\ & -\ 60 \\ \hline & 45 \leftarrow \text{s} \end{array}$$

B. Changing from hours to minutes and from minutes to hours.

Hours to Minutes: 1 h = 60 min

Examples :

4h = 4 × 60 min = **240 min.**
3h 45 min = 3h + 45 min
= 3 × 60 min + 45 min
= 180 min + 45 min
= 225 min.

Minutes to hours : 1 m = $\frac{1}{60}$ h

Examples :

540 min = (540 ÷ 60) h = **9 h.**
352 min = (352 ÷ 60) h
= **5h 52 min**
or
352 min = 300 min + 52 min
= (300 ÷ 60)h + 52 min
= 5h + 52 min
= **5h 52 min.**

$$\begin{array}{r|l} & 5 \leftarrow \text{h} \\ 60 & 352 \\ & -300 \\ \hline & 52 \leftarrow \text{min} \end{array}$$

C. Changing from seconds to hours, minutes and seconds.

2105s = (2105 ÷ 60) min
= 35 min + 5s = **35 minutes 5 seconds.**

```
       35  ← min
60 ) 2105
   - 180↓
     305
   - 300
       5  ← s
```

4931s = (4931 ÷ 60) min
= 82 min + 11s
= 60 min + 22 min + 11s
= 1h + 22 min + 11s
= **1 hour 22 minutes 11 seconds.**

```
      1  ← h
60 ) 82
   - 60
     22  ← min
```

```
       82  ← min
60 ) 4931
   - 480↓
     131
   - 120
      11  ← s
```

Fill in the blanks:

1. 5 min = ____________ s **2.** 16 h = ____________ min

3. 360s = ____________ min **4.** 660 min = ____________ h

5. 4 min 42 s = ____________ s **6.** 2h 45 min = ____________ min

Exercise 43

1. How many seconds are there in:

(a) 6 minutes **(b)** 15 minutes **(c)** $\frac{3}{5}$ minute **(d)** 2 hours

(e) 7 min 16 s **(f)** 10 min 35 s **(g)** $\frac{11}{15}$ min **(h)** $\frac{17}{30}$ min

2. How many minutes are there in:

(a) 8 hours **(b)** $\frac{1}{2}$ hours **(c)** $\frac{3}{4}$ hours **(d)** 5 hours 25 min

(e) 9 hours 12 minutes **(f)** $9\frac{1}{2}$h **(g)** $7\frac{5}{12}$h **(h)** $8\frac{13}{15}$h

3. How many minutes are there in:

(a) 480 seconds **(b)** 720 seconds **(c)** 120 seconds **(d)** 145 seconds
(e) 330 seconds. **(f)** 900 sec **(g)** 1140 sec **(h)** 1900 sec

4. How many hours are there in:

(a) 180 min **(b)** 420 min **(c)** 90 min **(d)** 205 min

(e) 135 min **(f)** 300 min **(g)** 1050 min

OPERATIONS ON MEASURES OF TIME

★ **Addition:**

Example 1.

	1		
	3	minutes	31 seconds
+	2	minutes	40 seconds
	6	minutes	11 seconds

Think !

31 seconds + 40 seconds
= 71 seconds
= 60 seconds + 11 seconds
= 1 minute 11 seconds

Example 2.

	1		
	8	hours	45 minutes
+	11	hours	28 minutes
	20	hours	13 minutes

Think !

45 min + 28 min = 73 min
= 60 min + 13 min
= 1 h + 13 min

Example 3:

	1		1		
	5	hours	30	minutes	45 seconds
+	2	hours	42	minutes	30 seconds
	8	hours	13	minutes	15 seconds

Think !

45 seconds + 30 seconds
= 75 seconds
= 1 minute + 15 seconds
30 mins. + 1 min. + 42 min.
=73 minutes
= 1 hour 13 minutes

Add:

1. 50 minutes + 10 minutes + 40 minutes = ____________.
2. 30 seconds + 15 seconds + 55 seconds = ____________.

3.

	8	minutes	14 seconds
+	15	minutes	23 seconds

4.

	12	hours	27 minutes
+	5	hours	58 minutes

5.

	1 hours	20 minutes	36 seconds
+	2 hours	45 minutes	55 seconds

6.

	3 hours	12 minutes	44 seconds
+	4 hours	48 minutes	16 seconds

Subtraction:

Example 1. Subtract 12 minutes 25 seconds from 35 minutes 12 seconds.

Solution.

	34	72
	~~35~~ minutes	~~12~~ seconds
−	12 minutes	25 seconds
	22 minutes	47 seconds

Think !

12 < 25, So we borrow
1 min = 60 seconds from 35 min.
60s + 12 s = 72s
72s − 25 s = 47s
Now subtract minutes
34 min − 12 min = 22 min

Example 2. Aparna completes a painting in 8 hours 26 minutes and Ruchira in 6 hours 45 minutes. How much more time does Aparna take to complete the painting?

Solution. You have to subtract 6 hours 45 minutes from 8 hours 26 minutes to find the required time.

	7	86
	~~8~~ hours	~~26~~ minutes
−	6 hours	45 minutes
	1 hour	41 minutes

Think !

26 < 45, So borrow 1 h = 60 min
from 8 hours
60 min + 26 min = 86 min
86 min − 45 min = 41 min
7h − 6h = 1h

Aparna takes 1 hours 41 minutes more to complete the painting.

Example 3.

Solution.

	4	84	91
	5 hours	25 minutes	31 seconds
−	2 hours	40 minutes	50 seconds
	2 hours	44 minutes	40 seconds

Think !

= 5 h + (24 min + 1 min) + 31s
= 5 h + (24 min + (60 + 31)s
= (4 hours + 1 hour) + 24 min + 91s
= 4 h + (60 + 24) m + 91s
= 4h 84 min 91 s

Now 4 hours 84 minutes 91 seconds
− 2 hours 40 minutes 50 seconds

2 hours 44 minutes 41 seconds

1. 1 hours 40 minutes − 50 minutes = ____________________ .

2. 4 minutes 25 seconds − 45 seconds = ____________________ .

Subtract:

3.
	20 minutes	15 seconds
−	17 minutes	35 seconds

4.
	4 hours	20 minutes
−	1 hours	50 minutes

5.
	3 hours	12 minutes	22 seconds
−	1 hours	15 minutes	45 seconds

6.
	5 hours	40 minutes	18 seconds
−	2 hours	16 minures	35 seconds

EXERCISE 44

Add:

1. 16 minutes 45 seconds
+ 12 minutes 22 seconds

2. 5 hours 20 minutes
+ 6 hours 40 minutes

3. 15 hour 27 minutes
+ 3 hours 42 minutes

4. 2 hours 15 minutes 55 seconds
+ 3 hours 58 minutes 22 seconds

5. 3 hours 2 minutes 3seconds
+ 4 hours 21 minutes 57 seconds

Subtract:

6. 18 minutes 29 seconds
− 16 minutes 31 seconds

7. 3 hours 40 minutes
− 2 hours 45 minutes

8. 5 hours 20 minutes
− 3 hours 58 minutes

9. 23 hours 4 minutes 58 seconds
− 1 hours 5 minutes 59 seconds

10. 16 hours 10 minutes 34 seconds
− 13 hours 20 minutes 39 seconds

11. Anil takes 40 minutes to walk to school and 35 minutes to reach home from school by bus. How long do the two journeys take?

12. Satish went to the beach on Sunday. He spent 1 hour 43 minutes fishing and 1 hour 20 minutes swimming. How much time did he spend in both the activities?

13. Doon Express took 26 hours 37 minutes 10 seconds in travelling from Dehra Dun to Lucknow. During this time the train had a total stoppage of 8 hours 49 minutes 55 seconds at different stations. For how much time was the train moving?

14. In a cricket match, the captain of the team played for 4 hours 28 minutes 13 seconds in the first innings and for 3 hours 49 minutes 35 seconds in the second innings. Find for how much more time he played in the first innings.

24-HOURS TIME NOTATION

You have already studied the 24 hour clock in class IV. The 24-hour time notation is represented by a '24 hour clock'.

Here, the time from **12 midnight to 12 midnight (1 day)** is given as: **0000 hours to 2400 hours,** where the first two digits refer to hours and the last two to minutes.

15 25
↑ ↑
hour minutes

3 : 25 p.m. | 15:25 |

The table given below shows the 24 hour time equivalents of 12 hours time.

12 HOUR TIME	24 HOUR TIME	12 HOUR TIME	24 HOUR TIME
12:01 midnight	0001 hours	12:00 noon	1200 hours
1:00 a.m.	0100 hours	1:00 p.m.	1300 hours
2:00 a.m.	0200 hours	2:00 p.m.	1400 hours
3:00 a.m.	0300 hours	3:00 p.m.	1500 hours
4:00 a.m.	0400 hours	4:00 p.m.	1600 hours
5:00 a.m.	0500 hours	5:00 p.m.	1700 hours
6:00 a.m.	0600 hours	6:00 p.m.	1800 hours
7:00 a.m.	0700 hours	7:00 p.m.	1900 hours
8:00 a.m.	0800 hours	8:00 p.m.	2000 hours
9:00 a.m.	0900 hours	9:00 p.m.	2100 hours
10:00 a.m.	1000 hours	10:00 p.m.	2200 hours
11:00 a.m.	1100 hours	11:00 p.m.	2300 hours
		12:00 midnight	2400 hours

Thus,

3:35 a.m. = 0335 hours　　　4:15 p.m. = 1615 hours

11:40 a.m. = 1140 hours　　　10:30 p.m. = 2230 hours.

So now, using the 24 hour time notation and our knowledge of addition and subtraction of time, we can easily find the **elapsed time,** i.e., the time between two given times or the starting time when the finishing time and duration of activity is given or the finishing time when starting time and duration of activity is given.

Example 1. How long is it from:

(a) 1652 hours to 2143 hours　(b) 4:35 a.m. to 6:20 p.m.

Solution. (a) The elapsed time is 2143 hours – 1652 hours.

21 43 hours	~~21~~ 20 hours	~~43~~ 103 minutes
– 16 52 hours	– 16 hours	52 minutes
	4 hours	51 minutes

The elapsed time is **4 hours 51 minutes.**

(b) To find the elapsed time between 4:35 a.m and 6:20 p.m., first convert the time to 24-hour time.

4 : 35 a.m. = 0435 hours

6:20 p.m. = 1820 hours

Elapsed time =	~~18~~ 17 hours	~~20~~ 80 minutes
	– 4 hours	35 minutes
	13 hours	45 minutes

The elapsed time is **13 hours 45 minutes.**

Example 2. Using a 24 hour clock, find the time:

(a) 4 hours 25 minutes after 20 40 hours

(b) 8 hours 15 minutes before 2 : 35 p.m.

Solution. (a) Add 4 hours 25 minutes to 20 40 hours to find the required time.

20 h 40 min + 4h 25 min = 25h 5 min = 24 h + 1h 5 min.

⇒ 1h 5 min after 12:00 midnight

= **1:05 a.m. (It becomes next day)**

(b) 2:35 p.m = 14 35 hours = 14 h 35 min

∴ Required Time = 14h 35 min – 8h 15 min
= 6 h 20 min

∴ Required time = 6 hours 20 minutes = **0620 hours (6:20 a.m.).**

EXERCISE 45

1. **Find the elapsed time in hours and minutes between:**
   - **(a)** 0815 hours to 1945 hours
   - **(b)** 10:00 a.m to 4:40 p.m.
   - **(c)** 1205 hours to 2302 hours
   - **(d)** 2:00 p.m to 1:15 a.m.
2. **What time would it be**
   - **(a)** 3 hours 42 minutes after 8:25 a.m?
   - **(b)** 6 hours 25 minutes after 4:12 p.m?
   - **(c)** 2 hours 35 minutes before 1:15 p.m?
   - **(d)** 3 hours 40 min after 1825 hours?
   - **(e)** 7 hours 50 minutes before 0810 hours?
   - **(f)** 12 hours before 1500 hours?
3. Manasi completed a 500-piece jigsaw puzzle in 3 hours 40 minutes. Meghana took 55 minutes less than Manasi to complete the same puzzle. Amrita took 1 hour 35 minutes more than Meghana. Find the time taken by Amrita to complete the puzzle.

4. A cyclist started her journey at 0715 hours. She reached her destination at 1145 hours. She rested for 35 minutes and took the same journey back, but the journey back took her 25 minutes longer than the original journey. What time did she reach home?

5. A flight is supposed to start at 1025 hours and reach its destination at 1320 hours. If it starts 15 minutes late and arrives 19 minutes late, how long does it take to reach its destination?

YEARS, MONTHS AND WEEKS

By studying the calendar, we know that:

7 days = 1 week, 12 months = 1 year.
About 52 weeks = 1 year, 4 weeks = 1 month.

Changing years to months and months to years

Years to months: 1 year = 12 months

5 years = (5 × 12) months = **60 months**

2 years 8 months = 2 years + 8 months
= (2 × 12) months + 8 months
= 24 months + 8 months
= **32 months.**

48 months = (48 ÷ 12) years
= **4 years.**

21 months = 12 months + 9 months
= **1 year 9 months.**

$$12 \overline{)21} \quad \text{quotient } 1 \leftarrow \text{Year}$$
$$-12$$
$$9 \leftarrow \text{Months}$$

Months to Years:

$1 \text{ month} = \frac{1}{12} \text{ year}$

Example 1. **The construction of a new tennis court in Madhavi's school was to be completed in 6 weeks 2 days. 4 weeks 3 days have passed since the construction started. How much more time is left?**

Solution. Remaining time = 6 weeks 2 days − 4 weeks 3 days.

Weeks	Days
5	9
~~6~~ weeks	~~2~~ days
− 4 weeks	3 days
1 week	6 days

Think !
2 days < 3 days, so borrow 1 week
= 7 days from 6 weeks.
7 days + 2 days = 9 days
9 days − 3 days = 6 days
5 weeks − 4 weeks = 1 week

Hence, **1 week 6 days** more are left for the completion of the tennis court.

Example 2. **(a) Aman is 4 years 6 months old. His sister Anamika is 3 years 7 months elder to him. How old is she?**

(b) Raghav went for an assignment to London, when he was 25 years 5 months old. Now he is 34 years 2 months old. How long has he been there?

Solution. (a) Anamika's age

= Aman's age + 3 years 7 months
= 4 years 6 months + 3 years 7 months
= 7 years 13 months = 7 years + 12 months + 1 month
= 7 years + 1 year + 1 month = 8 years 1 month.
∴ Anamika's age is **8 years 1 month.**

Years	Months
4	6
3	7
7	13
Also, 12 months = 1 year	

(b) Raghav has been in London for (34 years 2 months − 25 years 5 months)

Years	Months
33	14
~~34~~ years	~~2~~ months
− 25 years	5 months
8 years	9 months

Think !
2 months < 5 months, so we borrow
1 year = 12 months from 34 years.
12 months + 2 months = 14 months
14 months − 5 months = 9 months
33 years − 25 years = 8 years

∴ Raghav has been in London for **8 years 9 months.**

EXERCISE 46

Add:

1.

5 years	8 months
+ 11 years	9 months

2.

18 years	10 months
+ 4 years	5 months

3.

2 weeks	4 days
+ 1 week	8 days

Subtract:

4.

16 years	9 months
− 5 years	11 months

5.

8 weeks	3 days
− 4 weeks	4 days

6.

9 years	7 months
− 2 years	8 months

7. A painter is asked to complete two paintings in 5 weeks. He completed the first painting in 3 weeks 4 days. How much time is left for him to complete the second painting?

8. Neeta spent her holidays 6 weeks 2 days in Varanasi, Kolkata and Darjeeling. She stayed in Varanasi for 1 week 3 days and in Kolkata for 2 weeks 6 days. How many days did she spend in Darjeeling?

9. Sangita is 10 years old. Her brother is 3 years 4 months younger. How old is her brother?

10. In 2 years 9 months from now, Suresh will be 13 years old. How old is Suresh now?

11. Kathy is 8 years 4 months old. Her mother is 40 years 3 months old. How old was her mother when Kathy was born?

12. I started teaching at the age of 21 years 5 months. I am now 43 years 2 months old. How long have I been teaching?

13. Amit was 10 years 5 months old when his sister Anubhuti was born. His sister is now 39 years 3 months old. How old is Amit now?

14. Anjali went to a boarding school at the age of 6 years 2 months. Now that she is 14 years 7 months, she is leaving the school. How long did she stay in the boarding school.

MENTAL MATHS–9

1. Which unit is the best for measuring the length of time it takes to drive 250 km in a car ? (seconds, minutes, hours or days)

2. What time in the digital clock is 6 minutes after 11:54 p.m.?

3. The total number of months in $10\frac{1}{2}$ years = ____________.

4. The time passed between 8:00 a.m. and 1425 hours is ____________.

5. The date one week after March 9 is ____________ .

6. Minutes in $2\frac{1}{2}$ hours = ________________ .

7. Anita left home at 6:45 a.m. for Delhi. It took her 2 hours 15 minutes to reach Delhi. Show her reaching time on the given clock.

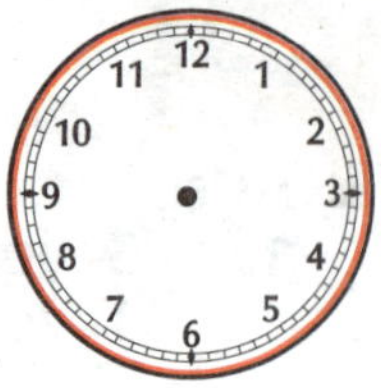

8. How much is 40 minutes before 0040 hours in 12 hour clock?

9. 8 years 8 months + 1 year 4 months = ________________ .

10. How many seconds in 1 hour ? ________________ .

11. 6:12 p.m. = ________________ hours..

12. 14 25 hours= ________________ p.m.

13. 11:25 p.m. = ________________ hours .

KNOW YOUR NUMBERS

THE NUMBER SIX

6

6 = 1 × 2 × 3 = 1 + 2 + 3

6 is the first **perfect number**. A perfect number is a number that equals the sum of its factors.

Insects like flies, moths, ants, beetles have six legs. A die with which you play 'Ludo' has six faces and the highest number on it is six. There are six feet in a ***fathom***.

A fathom is a unit of length used by sailors and equals 1·8288 metres.

Volley ball and Ice hockey are played with teams of 6 players.

Sex - and *hex*—means six. There are six sides in a hexagon and six musicians in a sextet.

Sextuplets are six children born together.

The cells in a honeycomb are hexagonal.

Six planets-*Mercury*, *Venus*, *Mars*, *Jupiter*, *Saturn* and *Uranus* can be seen with the naked eye from Earth.

PUZZLE

Jha earns more money than Sharma. Hem's salary is less than Raju's. Raju earns more than Sharma but less than Jha. Who gets the most salary?

14 TEMPERATURE

In the month of June, the headlines of most of the papers read, "The country under a severe heat wave, temperatures soaring as high as 50°C. In winters, it is the opposite, most of the places are in the grip of a cold wave and temperatures fall as low as 0°C or even below it.

So what is temperature and how do we measure it?

Temperature is the measure to describe how hot or cold something is.

The instrument used for measuring temperature is a **thermometer**. It is a thin tube with a coloured liquid called mercury in it.

A. **1.** A thermometer shows the temperature of the air. It shows how warm the air is.

★ The thermometer at the right shows a temperature of 20 degrees.

★ 25 degrees is written 25°. Find these temperatures on the thermometer. 30°, 32°, 50°, 54°, 70°, 76°.

2. When the air gets warmer, the red liquid expands, so it moves up the thin tube.

★ Where would the top of the coloured liquid be when the temperature is 10° warmer than 40° ?

3. When the air gets cooler, the red liquid contracts, occupies less space, so moves back down the tube.

★ Where would the top of the red liquid be when the temperature is 10° cooler than 40° ?

Write the correct temperature :

1. 10° cooler than 70°. ____________

2. 20° cooler than 30°. ____________

3. 10° warmer than 80° ____________

4. 40° warmer than 20°. ____________

Find the answer :

5. How many degrees cooler is 20° than 50° ? ____________

6. How many degrees warmer is 70° than 40° ? ____________

7. The thermometer showed 55° at noon. It was 13° cooler at night. What did the thermometer show then ? ____________

B. Temperature in the metric system is measured in degree Celsius (°C).

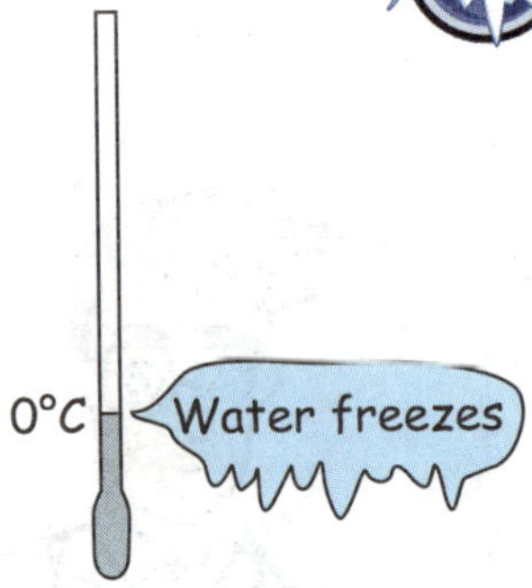

We call the point at which **water freezes 0°C** (zero degrees Celsius). We call the point at which **water boils 100°C.** Between 0°C and 100°C, this thermometer is marked off in intervals of ten degrees, labelled 10, 20, 30, and so on to 100. The smaller markings show intervals of two degrees.

You can get an idea of some other temperatures by looking at the thermometer at the right. Count from 30 to determine that **normal body temperature is 37°C**.

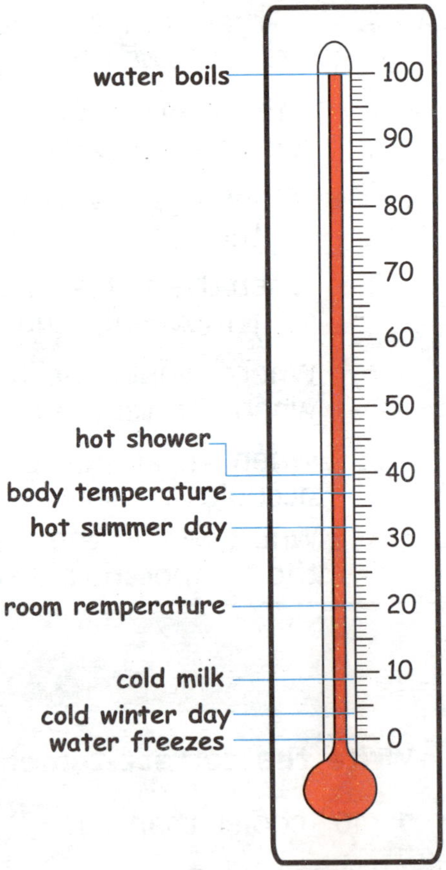

Skill Practice

Class Work

1. What is the temperature shown on the thermometer ?

(a) hot summer day________ **(b)** hot shower________

(c) room temperature ________ (d) cold milk ________

(e) water boiling temperature ________ (f) water freezing temperature ________

2. Circle the most likely temperature.

	a.	b.	c.
(i) planting a garden	a. 25°C	b. 0°C	c. 5°C
(ii) swimming in a pond	a. 21°C	b. 12°C	c. 32°C
(iii) wearing a woollen coat	a. 75°C	b. 28°C	c. 6°C
(iv) washing a car	a. 25°C	b. 5°C	c. 10°C

3. Find the difference between the two temperatures:

(a) between 10°C and 4°C ________ **(b)** between 20°C and 60°C ________

(c) between 30°C and 45°C ________ **(d)** between 40°C and 5°C ________

4. Here are temperatures of main cities on a day in August

Agra	37°C	Bangalore	30°C
Ahmedabad	35°C	Chennai	34°C
Bhopal	31°C	Kolkata	33°C
Coimbatore	35°C	Mumbai	32°C
Dehradun	30°C	Lucknow	36°C
Gangtok	19°C	Hyderabad	34°C
Simla	24°C	Jaipur	34°C
Delhi	35°C		

From the above chart, find

(a) Which city had the highest temperature ? ________

(b) Which city had the lowest temperature ? ________

(c) What is the difference between the temperatures of Agra and Gangtok ? ________

(d) Which cities had the same temperature ? ________

C. Two kinds of thermometers

A thermometer is used to measure temperature.

Two standard units of temperature are the Fahrenhiet degree **(F)** and the Celsius degree **(C)**, sometimes called centigrade.

In India, we use the centigrade or the celsius scale. Celsius thermometer was designed by the Swedish astronomer Celsius Anders (1701-44) and so the word celsius. In short it is 'cel'. Water freezes at 0°C also written as 0° cel and boils at 100°C also written as 100° cel.

In Fahrenheit unit, water freezes at 32°F and boils at 212°F.

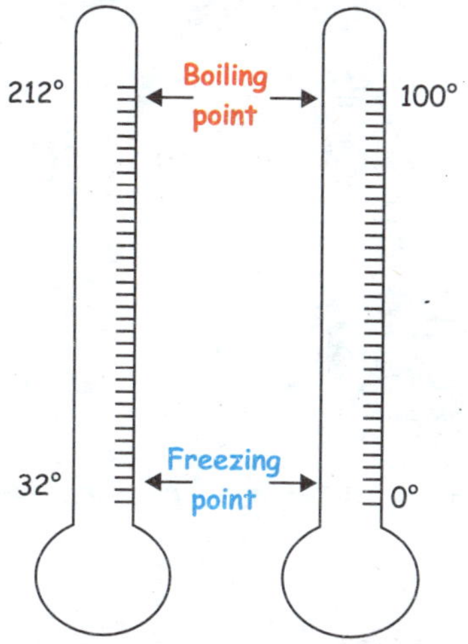

FAHRENHEIT CELSIUS
THERMOMETER

The normal temperature of the human body is 37°C or 98·4° F.

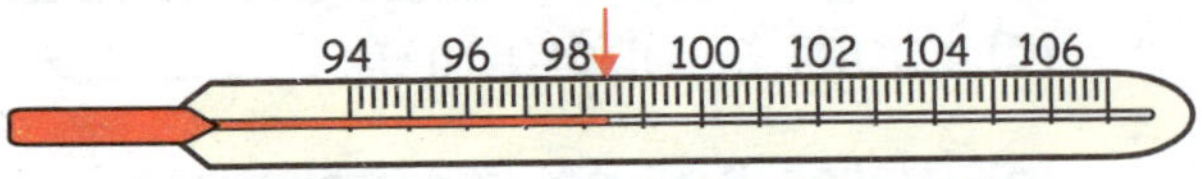

A CLINICAL THERMOMETER WITH FAHRENHEIT SCALE

★ How many degrees is it from the freezing point to the boiling point on the Celsius thermometer ? on the Fahrenheit thermometer ?

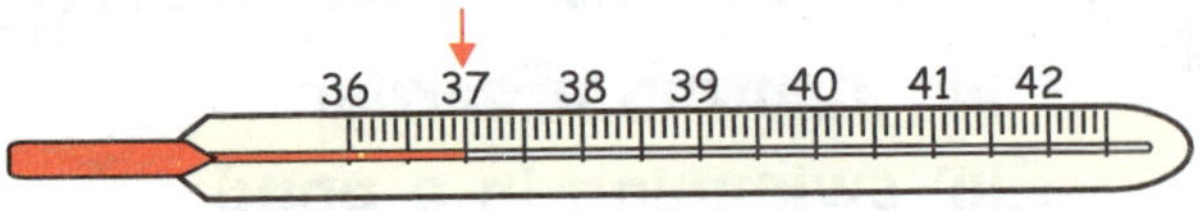

A CLINICAL THERMOMETER WITH CELSIUS SCALE

The thermometer used by a doctor to measure the temperature of a human body is called **Clinical thermometer.** Nowadays, thermometers with celsius scale are also used by doctors.

Before using a clinical thermometer it should always be adjusted by giving jerks so as to bring the level of mercury in the tube much below the degree marked 98° which is approximately the body temperature in normal conditions.

Conversion from one unit to another

- **To change from Fahrenheit to Celsius.**

Step 1 : Subtract 32,
Step 2 : Multiply by 5,
Step 3 : Divide by 9.

Example : 122° F

122 − 32 = 90
90 × 5 = 450
450 ÷ 9 = 50
∴ 122°F = 50°C

Temperature in °F → Subtract 32 and multiply the result by $\frac{5}{9}$ → Temperature in °C

- **To change from Celsius to Fahrenheit.**

Step 1 : Multiply by 9,
Step 2 : Divide by 5,
Step 3 : Add 32

Example : 80°C

80 × 9 = 720
720 ÷ 5 = 144
144 + 32 = 176
∴ 80°C = 176°F

Temperature in °C → divide by $\frac{9}{5}$ and add 32 → Temperature in °F

EXERCISE 47

1. Manish was not feeling well. His mother took his temperature with this thermometer.

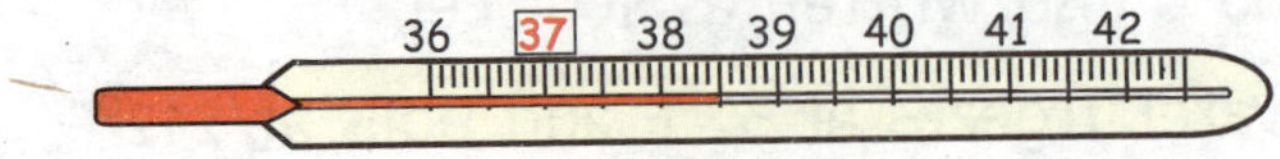

(a) What was his temperature ?

(b) Why is there a box round the 37 on the scale ?

(c) Why does the scale only go from 35°C to 42°C ?

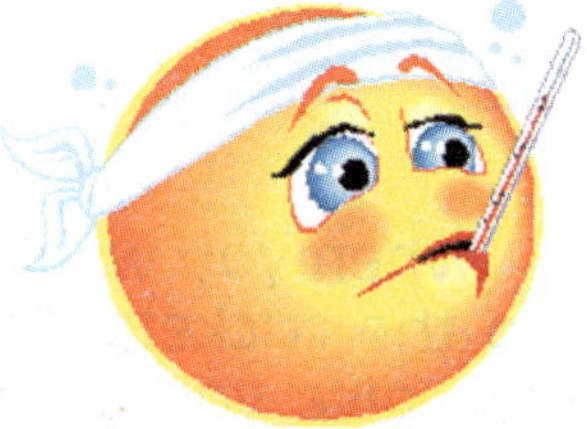

2. Can you read the temperature on these thermometers ?

(a) 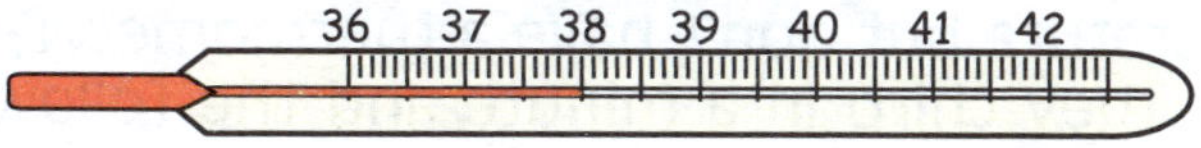

(b) 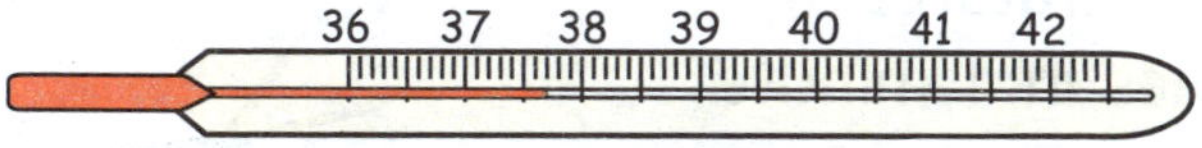

(c) 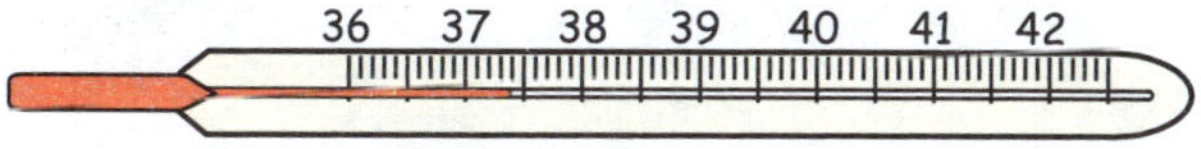

(d) 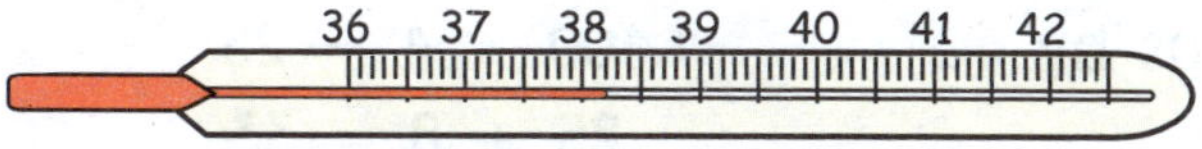

3. Express the following temperatures given in Celsius scale, in Fahrenheit scale.

(a) 30°C **(b)** 10°C **(c)** 40°C **(d)** 100°C

4. Express the following temperatures, given in Fahrenheit scale, in Celsius scale:

(a) 212°F **(b)** 284°F **(c)** 95°F **(d)** 455°F

5. (a) A comfortable room temperature is said to be 68°F, or ________°C.

(b) Your normal body temperature is about 98°F, or about ________ °C.

MENTAL MATHS–10

1. 100°C = ________ °F. **2.** 32°F = ________ °C.

3. The boiling point of water on the celsius scale is ________ °C.

4. The normal body temperature is ________ °F = ________°C.

5. To convert °F to °C we use the formula ________________ .

6. To which scale are the clinical thermometers marked?

7. The body temperature of a patient is 102°F. Does it show signs of fever? Yes/No

8. 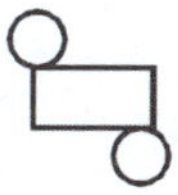 is the net of a ________________ .

9. 4 years 3 months =________________ months.

10. A quarter of an hour = ________________ minutes.

11. If the 5th day of a month is a saturday, what day will be the 19th of the same month?

12. In 2008, February has ________________ days.

MATHEMATICS FOR FUN

Try this !

When you want to know the temperature but don't have a thermometer, try listening to the crickets. The number of times they chirp in a minute and the following method can give you a good idea of the temperature.

Suppose the cricket chirps 100 times in 1 minute

Step 1. Divide the number of chirps by 4 $100 \div 4 = 25$

Step 2. Add 8 to the quotient. $25 + 8 = 33$

Step 3. Multiply the sum by $\frac{5}{9}$. $33 \times \frac{5}{9} = \frac{165}{9} = \frac{55}{3} = 18\frac{1}{3}$

$\therefore$ The temperature is about 18°C.

Find the approximate temperature if the number of chirps per minute is

1. 120 **2.** 160. **3.** 200 **4.** 220 5. 227.

15 PERCENTAGES

Report Card		
Name : Genie		
Class : V		Year : 2008
Subject	Term I	Term II
English	92%	94%
Hindi	95%	99%
Maths	100%	100%

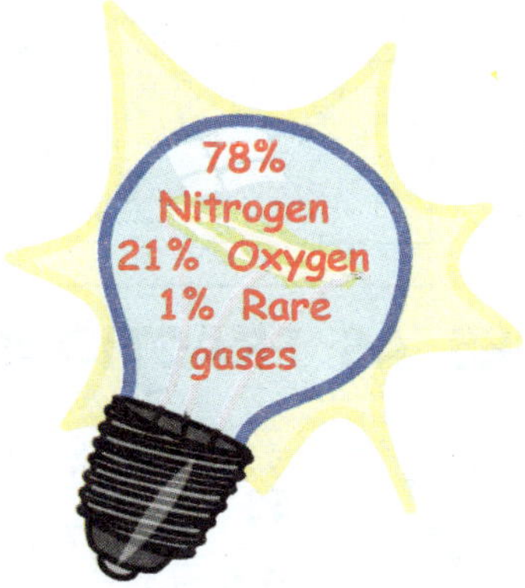

35% Off

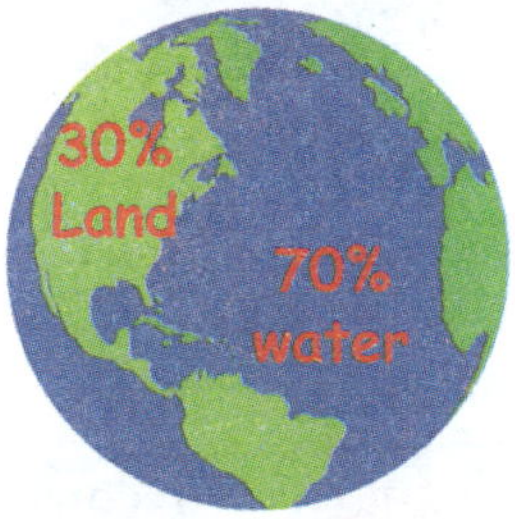

Percentages are often used to compare quantities. They are usually seen on your report card, sale banners, advertisements, statistical reports etc.

Percentage is a fraction with denominator 100.

You will be able to understand this with the help of the given example.

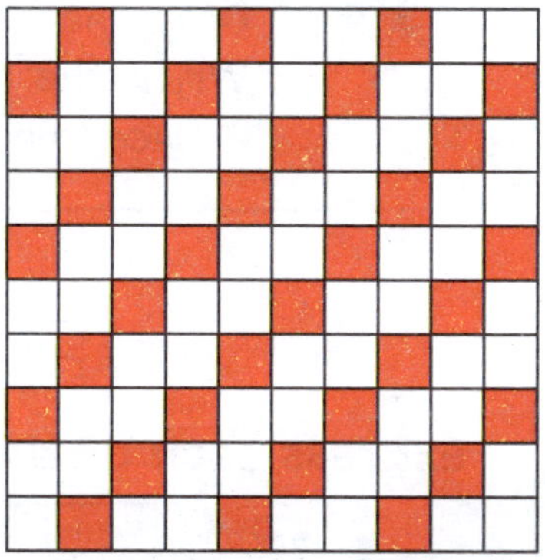

Look at this square. It is divided into 100 smaller squares. The shaded squares will always be a fraction of 100. There are 33 shaded squares in this picture.

33 out of 100 squares are shaded.

You can say that $\frac{33}{100}$ of the large square is shaded.

You can also say that 33 squares per hundred are shaded. When a fraction has 100 as its denominator, we call it a **percentage.** $\frac{33}{100}$ can be named as **33 per cent.** Per cent means per hundred. You can say that 33 per cent of the squares are shaded.

For per cent, we use the symbol %. Thus 33 per cent is written as 33%.

Per cent means per hundred. The symbol for per cent is %.

33%	=	$\frac{33}{100}$	=	0.33
Per cent		**Fraction**		**Decimal**

Wow! That's interesting one quantity can be written in so many different ways.

Write each of the following as percentage:

1. $\frac{37}{100} =$ **37%**   **2.** $\frac{40}{100} =$ ☐   **3.** $\frac{11}{100} =$ ☐   **4.** $\frac{25}{100} =$ ☐

5. $\frac{43}{100} =$ ☐   **6.** $\frac{58}{100} =$ ☐   **7.** $\frac{65}{100} =$ ☐   **8.** $\frac{100}{100} =$ ☐

EXPRESSING PERCENTAGES AS FRACTIONS:

Examples.

$13\% = \frac{13}{100}$ $9\% = \frac{9}{100}$ $41\% = \frac{41}{100}$ $57\% = \frac{57}{100}$ $80\% = \frac{\overset{4}{\cancel{80}}}{\underset{5}{\cancel{100}}}$ (Divide by 20) $= \frac{4}{5}$.

$5\% = \frac{\overset{1}{\cancel{5}}}{\underset{20}{\cancel{100}}} = \frac{1}{20}$ $14\% = \frac{\overset{7}{\cancel{14}}}{\underset{50}{\cancel{100}}} = \frac{7}{50}$ $36\% = \frac{\overset{9}{\cancel{36}}}{\underset{25}{\cancel{100}}}$ (Divide by 4) $= \frac{9}{25}$

Remember: To express a fraction in lowest terms, divide each term of fraction by HCF.

Rule: To convert a percentage to a fraction
- ★ Drop the '%' sign
- ★ Write the number over 100
- ★ Reduce the fraction formed to lowest terms.

Express each of the following percentages as a fraction in its lowest terms

1. 7% = ☐   **2.** 19% = ☐   **3.** 63% = ☐

4. $2\% = \frac{2}{100} = \frac{1}{50}$.   **5.** 10% = ☐   **6.** 25% = ☐

7. 32% = ☐   **8.** 54% = ☐   **9.** 95% = ☐

EXPRESSING PERCENTAGES AS DECIMALS

Examples.

$45\% = \frac{45}{100} = \mathbf{0.45}$ $27\% = \frac{27}{100} = \mathbf{0.27}$ $107\% = \frac{107}{100} = \mathbf{1.07}$

$24.7\% = \frac{24.7}{100} = \mathbf{0.247}$ $3.6\% = \frac{3.6}{100} = \mathbf{0.036}$ $8\% = \frac{8}{100} = \mathbf{0.08}$

Note: Dividing by 100 moves decimal point 2 places to the left .45.

Rule: To convert a percentage to decimal:

- ★ Drop the '%' sign
- ★ Divide by 100. Move decimal point two places to the left.

Skill Practice

Class Work

Write each of the following percentages as decimals:

1. 25% $= \frac{25}{100}$ $= 0.25$	**2.** 5% = =	**3.** 1% = =	**4.** 75% = =
5. 90% = =	**6.** 125% = =	**7.** 6.2% = =	**8.** 28.3% = =

EXPRESSING FRACTIONS AND DECIMALS AS PERCENTAGES

★ Fractions to Percentages

Method: Make the denominator of each fraction 100.

Examples.

$\frac{4}{5} = \frac{4 \times 20}{5 \times 20} = \frac{80}{100} = \mathbf{80\%}$.

$\frac{1}{2} = \frac{1 \times 50}{2 \times 50} = \frac{50}{100} = \mathbf{50\%}$.

Can you express $\frac{1}{6}$ as a percentage with the above method ?

Let us see what we get when we multiply the above given fractions by 100%.

$\frac{4}{\cancel{5}_1} \times \cancel{100}^{20}\% = \mathbf{80\%}$

$\frac{1}{2} \times 100\% = \mathbf{50\%}$

Oh Yes! we get the same answer.

$\frac{1}{6} = \frac{1}{\cancel{6}_3} \times \cancel{100}^{50}\% = \frac{50}{3}\% = \mathbf{16\frac{2}{3}\%}$

Is not this Easy ?

Rule: To convert a fraction to percentage

- ★ Multiply by 100 %.
- ★ Simplify.

★ Decimals to Percentages

$0.54 = \frac{54}{100} = \mathbf{54\%}$ $\qquad 0.06 = \frac{6}{100} = \mathbf{6\%}$ $\qquad 0.375 = \frac{375}{1000} = \frac{37.5}{100} = \mathbf{37.5\%}$

Did you notice that in all these cases the decimal point has shifted two places to the right. 0.54 0.06 0.375

What is the **multiplier** that shifts the decimal point in a **multiplicand,** two places to the right ? **10, 100, 1000 or any other number!**

Rule : To convert decimal to percentage, multiply by 100%.

0.39 = 0.39 × 100% = **39%**

8 = 8 × 100% = **800%**

0.007 = 0.007 × 100 = **0.7%**

4.8 = 4.8 × 100 = **480%**

Skill Practice — **Class Work**

1. Express each of the following as a percentage.

a. 0.21 = **0.21 × 100% = 21%**	**d.** 0.138 = 138%
b. 0.72 = 72%	**e.** 1.8 = 180%
c. 0.06 = 6%	**f.** 1 = 100%

2. Circle the correct decimal equivalent of 9%:

(a) 9 **(b)** 0.9 **(c)** 0.09 **(d)** 0.009

3. Circle the correct percentage equivalent of 3.05.

(a) 3050% **(b)** 30.50% **(c)** 305% **(d)** 3.05%

4. Circle the percentage equivalent of $\frac{1}{5}$.

(a) 25% **(b)** 20% **(c)** 10% **(d)** 40 %

5. Circle the fraction equivalent of 40%.

(a) $\frac{4}{100}$ **(b)** $\frac{1}{25}$ **(c)** $\frac{2}{10}$ **(d)** $\frac{2}{5}$

6. Complete the table for the shaded squares drawn below:

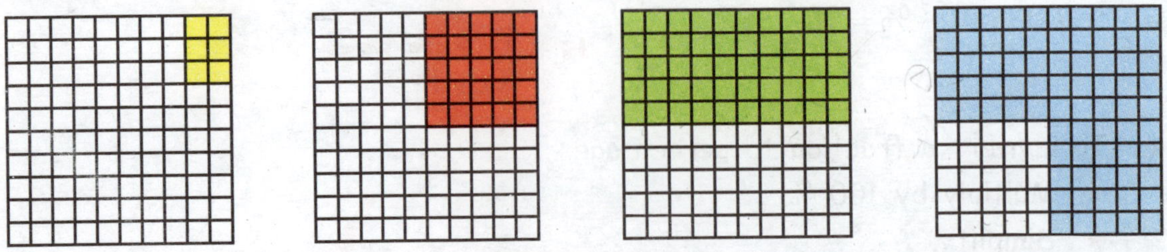

(a) (b) (c) (d)

Shaded squares	fraction	decimal	per cent
(a) 6	$\frac{6}{100}$	0.06	6%
(b) 25	$\frac{25}{100}$	0·25	25%
(c) 50	$\frac{50}{100}$	0·50	50%
(d) 75	$\frac{75}{100}$	0·75	75%

7. Fill in the blanks:

(a) The word 'cent' means 100

(b) 48 per cent means $\frac{48}{100}$

(c) $\frac{69}{100}$ is the same as 69%

(d) The symbol % is the sign for percent

TO EXPRESS A QUANTITY AS A PERCENTAGE OF ANOTHER QUANTITY

Example 1. What per cent of 20 is 15 ?

Solution. The question means if a quantity is 15 parts out of 20, how much is it out of 100.

$$\frac{15}{20} = \frac{15}{\cancel{20}_{1}} \times \cancel{100}^{5}\% = \mathbf{75\%}$$

Step 1: Write 15 as a fraction of 20
Step 2: Change the fraction to %.

Thus, 15 is **75%** of 20.

Look at the following pictures for better understanding:

$\frac{1}{2}$

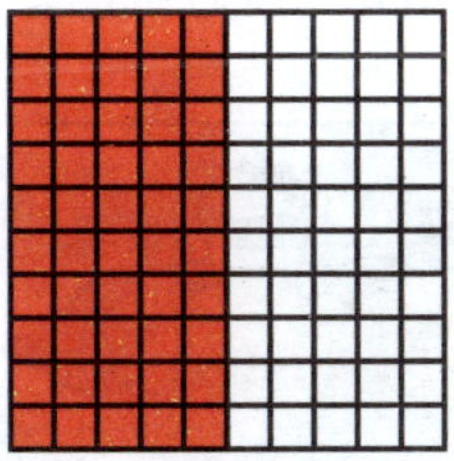

$\frac{50}{100}$

$$\frac{1}{2} = \frac{50}{100} = \mathbf{50\%}$$

One part out 2 equal parts = 50 parts out of 100 equal parts.

Example 2. What percentage of 3 m is 12 cm ?

Solution. 3 m = 3 × 100 cm = 300 cm

$$\frac{12\text{ cm}}{3\text{m}} = \frac{12\text{ cm}}{300\text{ cm}} = \frac{\cancel{12}^{4}}{\cancel{300}_{\cancel{3}_{1}}} \times \cancel{100}^{1}\% = \mathbf{4\%.}$$

Maths Alert !
Change to the same units.

What per cent of:

1. 4 is 3 ? $= \frac{3}{4} \times 100\% = 75\%$	**2.** 18 is 9 ?	**3.** 50 is 1 ?
4. 25 is 24 ?	**5.** 10 is 2 ?	**6.** 24 is 12 ?

FINDING PERCENTAGE OF A NUMBER

Examples:

$$10\% \text{ of } 70 = \frac{10}{100} \text{ of } 70 = \frac{10}{100} \times 70 = \mathbf{7}.$$

$$50\% \text{ of } 18 = \frac{50}{100} \text{ of } 18 = \frac{50}{100} \times 18 = \mathbf{9}.$$

$$5\% \text{ of Rs } 20 = \frac{5}{100} \text{ of Rs } 20 = \text{Rs } \frac{5}{100} \times 20 = \textbf{Re 1}.$$

$$7\% \text{ of 1 litre} = \frac{7}{100} \text{ of 1 litre} = \frac{7}{100} \text{ of } 1000\ ml = \frac{7}{100} \times 1000\ ml = \mathbf{70}\ ml.$$

Find the value of:

1. 2% of 200 $= \frac{2}{100}$ of 200 $= \frac{2}{100} \times 200 = 4$	**2.** 50% of 480	**3.** 40% of 500	**4.** 15% of 2 kg (in g)

EXERCISE 48

1. **Express the first quantity as a percentage of the second quantity:**

   (a) 3; 5 **(b)** 400 g; 2 kg **(c)** 6 mm; 3 cm **(d)** 24 min; 2h

   (e) 500 *ml*; 2*l* **(f)** 6 months; 1 year **(g)** 90°; 360°

2. Karishma scored 44 marks out 55 marks in Science. What percentage of marks did she get ?

3. Anil said that $\frac{3}{4}$ of his friends wear glasses. What per cent of his friends wear glasses.

4. **Find the value of:**

   (a) 30% of 120 **(b)** 22% of 50 **(c)** 15% of 2kg (in g)

   (d) 25% of 1 min (in s) **(e)** 18% of 2 *l* (in *ml*) **(f)** 30% of 1 day (in h)

   (g) 5% of 4 m (in cm) **(h)** $33\frac{1}{3}$% of Rs 6000 **(i)** $16\frac{2}{5}$% of 6 *kl*

5. In a school, 40% of the children come to school by bus. If there are 600 children in all in a school, how many children come by bus ?

MENTAL MATHS–11

1. A kitemaker made 7800 kites in 13 days. The number of kites made in 1 day are ____________

2. 0.45 as % = ____________ %.

3. 25% as a fraction = ____________ .

4. $\frac{100}{100}$ as a % = ____________% .

5. 32.5% as a decimal = ____________ .

6. 20% of Rs 5 = ____________

7. 7 as a per cent of 10 = ____________ .

8. A man spends 50% of his income; what fractions of his income does he save ?

9. The percentage of the letter 'D' in the word ' INDIA' is ____________ .

10. The percentage of the letter 'S' in the word 'CLASS' is ____________ .

16 MONEY

UNITARY METHOD

For your birthday party, your mother needs to buy 16 key chains as return gifts. When she asks the shopkeeper for the price of 1 key chain, he replies "Rs 300 for a dozen" (1 dozen = 12). To know the price of 16 key chains, she needs to know the price of one key chain. How does she find it ?

12 key chains cost Rs 300

∴ 1 key chain costs Rs 300 ÷ 12 = Rs 25

16 key chains cost Rs 25 × 16 = Rs 400

> One key chain will cost less, so divide

> 16 key chains will cost more, so multiply

```
     25
12)300
   -24↓
     60
    -60
      0
```

Thus, if the price of any number of articles is given, the price of one article **(unit rate)** will be **less** and can be found by **dividing** the total price by the number of articles.

Now that you know the unit rate, you can find the price of any number of articles by multiplying the unit rate by the number of articles.

> This method of finding the value of 1 item first and then finding the value of the required number of items is called **unitary method**. The word **unitary** is derived from the word **"unit"** (one).

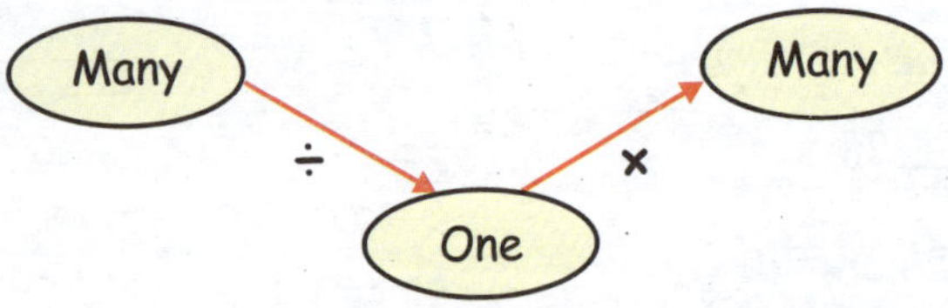

Example. If 5 chairs cost Rs 630, what is the cost of 12 such chairs?

Solution. 5 chairs cost Rs 630

> 1 chair costs less, so divide

1 chair costs Rs 630 ÷ 5 = $\frac{\text{Rs } 630}{5}$

12 chairs cost Rs $\frac{\overset{126}{\cancel{630}}}{\underset{1}{\cancel{5}}} \times 12$ = Rs 126 × 12 = **Rs 1,512.**

> 12 chairs cost more so multiply cost of 1 chair by 12.

Example 2. If 20 ice cream cones cost Rs 248, find the cost of 8 ice cream cones.

Solution. 20 ice cream cones cost Rs 248

1 ice cream cone costs Rs $\frac{248}{20}$

8 ice cream cones cost Rs $\frac{248}{20} \times 8$ = Rs $\frac{496}{5}$ = **Rs 99.20.**

Note: The unknown quantity is always placed on the right hand side.

Solve:

Find the cost of:

1. 15 candles if 4 candles cost Rs 8.

2. 8 bananas if one dozen bananas cost Rs 36.

Which is a better buy ?

12 for Rs 60

Super Pencils

12 pencils cost Rs 60

1 pencil costs Rs 60 ÷ 12 = **Rs 5**

20 for Rs 80

Smooth Pencils

20 pencils cost Rs 80

1 pencil costs Rs 80 ÷ 20 = **Rs 4**

To decide as to which box of pencils would be a better buy, you have to compare the cost of 1 pencil from each box, *i.e.*, find the unit rate and compare.

The "**Smooth Pencils**" box would be a better buy as the unit price is less than the other box.

Skill Practice

Class Work

Which is a better buy ?

1. 20 stickers for Rs 60 OR	12 stickers for Rs 48	2. 10 bananas for Rs 15 OR	5 bananas for Rs 8
Ans. :		**Ans. :**	

Home Work

EXERCISE 49

1. If 24 ball pens cost Rs 56, find the cost of 18 such pens. 
2. If a man earns Rs 3,900 in 6 weeks, how much does he earn in 10 weeks ?
3. If 3 litres of petrol cost Rs 135.45, what is the cost of 10 litres of petrol ? 
4. 5 shaving blades cost Rs 6.25, find the cost of 12 blades.
5. If one dozen oranges cost Rs 24.60, what would be the cost of 5 oranges? 
6. 5 gross pens cost Rs 2,160, find the cost of 43 pens.
   (1 gross = 12 dozen = 144)
7. 8 scores of bananas cost Rs 192, find the cost of 85 bananas. (1 score = 20) 
8. If 2 dozen notebooks cost Rs 288, find the cost of 1 score of notebooks. 
9. **Which is a better buy**—3 kg apples for Rs 105 or 5 kg apples for Rs 180?
10. **Which is a better buy**—A 2-litre pack of juice for Rs 130 or 5-litre pack of juice for Rs 310?

PROFIT AND LOSS

A shopkeeper bought a mobile phone for Rs 2,800 and sold it to you for Rs 3,100. Did he **earn** or **lose** money ?

Before answering this question, let us understand two very simple terms.

> **Cost Price:** The amount of money which you pay to buy an article is called the cost price, written as C.P.
>
> **Selling Price:** The amount which you receive from the customer by selling an article is called the selling price, written as S.P.

Now we come back to our question. You can very well see that Rs 3,100 (the shopkeeper's selling price) > Rs 2,800 (the shopkeeper's cost price), *i.e.*, Money he receives > Money he paid.

So he earns money. The money earned is called **profit**.

> So, when the selling price (S.P.) of an article is greater than its cost price (C.P.), a profit is made.

How much money did he earn as profit ? Rs 3,100 − Rs 2,800 = **Rs 300.** Therefore, **Profit = Selling Price − Cost Price.**

On the other hand, Mr. Verma bought a table for Rs 1,600 and he had to sell it for Rs 1,350.

Here, we see that,

Rs 1,350 (Selling Price) < Rs 1,600 (Cost Price)

i.e., Money he received < Money he paid.

> So, he loses money. The money lost is called a **loss.** So, when the selling price (S.P.) of an article is less than its cost price (C.P.), a loss is made.

How much money did he lose ? Rs 1600 − Rs 1350 = **Rs 250.**

Hence, **Loss = Cost Price − Selling Price**

> **To sum up**
>
> 1. **When S.P. > C.P., there is a profit. Profit = S.P. − C.P.**
> 2. **When S.P. < C.P., there is a loss. Loss = C.P. − S.P.**

Example 1. A toy car is bought for Rs 153.25 and sold for Rs 125.50. Find the profit or loss made.

Solution. Cost Price (C.P.) = Rs 153.25, Selling Price (S.P.) = Rs 125.50

C.P. > S.P., so it is a loss

Loss = Cost Price − Selling price = Rs 153.25 − Rs 125.50 = **Rs 27.75**

Example 2. A shopkeeper buys a television set for Rs 7,200, spends Rs 120 on its transportation and sells it for Rs 9,600. Find the profit or loss made.

Solution. Cost Price (C.P.) = Rs 7,200, Selling Price (S.P.) = Rs 9,600

Cost of transportation = Rs 120

∴ Total Cost Price = Rs 7,200 + Rs 120 = Rs 7,320

S.P. > C.P., so there is a profit.

Profit = S.P. − C.P. = Rs 9,600 − Rs 7,320 = **Rs 2,280.**

> **Maths Alert !**
> Overheads such as repair cost, transportation charges, octroi etc are added to the Cost Price.

Find the profit or loss in each of the following:

Cost Price (C.P.)	Selling Price (S.P.)	Profit/Loss
1. Rs 70	Rs 80	S.P. > C.P, so Profit is made Profit = Rs 80 – Rs 70 = Rs 10
2. Rs 510	Rs 548	
3. Rs 475	Rs 462	
4. Rs 125.50	Rs 128.15	

TO FIND SELLING PRICE OR COST PRICE

To find selling price when,
- ★ Cost price and profit are given
- ★ Cost price and loss are given

S.P. = C.P. + Profit
S.P. = C.P. – Loss

Example 1. Find the selling price of a pair of shoes, when its cost price is Rs 762 and the profit made on it is Rs 48.

Solution. Cost Price = Rs 762, Profit = Rs 48

∴ Selling Price = Cost Price + Profit = Rs 762 + Rs 48

= **Rs 810.**

Example 2. During summers, a warm jacket which was bought by the shopkeeper for Rs 1,360 was sold at a loss of Rs 435. What was the selling price of the jacket ?

Solution. Cost Price = Rs 1,360, Loss = Rs 435

∴ Selling Price = Cost Price – Loss = Rs 1,360 – Rs 435 = **Rs 925.**

To find Cost Price when,
- ★ Selling Price and Profit are given.
- ★ Selling Price and Loss are given.

C.P. = S.P. – Profit
C.P. = S.P. + Loss

Example 3. Rajni bought a tricycle for her daughter Aadhya for Rs 750. The shopkeeper earned a profit of Rs 125. At what price did he buy it ?

Solution. Selling Price = Rs 750, Profit = Rs 125

∴ Cost Price = Selling Price – Profit = Rs 750 – Rs 125 = **Rs 625.**

Example 4. Find the cost price of a pair of slippers sold at Rs 345 for a loss of Rs 25.

Solution. Selling Price = Rs 345, Loss = Rs 25

∴ Cost Price = Selling Price + Loss = Rs 345 + Rs 25 = **Rs 370.**

Fill in the blanks:

	Cost Price	Selling Price	Profit or Loss
1.	Rs 400		Profit = Rs 25
2.	Rs 180.30		Loss = Rs 15.45
3.		Rs 765	Profit = Rs 45
4.		Rs 1045.25	Loss = Rs 46.85

EXERCISE 50

1. Find the profit or loss in each of the following cases :

(a) C.P. = Rs 150, S.P. = Rs 275 **(b)** C.P. = Rs 185.75, S.P. = Rs 172.50

(c) C.P. = Rs 200, S.P. = Rs 315.50 **(d)** C.P. = Rs 65.85, S.P. = Rs 55.25

2. A shopkeeper bought a refrigerator for Rs 12,000 and sold it for Rs 13,510. Find his profit or loss.

3. Rachana bought a saree for Rs 1150 and sold it to her friend for Rs 1070. Find her profit or loss.

4. A dealer bought 20 TV sets for Rs 3150 each, paid Rs 360 as the cartage for all and sold these at Rs 3500 each. Find his total profit or loss.

5. A fruit vendor bought 10 kg of mangoes for Rs 350. He sold them at Rs 42 per kg. How much profit did he make for 1 kg of mangoes ?

6. Pens were purchased by a stationery shop owner at 100 pens for Rs 350. He sold them at 8 for Rs 40. Find the profit made on each pen.

(Hint. Find the unit cost in both the cases, compare)

7. Fill in the blanks:

	Cost Price	Selling Price	Cartage, Octroi, etc.	Profit or Loss
(a)	Rs 170	Rs 200	Rs 10	—
(b)	Rs 508	Rs 615	—	Rs 95 (Profit)
(c)	Rs 860	Rs 710	Rs 20	—
(d)	Rs 325	—	Rs 15	Rs 30 (Loss)
(e)	—	Rs 56.10	Rs 3.15	Rs 4.28 (Loss)
(f)	Rs 1600	—	Rs 18	Rs 162 (Profit)
(g)	—	Rs 217.68	Rs 7.13	Rs 21.75 (Profit)

8. A bicycle is sold for Rs 2140 by making a profit of Rs 153. Find the cost price of the bicycle.

9. The cost price of a chair is Rs 312 and is sold by making a profit of Rs 58. What is the selling price of the chair ?

10. A book is bought for Rs 97.50 and sold by incurring a loss of Rs 12.35. Find the selling price of the book.

11. The selling price of a fan is Rs 1875. Loss incurred is Rs 176. Find the cost price of the fan.

PROFIT % AND LOSS %

I bought books from two book-sellers. Let us see how much profit each of them made.

It seems as if both the book shops have made the same profit, but it is not so.

In the first case, Rs 10 is a profit on a cost price of Rs 40.

In the second case, Rs 10 is a profit on a cost price of Rs 50.

How can we compare profits on different cost prices?

In the previous chapter you have learnt to compare costs by finding the unit cost. Here, we shall find the profit in both the cases on the some cost price, *i.e.*, profit per hundred rupees or **Profit %.**

Great Book Shop

On Rs 40, the profit is Rs 10

$\therefore$ On Re 1, the profit is Rs $\frac{10}{40}$

$\therefore$ On Rs 100, the profit is $\frac{10}{40} \times 100$

$= \text{Rs } 25$

$\therefore$ Profit per Rs 100 = Rs 25

or

Profit % = **25%.**

Lovely Book Shop

On Rs 50, the profit is Rs 10

$\therefore$ On Re 1, the profit is Rs $\frac{10}{50}$

$\therefore$ On Rs 100, the profit is Rs $\frac{10}{50} \times 100$

$= \text{Rs } 20$

$\therefore$ Profit per Rs 100 = Rs 20

or

Profit % = **20%.**

This shows that "Great Book Shop" made a higher profit than "Lovely Book Shop".

The profit % or loss % is always calculated on the cost price as you have seen above.

$$\textbf{Profit \%} = \frac{\textbf{Profit}}{\textbf{Cost Price}} \times 100 \qquad \textbf{Loss \%} = \frac{\textbf{Loss}}{\textbf{Cost Price}} \times 100$$

Example 1. **A book-seller buys books at Rs 45 each and sells them at Rs 63 each. Find his profit per cent on each book.**

Solution. C.P. of each book = Rs 45,

S.P. of each book = Rs 63

Profit = Rs 63 − Rs 45 = Rs 18

$$\text{Profit per cent} = \frac{\text{Profit}}{\text{C.P.}} \times 100 = \frac{18}{45} \times 100\% = \mathbf{40\%}.$$

Example 2. **A fruit vendor bought a box of 110 apples for Rs 500. He found 30 apples were rotten and he sold the rest at Rs 5 each. Find his percentage profit or loss.**

Solution. C.P. of 110 apples = Rs 500

30 apples were rotten.

He sold (110 − 30) or 80 apples at Rs 5 each.

S.P. of 80 apples = Rs 80 × 5 = Rs 400

Since S.P. < C.P., there is a loss.

Loss = Rs 500 − Rs 400 = Rs 100

$$\text{Percentage loss} = \frac{\text{Loss}}{\text{C.P.}} \times 100 = \frac{100}{500} \times 100\% = \mathbf{20\%}.$$

Find the percentage profit or loss as the case may be:

C.P. = Rs 70, S.P. = Rs 77	C.P. = Rs 250, S.P. = Rs 200	C.P. = Rs 500, S.P. = Rs 600

EXERCISE 51

1. Satpal bought a thing for Rs 500 and sold it for Rs 550. Find his percentage profit.
2. A vendor bought sweets for Rs 50 and sold these for Rs 80 to school children. Find his percentage profit.
3. Meena bought a saree for Rs 120 and found it defective. She sold it for Rs 100. Find her percentage loss.
4. Mr. Bakshi sold his washing machine for Rs 5,940. He had purchased it for Rs 5,500. What was the percentage profit ?
5. I bought music system for Rs 5,600 and sold it for Rs 5,320. What was my percentage loss ?
6. Rama purchased 20 pens at Rs 27 each. He sold these pens at Rs 30 a pen. Find his percentage profit or loss.

MENTAL MATHS–12

1. If 1 dozen ball pens cost Rs 60, how much will 1 pen cost ?
2. Cost of 20 fruit packs = Rs 100, cost of one pack = Rs ______
   Cost of 15 packs = Rs ______.
3. C.P. = Rs 400, S.P. = Rs 500, profit = Rs ______.
4. S.P. = Rs 540, loss = Rs 60, C.P. = Rs ______.
5. C.P. = Rs 4, S.P. = Rs 5, profit = Rs ______, profit % = ______.
6. Purchase price = Rs 70, transportation charges = Rs 10, C.P. = Rs ______.
7. Purchased a shawl for Rs 500, sold for Rs 400, loss = Rs______, loss % = ______.
8. 25% of Rs 1,200 = Rs ______.
9. $\frac{3}{4}$ of Rs 280 ______
10. $\frac{2}{5}$ of 1 metre =______.

KNOW YOUR NUMBERS

THE NUMBER SEVEN

7

7 IS A PRIME NUMBER

There are seven days in a week. There are seven colours in a rainbow. Music has seven notes. There are seven seas in the world. There are seven ancient wonders of the world. Only one of them exists today : the Pyramids of Egypt. Water polo is played with teams of seven players. You might have read the story of the seven sisters, the seven dwarfs. *Saat (seven) phare are* taken on the occasion of a Hindu marriage. Seven is supposed to be a lucky number.

According to the Bible, God created the world in 7 days.

17 PERIMETER AND AREA

PERIMETER

A hockey player, Preeti, was asked to run 5 times around a hockey field as a penalty for coming late for training.

Ranjana wants to put a lace around her square table cover. How much lace does she need ?

All these types of questions referring to the distance around a plane figure are answered by "Perimeter".

What is Perimeter ?

The word perimeter is derived from two Greek words **peri** meaning around and **metron** meaning measure.

Thus perimeter is the distance around a plane figure or the length of the boundary of a plane figure.

We measure the perimeter of a plane figure by adding the lengths of its sides.

- If the hockey field is a rectangular field 110 m long and 60 m wide, then Preeti has to run 5 times the distance around the field
  = 5 × Perimeter of the field
  = 5 × (110 m + 60 m + 110 m + 60 m)
  = 5 × 340 m = 1700 m = **1.7 km**

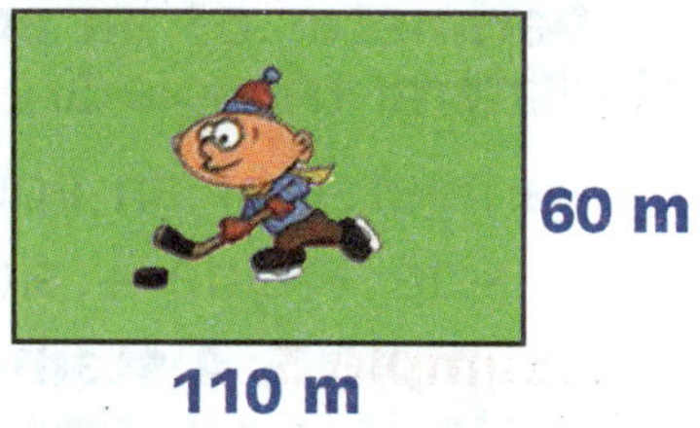

- If each side of the square table cover is 3 m, then the length of lace needed = Perimeter of table cover = 3 m + 3 m + 3 m + 3 m = **12 m.**

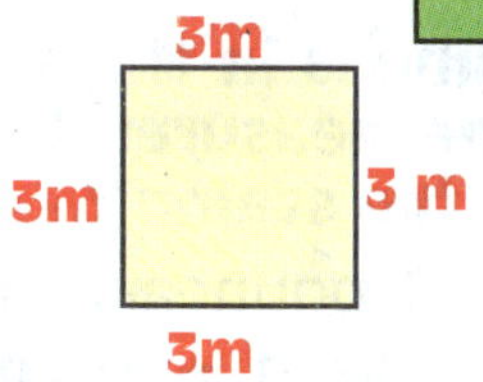

Skill Practice — **Class Work**

Find the perimeter of the following figures:

1.

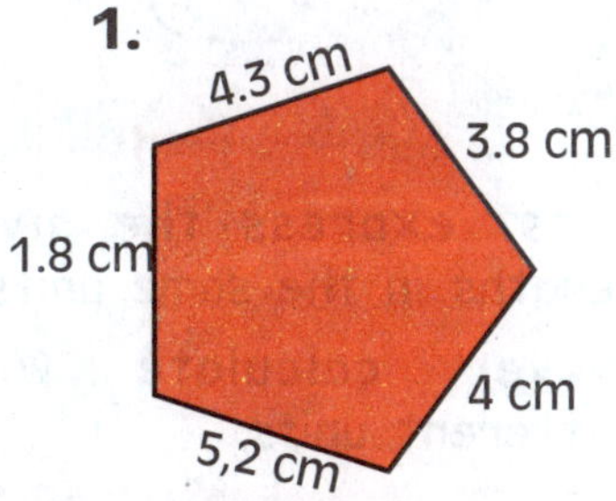

P = ______________

= ______________

2.

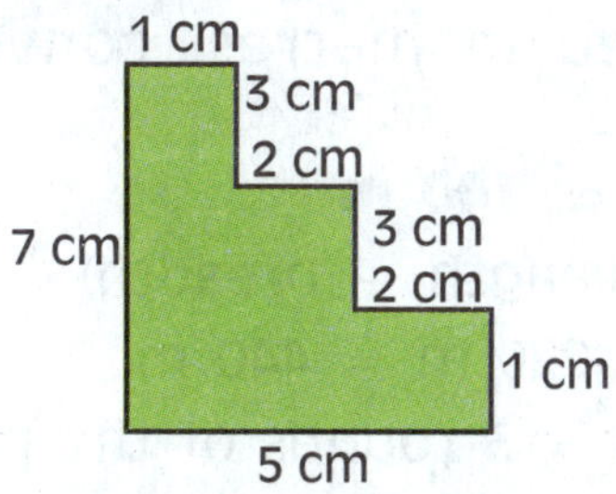

P = ______________

= ______________

3.

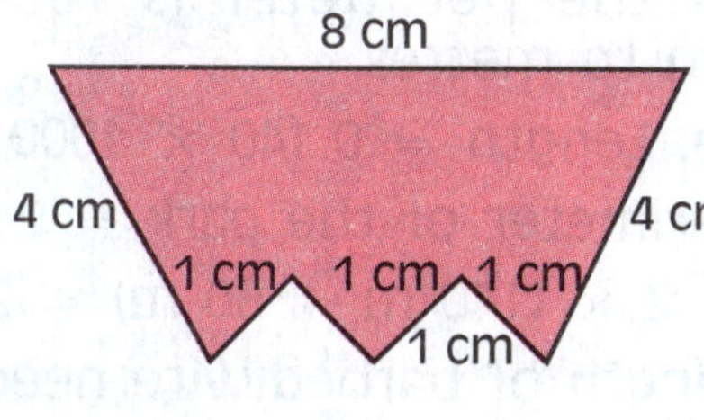

P = ______________

= ______________

Perimeter of a rectangle : A rectangle is a plane figure whose opposite sides (facing each other) are of equal lengths. So, we have two equal lengths and two equal breadths.

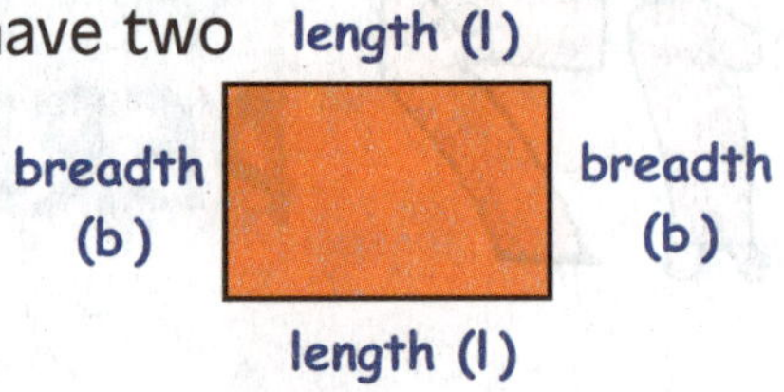

Perimeter of a rectangle

= length + breadth + length + breadth

= length + length + breadth + breadth

= 2 × length + 2 × breadth or 2 × (length + breadth)

Example 1. Find the perimeter of a carpet whose length = 3 m and breadth = 1.8 m.

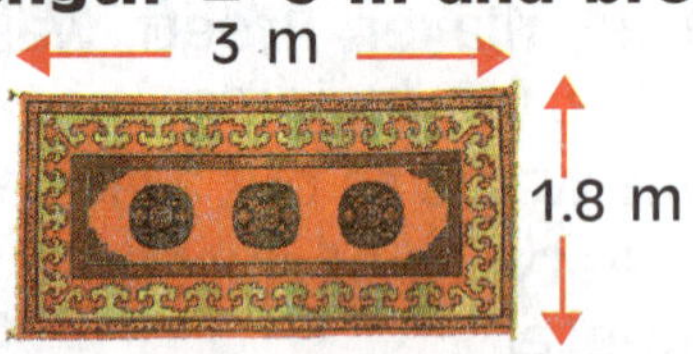

Solution. Perimeter of the carpet

= 2 × (length + bredth)

= 2 × (3m + 1.8 m) = 2 × 4.8 m = **9.6 m.**

Perimeter of a square : A square is a plane figure whose all four sides are of equal lengths.

Perimeter of a square = side + side + side + side

= 4 × side

Example 2. Find the perimeter of a square book on animals of side 20 cm.

Solution. Perimeter of the book = 4 × side = 4 × 20 cm = **80 cm.**

Perimeter of an equilateral triangle: All the three sides of an equilateral triangle are equal.

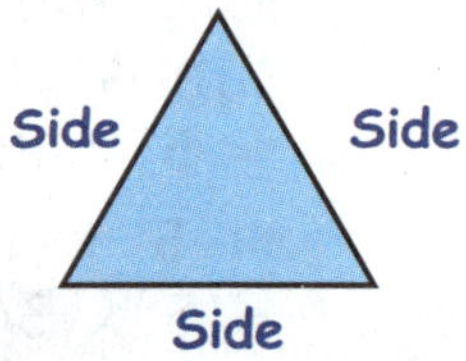

Perimeter of an equilateral triangle

= side + side + side **= 3 × side.**

Example 3. A stamp is in the shape of an equilateral triangle. Its each side measures 3 cm. What is the length of the boundary of the stamp ?

Solution. Length of boundary of the stamp

= 3 × side = 3 × 3 cm **= 9 cm.**

Example 4. The length of a rectangular park is 0.140 km and its breadth is 80 m. What is the perimeter of the park in metres ? Three rounds of barbed wire are needed to fence the park. Find the cost of fencing the park if the barbed wire costs Rs 2.20 per metre.

Solution. Length = 0.140 km, breadth = 80 m

As the perimeter is required in metres, convert length to metres.

So, Length = 0.140 × 1000 m **= 140 m**

Perimeter of the park = 2 × (length + breadth)

= 2 × (140 m + 80 m) = 2 × 220 m = 440 m.

Length of barbed wire needed = 3 rounds of the park

= 3 × perimeter of park = 3 × 440 m **= 1320 m**

Now, cost of 1m = Rs 2.20

∴ Cost of 1320 m of barbed wire = Rs 2.20 × 1320 = **Rs 2904.**

> **Maths Alert!**
> First express the given lengths in the same units.
> Never calculate with different units.

Example 5. **The perimeter of a rectangle is 42 cm and its length is 10 cm. Find its breadth.**

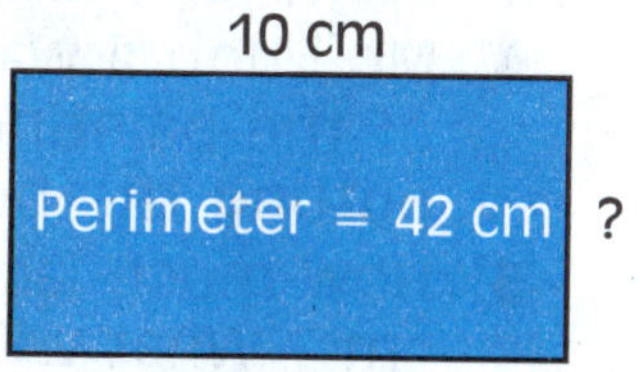

Solution. Perimeter = 2 × (length + breadth)

Therefore, length + breadth = Perimeter ÷ 2

10 cm + breadth = 42 cm ÷ 2 = 21 cm

breadth = 21 cm − 10 cm = **11 cm.**

Example 6. **What is the length of the side of a square park whose perimeter is 60 m?**

Solution. Perimeter = 4 × side,

∴ side = perimeter ÷ 4

Hence, length of a side of the square park = 60 m ÷ 4 = **15 m.**

1. Measure the lengths of sides of each figure with your ruler and find the perimeter for each figure:

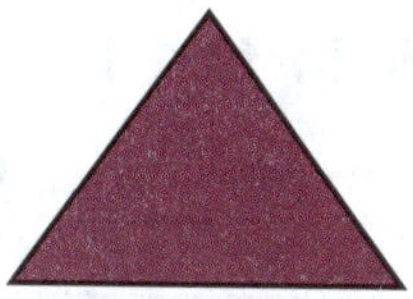

Length = ________ Side = ________ Sides = ________

Breadth = ________ Perimeter = ________ Perimeter = ________

Perimeter = ________ = ________ = ________

= ________

2. The perimeter of the figure drawn alongside = ________

____

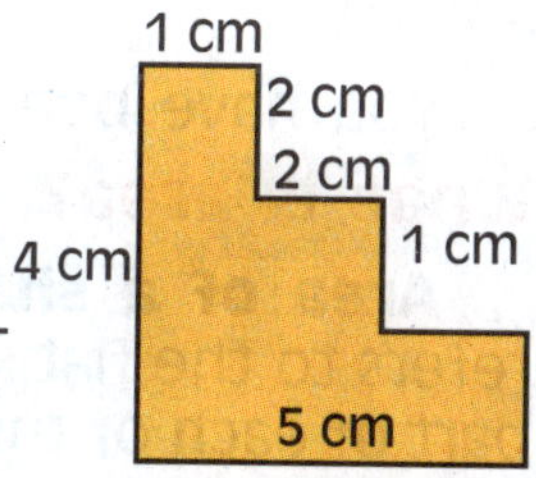

EXERCISE 52

1. Find the perimeter of a square with side length :

(a) 14 cm **(b)** 6 mm **(c)** 4.7 m **(d)** 2.17 km **(e)** 8.9 cm **(f)** 0.19 m

2. Find the perimeter of a rectangle given:

		(a)	(b)	(c)	(d)	(e)	(f)
(a)	**Length**	8 cm	16 m	7 mm	6.4 cm	1.5 m	12.1 km
	Breadth	5 cm	14 m	2.5 mm	36 mm	70 cm	3.2 km

3. How many metres of fencing is required to enclose a rectangular vegetable garden 15 m long and 6.5 m wide ?

4. Mr. Chopra wants to put a wooden border around a painting. If the length of the painting is 40 cm and its breadth is 25 cm, what will be the cost of putting the border at Rs 40 per m?

5. How much lace will be needed to put around a square cushion of side 65 cm ? What will be the cost of the lace required, if the lace costs Rs 8 per m?

6. A rope of length 24 m was used to fence off a square patch of land for a garden. Find the length of the side of the square garden.

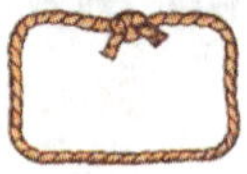

7. The following diagram shows a rectangle and a square with dimensions given.

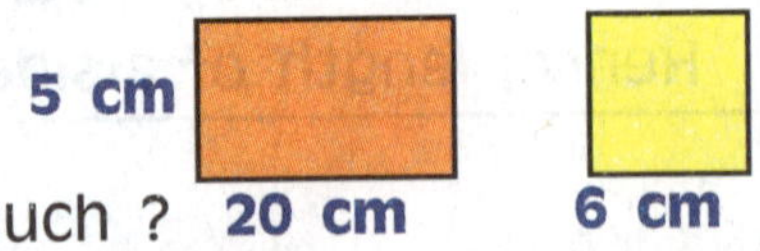

(a) What is the perimeter of each figure ?

(b) Which figure has a greater perimeter and by how much ?

8. Mahesh jogs 5.2 km everyday. He jogs around a rectangular field 700 m long and 600 m wide. How many rounds does he take of the whole field ?

[**Hint** : Find the perimeter of the field.

Then, Number of rounds = Total distance jogged ÷ Perimeter]

9. **Fill in the blanks in the given table:**

	(a)	(b)	(c)	(d)
Length	14 m		20 mm	
Breadth		8 cm		12 cm
Perimeter	58 m	34 cm	50 mm	40 cm

10. **Find the length of a side of a square whose perimeter is:**

(a) 12 cm **(b)** 10.4 cm **(c)** 36 m **(d)** 7.2 m

AREA

You have been introduced to the concept of area in Class IV. Let us revise.

What is area?

Area of a shape is the amount of space it covers. It refers to the flat surface taken by a plane figure. The coloured part of each of the shapes shows the amount of space covered by each of them on this sheet of paper, *i.e.*, it shows their area. Which shape has greater area ? To answer this question we need to know as to which shape covers more surface on paper. Obviously! shape A.

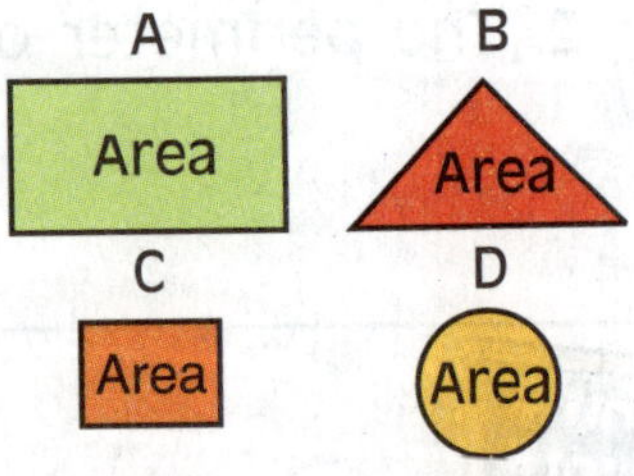

Area is found by calculating how many square units are needed to cover the given shape.

Units of measurement of area

The units used to measure area are based on the units of length, *i.e.*, mm, cm, m and km.

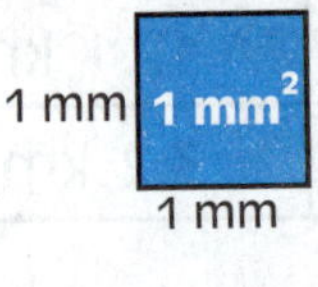

This is an area equal to 1 square millimetre (mm^2)

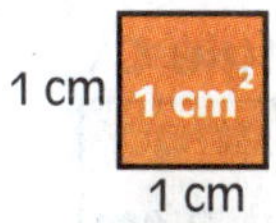

This is an area equal to 1 square centimetre (cm^2)

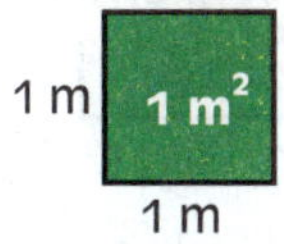

and
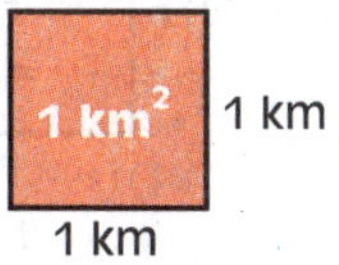

represents other units of measuring area such as **1 square metre (1 m^2)** and **1 square kilometre (km^2).**

The unit used for measuring area depends on the size of the area being measured. **For example,** the areas of the following are usually measured as under:

A stamp  in $\mathbf{mm^2}$

A 100 rupee note  in $\mathbf{cm^2}$

The floor of your classroom in $\mathbf{m^2}$

A country  in $\mathbf{km^2}$.

Find the areas of the following shaded shapes by counting the number of squares they cover. Give your answer in square units. Here we have some half squares also. **Count 2 half squares as 1 unit².**

AREA OF A RECTANGLE

Here are three rectangles, covered with small squares. Each small square stands for 1 square centimetre, *i.e.*, each small square is 1 cm on each side.

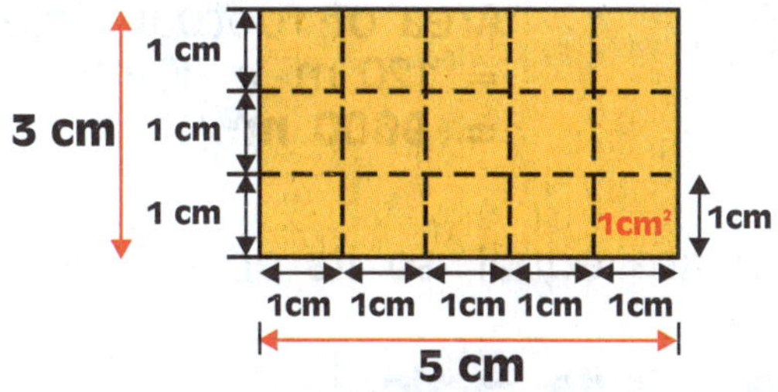

A

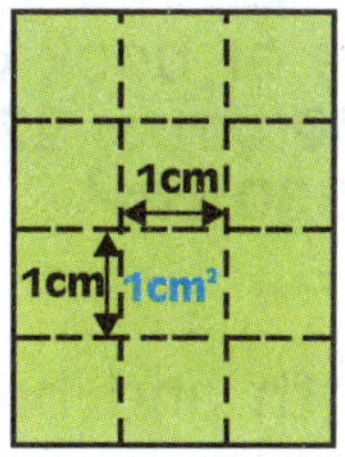

B

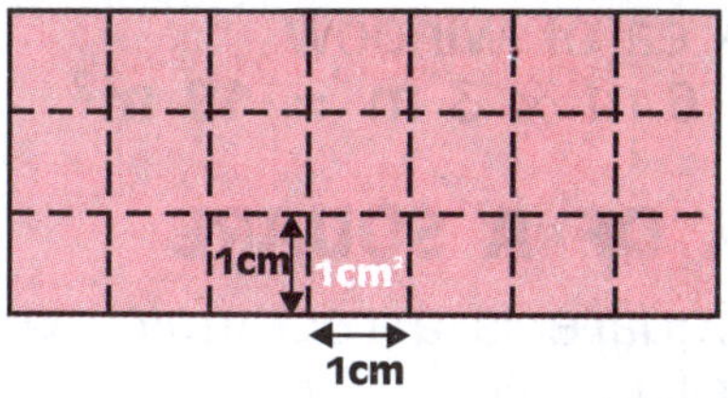

C

Count the number of small squares that cover each rectangle. This will give you the area of the rectangle in square centimetres. Also write the length and breadth of each rectangle in centimetres. What do you observe ?

Rectangle A	Rectangle B	Rectangle C
Number of small squares = 15	Number of small squares = 12	Number of small squares = 21
Area = 15 cm^2 (or 15 sq cm)	Area = 12 cm^2 (or 12 sq cm)	Area = 21 cm^2 (or 14 sq cm)
Length = 5 cm	Length = 3 cm	Length = 7 cm
Breadth = 3 cm	Breadth = 4 cm	Breadth = 3 cm

From the above examples, you can see that there is a quick way to find the area in each case without having to count the smaller squares. Since $5 \times 3 = 15$; $3 \times 4 = 12$ and $7 \times 3 = 21$, we can say that the area of each of the rectangles can be found out by multiplying the measures of length and breadth of the rectangle.

Rectangle A

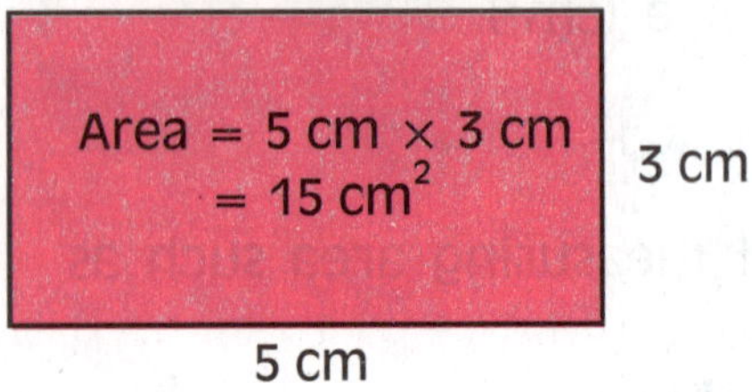

Rectangle B

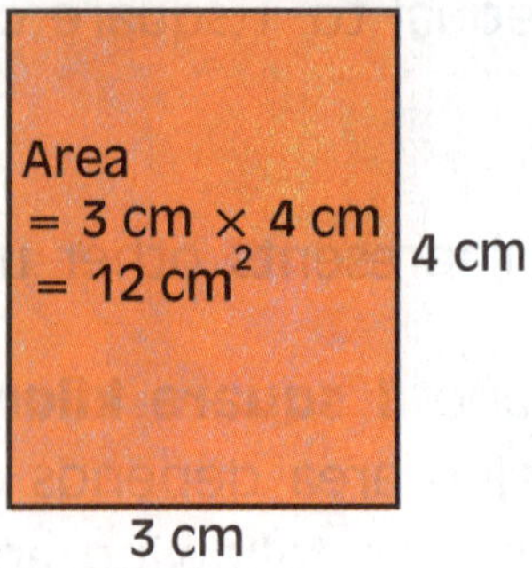

Rectangle C

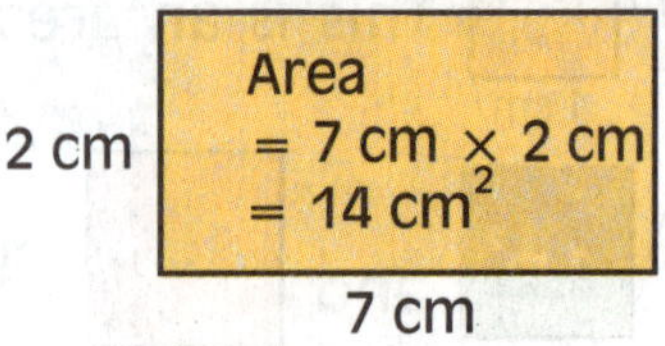

Conclusion:

Area A of a rectangle of length l and breadth **b** is $l \times$ **b.**

i.e., **Area of a rectangle = length × breadth**

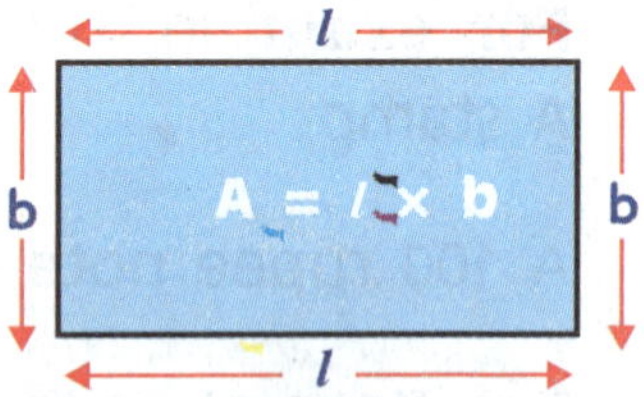

Examples:

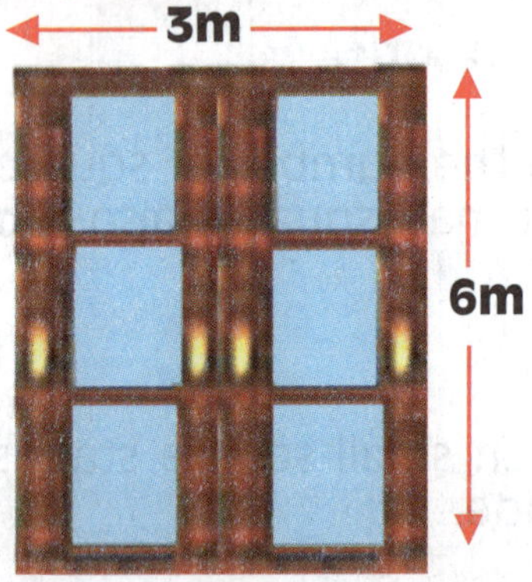

Area of window
= 6 m × 3 m = **18 m²**

Area of book cover
= 30 cm × 22 cm
= 660 cm²

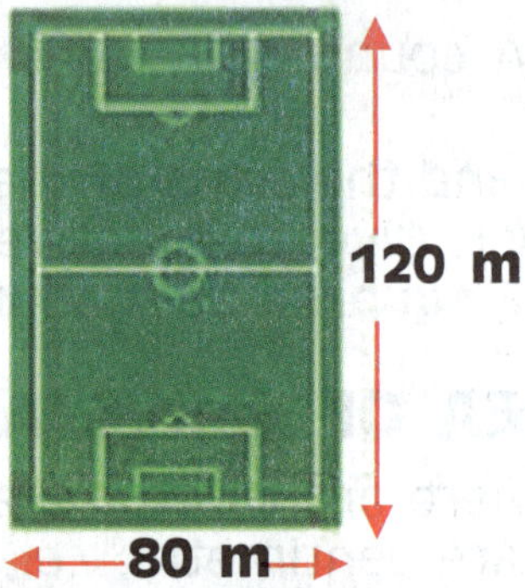

Area of football field
= 120 m × 80 m
= 9600 m²

AREA OF A SQUARE

A square is a rectangle whose length and breadth are equal, so its area = length × length

Therefore, **Area of a square = side × side = side²**

Examples :

Area of square card
= 15 cm × 15 cm
= **225 cm².**

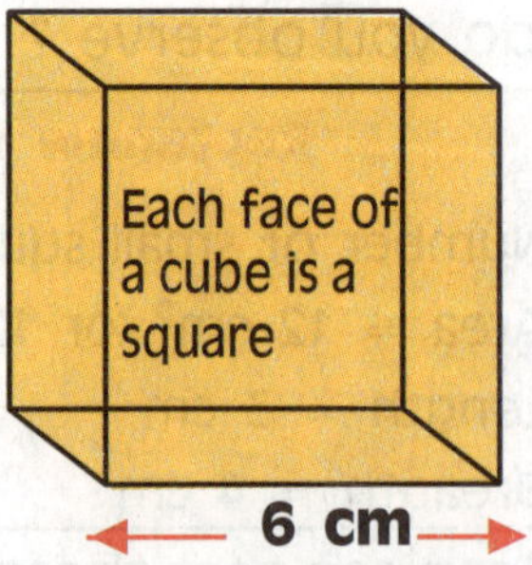

Area of 1 face of a cube
= 6 cm × 6 cm = **36 cm².**

Area of square carpet
= 2.4 m × 2.4 m
= **5.76 m².**

Skill Practice

Class Work

Find the area of each of the rectangles and squares drawn below:

1.

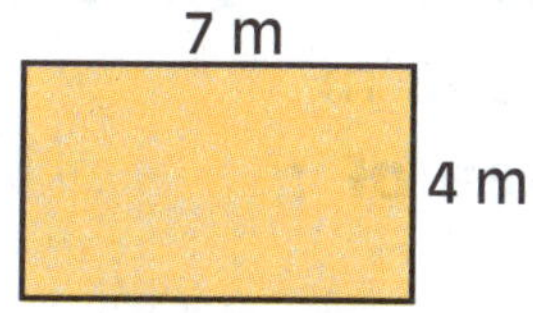

Area = ____________

= ____________

2.

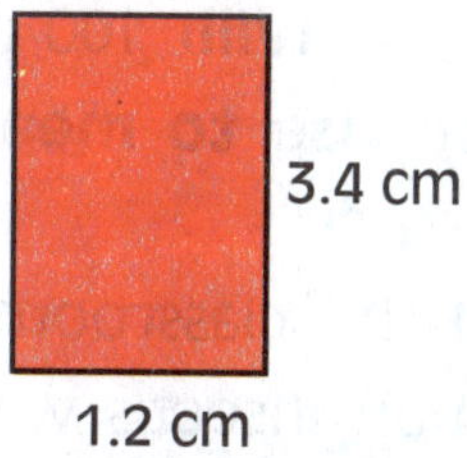

Area = ____________

= ____________

3.

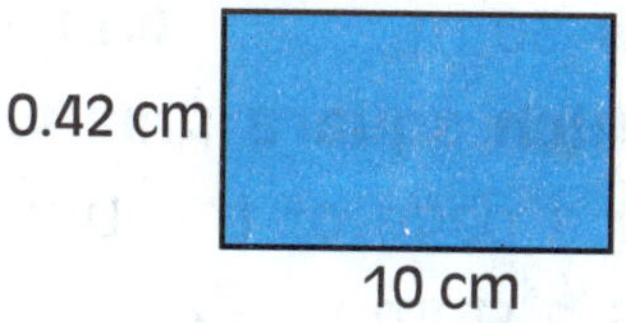

Area = ____________

= ____________

4.

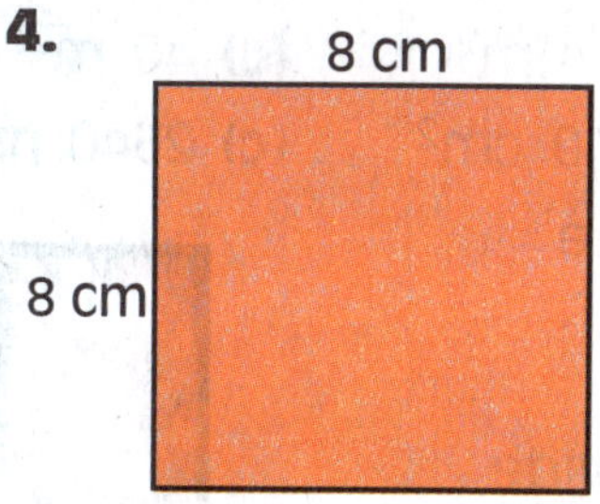

Area = ____________

= ____________

5.

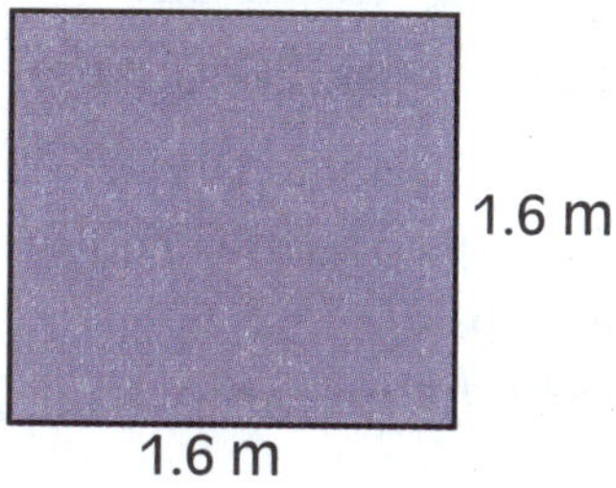

Area = ____________

= ____________

6.

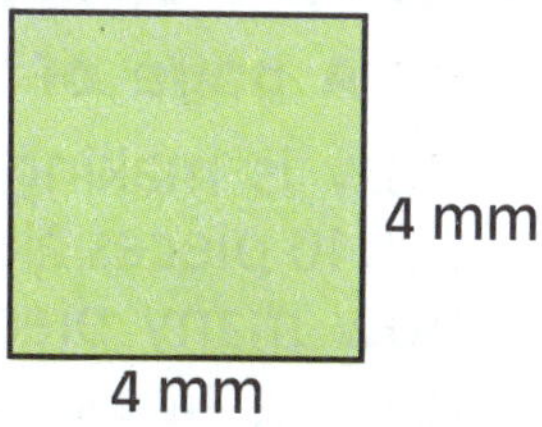

Area = ____________

= ____________

Home Work

EXERCISE 53

1. Copy and complete the table for the area of rectangles :

	a	b	c	d	e
Length	20 cm	14 km	13 mm	12.5 m	25 mm
Width	8 cm	10 km	3 mm	10 m	60 mm
Area					

2. Calculate the area of each of the objects shown below

(a)

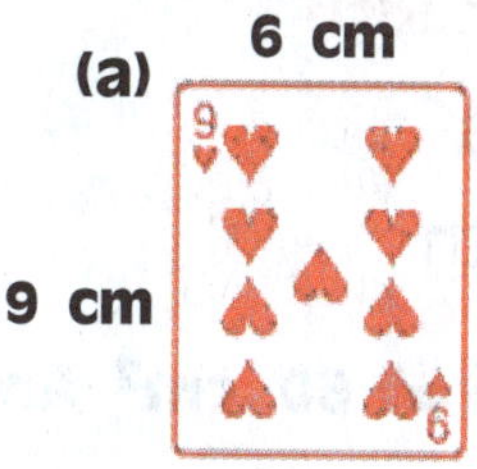

Playing card

(b)

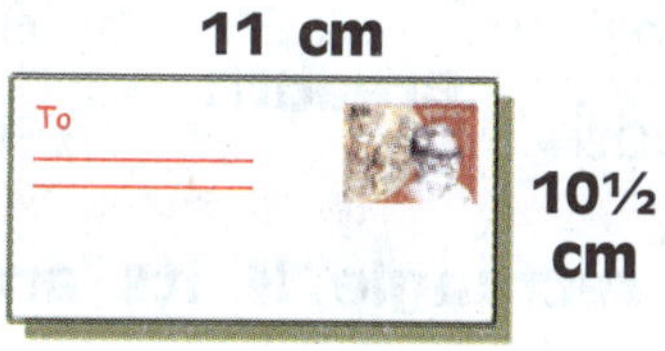

Envelope

(c)

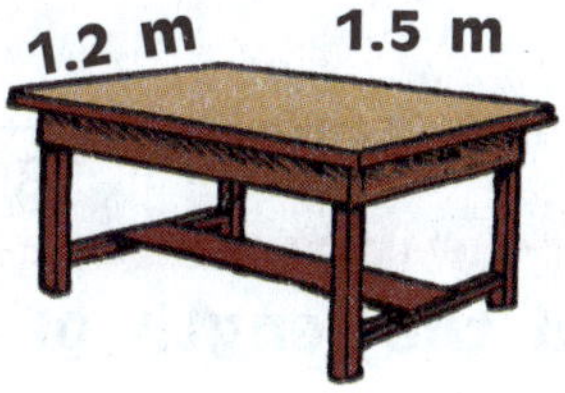

Desktop

(d)

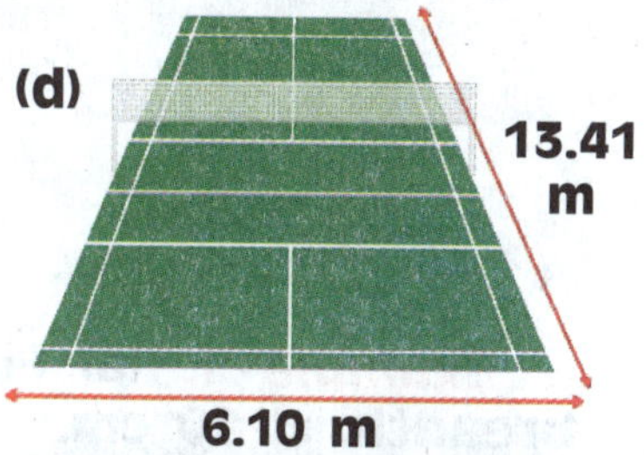

Badminton Court

13.41 m × 6.10 m

3. **Find the area of each of the rectangles described below:**

(a) 4 cm by 3 cm **(b)** 8 cm by 6 cm **(c)** 20 cm by 15 cm

(d) 28 cm by 21 cm **(e)** Length = 1 m 75 cm, Width = 80 cm

4. **Find the area of a square whose one side is**

(i) 3 cm **(ii)** 5 cm **(iii)** 13 cm **(iv)** 1.2 cm **(v)** 2.5 cm

(vi) 30 m **(vii)** 13.4 km **(viii)** 100 m **(ix)** 12.16 m

5. **Which square unit would you use to measure the area of :**

(a) a page of this book (**Ans.** sq cm)

(b) a stamp **(c)** the classroom floor **(d)** Delhi

(e) the blackboard **(f)** an insect's wing **(g)** the Indian Ocean

(h) a sports field

6. **Choose the best estimate for the area. Write *a*, *b* or *c*.**

(i) An envelope **(a)** 200 cm^2 **(b)** 200 mm^2 **(c)** 200 m^2

(ii) An identity card **(a)** 40 cm^2 **(b)** 40 mm^2 **(c)** 40 m^2

(iii) A page of a newspaper **(a)** 2640 mm^2 **(b)** 2640 cm^2 **(c)** 2640 m^2

7. Mary is making a quilt from square pieces of cloth. She fits 10 pieces along one side and 15 pieces along the other. How many pieces does she need ?

8. What is the area of the front of a watch that is 25 mm wide and 23 mm high ?

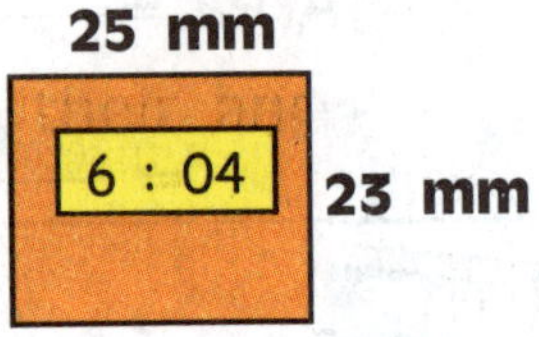

Front of Watch

9. Compare the areas of a square 15 cm on each side and a rectangle 11 cm by 8 cm. Which figure has the greater area and by how much ?

10. Which of the two—a rectangle of length 16 cm and breadth 12 cm or a square of side 14 cm - has more area and by how much more ?

11. A rectangular field measures 400 m by 100 m and a square field 200 m by 200 m.

(i) Do the two fields have the same area ?

(ii) Which field requires more fencing and by how much more ?

[**Hint :** Length of fencing = perimeter = sum of all sides]

Area and measure of length or breadth being given, to find the other

Since, **Area of a rectangle = length × breadth**

$$\therefore \textbf{Length} = \frac{\text{Area}}{\text{Breadth}}, \quad \textbf{Breadth} = \frac{\text{Area}}{\text{Length}}$$

Example 1. **(a) Find the length of a rectangle if its area = 60 cm^2 and its breadth = 10 cm.**

(b) Find the breadth of a rectangle if its area = 65 cm^2 and its length = 13 cm.

Solution.

(a) Area = 60 cm^2, Breadth = 10 cm

$\therefore$ Length = Area ÷ Breadth = 60 cm^2 ÷ 10 cm = **6 cm.**

(b) Area = 65 cm^2, Length = 13 cm

$\therefore$ Breadth = Area ÷ Length = 65 cm^2 ÷ 13 cm = **5 cm.**

Example 2. The perimeter of a square is 12 cm. Find its area.

Solution. To find the area, we need the side of the square.

Side of the square = Perimeter ÷ 4

= 12 cm ÷ 4 = 3 cm

$\therefore$ Area = side × side = 3 cm × 3 cm = **9 cm^2.**

Skill Practice

Class Work

Find:

1. Length

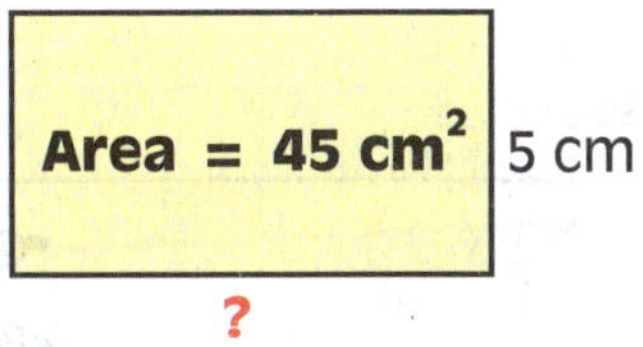

2. Breadth

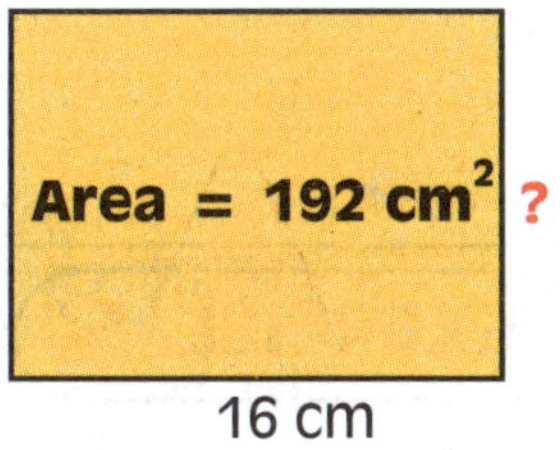

3. Breadth

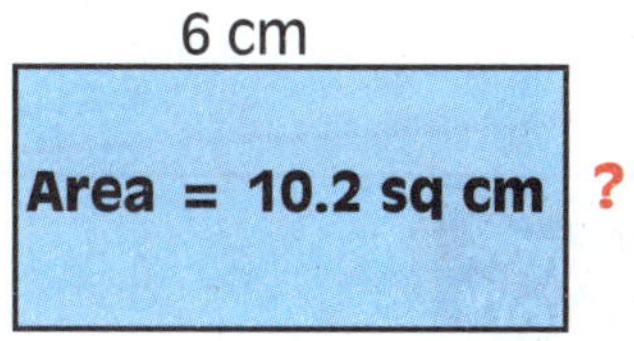

4. A room is 16 m long and 10 m wide. Find its perimeter and area of the carpet required to cover the floor.

Ans.

5. The perimeter of a square is 36 cm. Find its area.

Ans.

6. The area of a rectangle is 20 cm^2. Its length is 5 cm. Find its breadth. Also find the perimeter of the rectangle.

Ans.

EXERCISE 54

1. Find the length of the rectangle whose:

(a) Area = 320 sq. cm and breadth = 16 cm

(b) Area = 1800 sq. cm and breadth = 40 cm

(c) Area = 216 sq. cm and breadth = 12 cm

(d) Area = 10.2 sq. cm and breadth = 1.7 cm

2. Find the breadth of the rectangle whose:

(a) Area = 187 cm^2 and length = 17 cm

(b) Area = 315 mm^2 and length = 21 mm

(c) Area = 5.5 m^2 and length = 0.25 m

(d) Area = 10000 m^2 and length = 100 m

3. Find the perimeter of a rectangle whose:

(a) Area = 1050 sq. cm, length = 35 cm

(b) Area = 18 cm^2, breadth = 6 cm

(c) Area = 4800 m^2, breadth = 60 m

(d) Area = $8\frac{2}{3}$ sq mm, length = $5\frac{7}{9}$ mm.

(e) Area = 10.2 sq cm, breadth = 17mm

(f) Area = 81.34 sq m, length = 8.3 cm

(g) Area = 0.0225 km^2, breadth = 50 m [**Hint:** breadth = $\frac{50}{1000}$ km = 0.05km]

18 VOLUME

VOLUME

In your previous classes you have learnt about solid figures.

Some solid figures are shown above. **Does each of these figures enclose a portion of space ?** Do you understand what we mean by the interior of a solid region? The interior of a balloon contains air. The interior of a milk bottle contains milk. You know a cup or a glass can only take so much of liquid and then overflows. Your box can hold up to a certain limit only. Beyond it you cannot fill it.

The amount a container can hold is called its **capacity.** Finding out the volume means measuring the space occupied by the interior of a solid figure.

> **The volume of a solid is the amount of space enclosed by it or the amount of space it takes up.**

The space covered by a 1 cm square is used as a unit of area measure. It is derived from the unit of length, *i.e.*, cm.

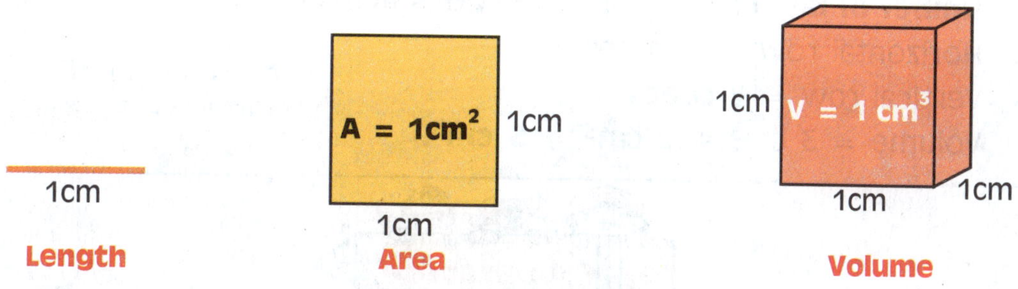

Similarly, the amount of space enclosed by a 1 cm cube (a cube whose each edge is 1 cm) is used as a unit of volume measure. It is also derived from the unit of length, i.e., cm.

The other units used for measuring volume are **cubic millimetre (mm^3)** and **cubic metre (m^3).**

The unit used for measuring volume depends on the size of the solid being measured.

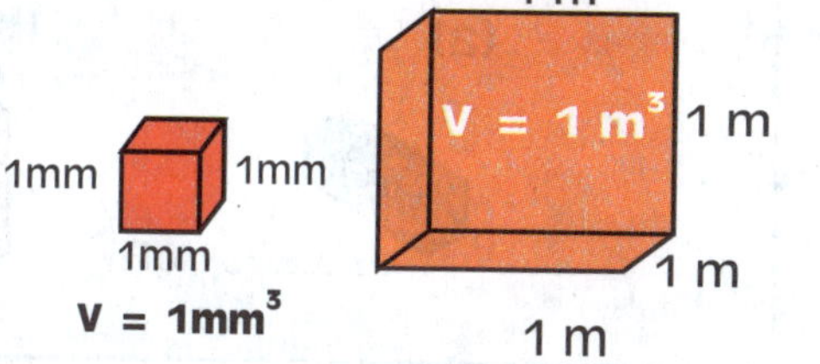

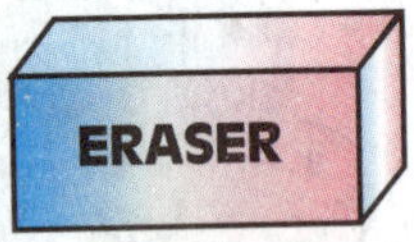

mm^3

cm^3

m^3

We say that the 1 cm cube is a unit cube and has a volume of 1 **cubic centimetre (1 cu cm).** We also write it as **1 cm³.** To measure the volume of a solid region we find out the number of unit cubes that make up the solid or fill the solid.

Example 1. Find the volume of these solids by counting the number of cubes in each solid. The volume of each small cube is 1 cm³.

(a)

(b)

(c)

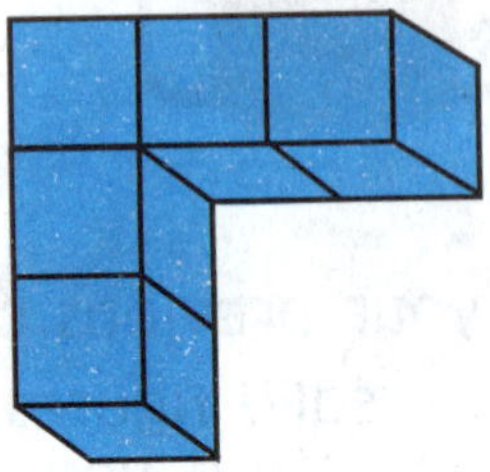

Solution.

(a) There are 4 cubes in the given solid, so, Volume = **4 cm³.**

(b) 1st row = 8 cubes

2nd row = 8 cubes

3rd row = 8 cubes

$\therefore$ Volume = $8\ cm^3 + 8\ cm^3 + 8\ cm^3$ = **24 cm³.**

Remember to count the hidden cubes

We can also say that, there are 3 layers of cubes and each layer consists of 8 cubes, so

Volume = Number of layers × Number of cubes in each layer = (3 × 8) cm³ = 24 cm³.

(c) Horizontal row = 3 cubes

Vertical row = 2 cubes

Volume = $3\ cm^3 + 2\ cm^3$ = **5 cm³.**

Do not count the common cube again

1. Find the volume of each of the solids given below by counting the number of cubes in each solid. The volume of each small cube is 1 cm³.

(a)

------- cm^3

(b)

------- cm^3

(c)

------ cm^3

(d)

------ cm^3

(e)  _______ cm³

(f)  _______ cm³

(g)  _______ cm³

(h)  _______ cm³

2. Fill in the blanks with the correct units :

	Length	Area	Volume
(a)	cm	cm^2	
(b)	mm		
(c)	m		

VOLUME OF A CUBOID

You know what a cuboid is. It is a rectangular solid. The following have the shape of a cuboid.

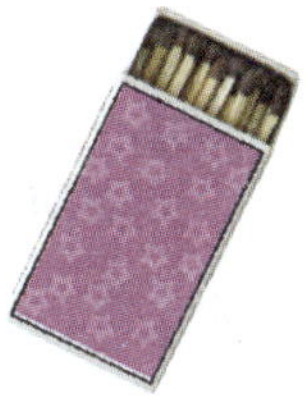

A match box

A biscuit gift pack

A Pencil box

You can find the volume of a cuboid by finding how many centimetre cubes can be placed in it.

Study the following examples:

See how a cuboid is built from 1-centimetre cubes and how we find its volume.

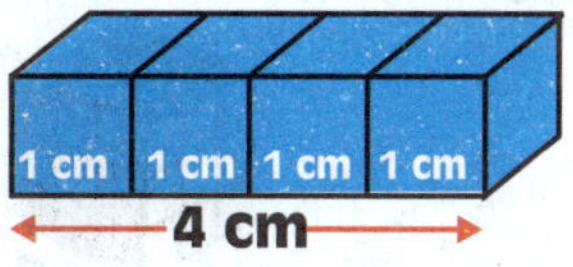

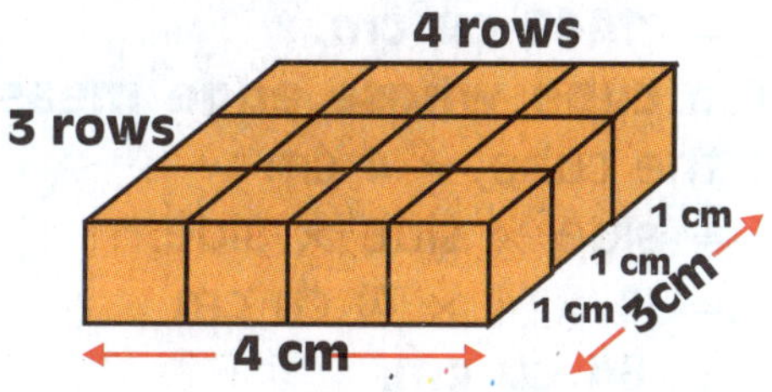

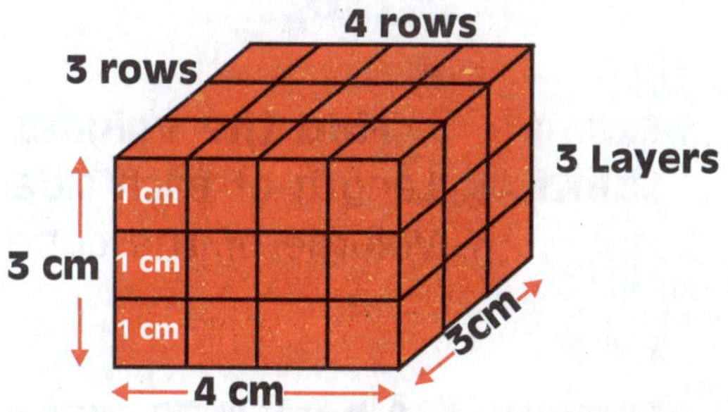

4 units per row		12 units per layer		36 units in all		
length (rows)	×	breadth (Unit cubes in each row)	×	height (Layers)	=	**volume**
↓		↓		↓		
4 cm	×	3 cm	×	3 cm	=	**36 cm^3**

The following cuboids are built from 1-centimetre cubes. Complete the table given below using these cuboids.

(a)

(b)

(c)

	Measurements in centimetres			Count	Volume
	Length (Rows)	Breadth (Unit cubes in each row)	Height (Layers)	Number of 1 cm cubes	Length × Breadth × Height
(a)					
(b)					
(c)					

From the table you will notice that

Volume of a cuboid = length × breadth × height

If all the edges of a cuboid are equal, then it is called a cube.

So, volume of a cube = side × side × side = side^3

Example 1. Find the volume of a cuboid whose length, breadth and height are 15 cm, 12 cm and 8 cm respectively.

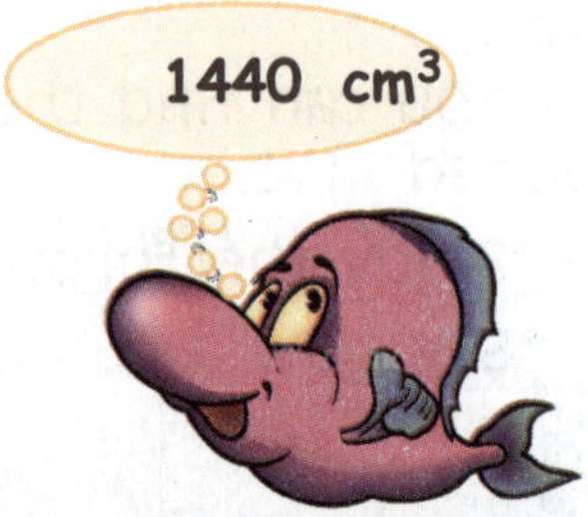

Solution. Length of the cuboid = 15 cm
Breadth of the cuboid = 12 cm
Height of the cuboid = 8 cm
Volume of the cuboid = length × breadth × height
= (15 × 12 × 8) cu cm
= 1440 cu cm.

Example 2. Find the volume of a cube whose edge measures 4 cm.

Solution. Length of each side of the cube = 4 cm
Volume of the cube = side × side × side
= (4 × 4 × 4) cu cm
= 64 cu cm.

Example 3. A book is 20 cm by 18 cm and 50 mm thick. How many cubic centimetres of space does the book occupy ?

Solution. Length = 20 cm
Breadth = 18 cm
Height = 50 mm = (50 ÷ 10) cm = **5 cm**
Volume of the book = (20 × 18 × 5) cu cm
= 1800 cu cm.

Skill Practice **Class Work**

Find the volume of each solid region. Show the unit with your answer.

1.

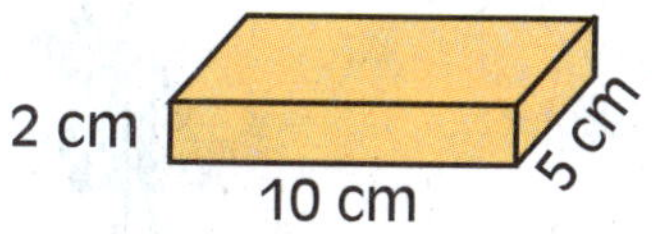

Volume = ____________

= ____________

2.

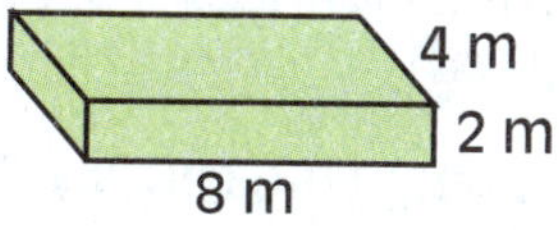

Volume = ____________

= ____________

3.

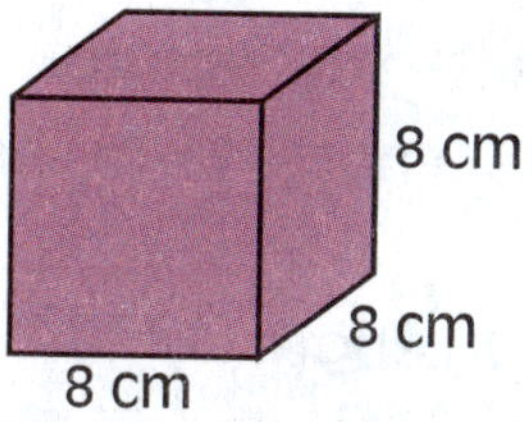

Volume = ____________

= ____________

4. l = 5 mm, b = 1.6 mm, h = 10 mm

Volume = ____________

= ____________

5. $l = \frac{4}{5}$ m, $b = \frac{7}{8}$ m, $h = \frac{10}{21}$ m

Volume = ____________

= ____________

Home Work

EXERCISE 55

1. Find the volume of each solid and write the unit with your answer.

(a)

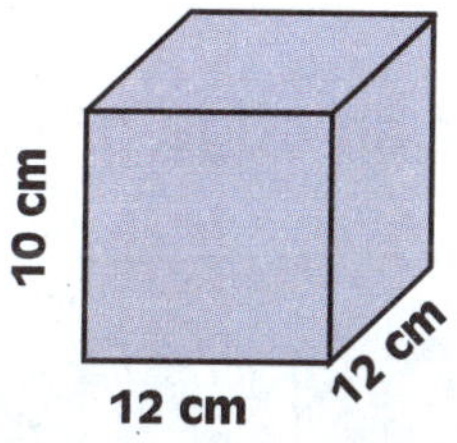

(b)

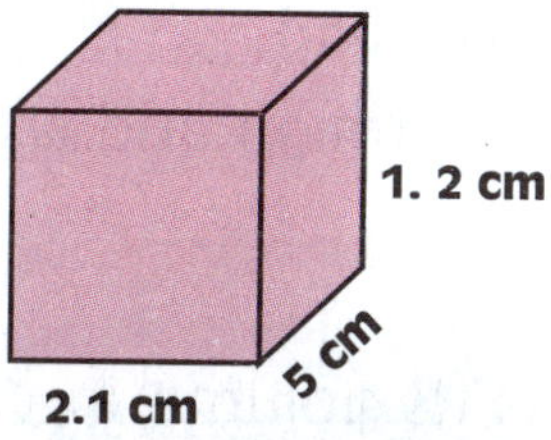

(c)

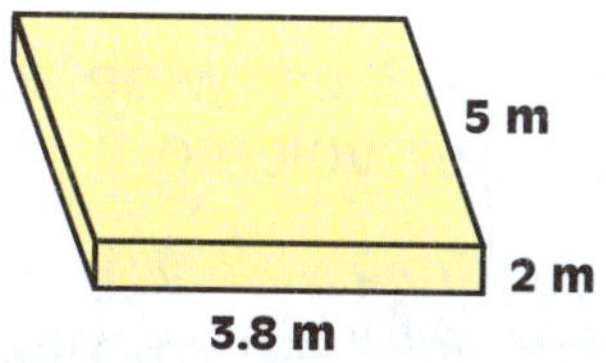

(d)

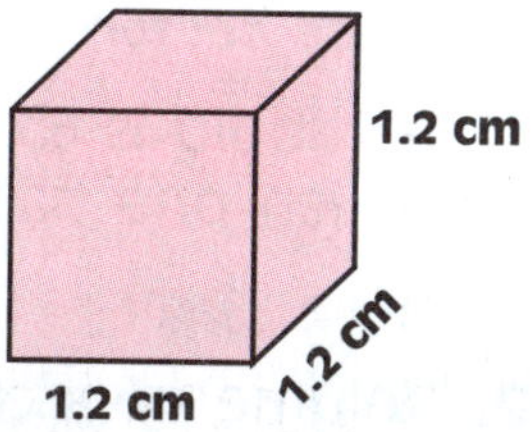

2. For a cuboid, copy and complete the table:

	Length	Breadth	Height	Volume
(a)	4 cm	2 cm	3 cm	
(b)	0.05 cm	3 cm	4 cm	
(c)	10 cm	2.6 cm	2 cm	
(d)	1.2 cm	0.8 cm	5 cm	
(e)	5 m	1.5 m	1.2 m	

3. Find the volume of the following cubes:

(a) edge 3cm **(b)** edge 5 cm **(c)** 1.1 cm

(d) edge 70 mm **(e)** edge 80 mm **(f)** 4.3 m

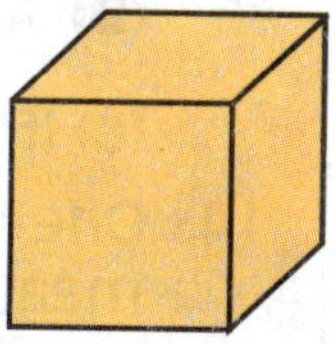

4. A 100-centimetre cube is the same size as a 1-metre cube.
How many cubic centimetres make 1 cubic metre ?

5. Which has the greater volume : box that measures 10 cm by 6 cm by 4 cm or one that measures 6 cm by 6 cm by 7 cm ?

6. A heap of wood is a pile 5 dm wide, 9 dm long, and 3 dm high. What is the volume of the heap of wood?

7. **Find the volume of the rectangular solids whose dimensions are as follows:**

(i) Length 15 m, breadth 12 m, thickness 75 cm;

(ii) Length $12\frac{1}{2}$ m, breadth 5.2 m, thickness 80 cm.

8. Calculate the volume of air in a six-room house, if the height of each room is 3 m, and the lengths and widths of the rooms, in metres, are as given by the following table:

Room	A	B	C	D	E	F
Length	9	4	5	6	8	3
Width	5	4	6	4	4	2

9. **Sameer has 80 one-centimetre cubes. Which of these cuboids could he build ?**

(a) 6 cm long, 5 cm wide, 4 cm high
(b) 8 cm long, 5 cm wide, 2 cm high
(c) 10 cm long, 4 cm wide, 5 cm high
(d) 6 cm long, 7 cm wide, 2 cm high
(e) 7 cm long, 5 cm wide, 2 cm high
(f) 10 cm long, 4 cm wide, 2 cm high

10. Write the length, breadth and height of six cuboids that can be made from 36 one-centimetre cubes. [**Hint:** Find all factors of 36 and then various possible combinations of these factors so that the product may remain 36.]

11. A cuboid is 6 cm long, 3 cm wide and 3 cm high and a cube has an edge of 6 cm. Which one has greater volume ?

12. Volume of a cuboid is $\frac{1}{8}$ cu m. What is its volume in cu cm ?

[**Sol.** 1 cu m = 1 m × 1 m × 1 m = 100 cm × 100 cm × 100 cm = 1000000 cu cm

$\therefore \frac{1}{8}$ cu m $= \frac{1}{8} \times 1000000$ cu cm = 125000 cu cm.]

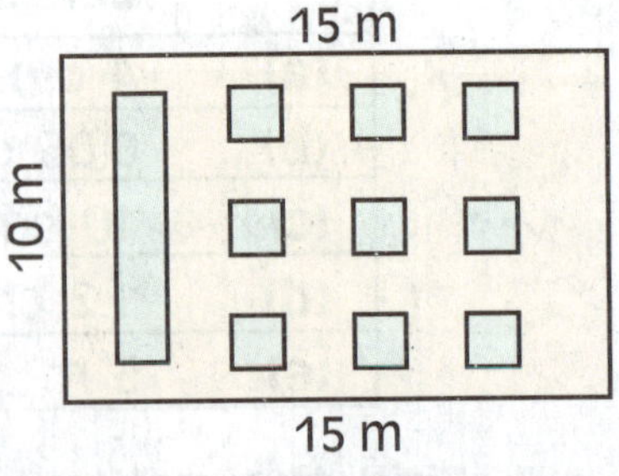

13. **The floor of a class room is shaped as shown at the right,**

(a) What is the perimeter of the floor ?

(b) What is the floor area of the room ?

(c) If the height of the room is 8 m, how many cubic metres of space is in the room ?

14. One centimetre, one square centimetre and one cubic centimetre are different measurements. Why ?

15. A trench was 30 m long, 2 m wide and 2 m deep. How many cubic metres of earth was removed to dig the trench ?

16. Complete the following table:

	Length	Breadth	Height	Volume
(a)	4 cm	2 cm	?	24 cm^3
(b)	5 cm	3 cm	?	30 cm^3
(c)	?	70 mm	5 mm	2100 mm^3
(d)	8 cm	?	0.5 cm	2.4 cm^3
(e)	?	4 mm	0.05 mm	1.04 mm^3

Height = Volume ÷ (Length × breadth)
= 24 cm^3 ÷ (4 cm × 2 cm)
= (24 ÷ 8) cm
= 3 cm.

[**Hint:** To find the unknown dimension, divide the volume by the product of the given dimensions.]

17. Fill in the blanks:

(a) The amount of space a solid occupies is called ....

(b) The volume of a solid pipe and a hollow pipe of the same shape and size are ....

(c) If a rectangular box is filled in by fourteen 1-cm cubes, its volume is ....

(d) The volume of a 1-cm cube is ....

(e) The volume of a rectangular box (or a cuboid) = ...

18. Choose the most appropriate unit (mm^3, cm^3, m^3) to measure the volume of:

(a) A room, **(b)** 100 gm butter pack, **(c)** a school bag, **(d)** your assembly hall, **(e)** 20 grains of sand, **(f)** a match box, **(g)** a railway compartment and **(h)** a brick

CAPACITY

The **volume** is the amount of space a solid takes up or encloses. **For example,** the amount of space in a swimming pool (in m^3).

The **capacity** is the amount of water that a given solid can hold.

Thus, the capacity of the swimming pool is the amount of water needed to fill the swimming pool.

The capacity of a gas cylinder is the amount of gas it can hold. The capacity of a fuel tank in a vehicle is the amount of petrol or diesel needed to fill it full. Standard unit of capacity is **litre (*l*).**

Liquids are often measured in litres, or millilitres.

1 litre = 1000 *ml* = 1000 cm^3

1 *ml* = 1 cm^3

Remember : 1 cm^3 = 1 *ml*, 1 m^3 = 1000 *l*

$$1m^3 = \frac{100\text{ cm} \times 100\text{ cm} \times 100\text{ cm}}{1000}\text{ litres}$$

$$= 1000\ l$$

Example 2. A rectangular tank measures 2 m by 3 m by 4 m and is full of water. What is its capacity in litres ?

Solution. The tank measures 2 m by 3 m by 4 m *i.e.*, 200 cm by 300 cm by 400 cm

Volume of the tank = (200 × 300 × 400) cm^3

Since 1 litre = 1000 cm^3, so capacity in litres = $\frac{200 \times 300 \times 400}{1000}$ litres = **24000 litres.**

Example 3. A machine for making ice freezes 5.76 litres of water into ice bricks measuring 4 cm by 3 cm by 2 cm. How many bricks will be made ?

Solution. Volume of 1 ice brick = $(4 \times 3 \times 2)$ cm^3 = 24 cm^3

Volume of water = 5.76 litres = 5.76×1000 cm^3 = 5760 cm^3

No. of bricks made $5760 \div 24 = \frac{5760}{24} = $ **240.**

EXERCISE 56

1. Find the volume of each of the following tanks and their capacity in litres.

(a) height = 3 m, length = 6 m, width = 4 m

(b) height = 1.5 m, length = 2 m, width = 3 m

(c) height = 0.16 m, length = 0.25 m, width = 0.2 m

(d) height = 16 cm, length = 50 cm, width = 15 cm

2. An oil can is 40 cm × 30 cm × 20 cm.

(a) Find the volume of the oil can.

(b) How many litres of oil can the can hold ?

(c) A litre of oil costs Rs. 15.80. For how much can one buy one can ?

3. A fish tank is 30 cm long, 15 cm broad and 10 cm high.

(a) Calculate its volume in cm^3.

(b) Write down its volume in millilitres and in litres.

4. A petrol tank is 50 cm long, 40 cm broad and 30 cm high.

(a) What is the capacity of the tank in (i) cm^3 (ii) litres ?

(b) Find the cost of filling it with petrol at Rs 29.50 per litre.

5. A container in the shape of a cube has sides one metre long. Calculate its capacity in (a) cubic centimetres, (b) millilitres and (c) litres.

MENTAL MATHS–13

1. What is the perimeter of this shape ?
2. What is the volume of this figure ?

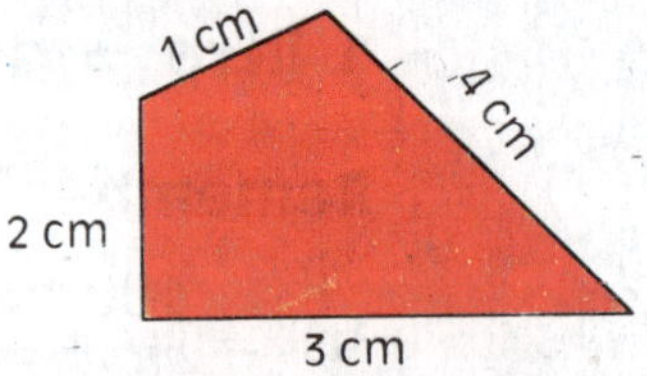

3. The perimeter of a square is 40 cm. Find its side.
4. The area of the carpet required to cover a floor of length 4 m and breadth 2 m is ....
5. Find the volume of the given match box.
6. Find the volume of a cube whose one edge is 4 cm.
7. 1000 cm^3 = ... litres.

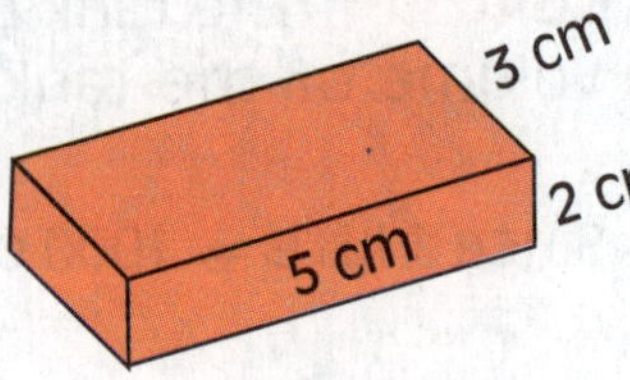

8. Express 4 litres in cm^3.

9. Roshan rotated the shape E by 360° in the clockwise direction. What was the final result ?

10. The cost of 8 pens is Rs 24. What is the cost of 10 pens ?

11. A man bought an article for Rs 25 and sold it for Rs 36. What did he have—profit or loss ? Find.

12. Express the shaded portion as:

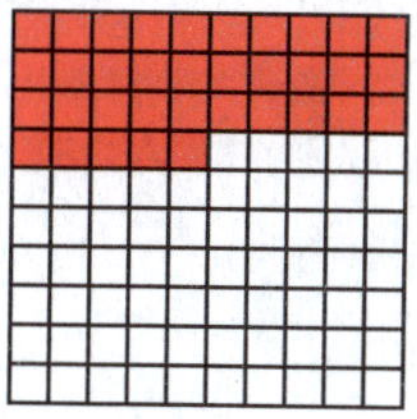

__________ Fraction   __________ Decimal   __________ Per cent

MIXED REVIEW – 3

1. **Add:** 1609275, 54832, 9870527, 6975, 88926.
2. Multiply 7084 by 3102. Divide the product by 121.
3. Subtract 19,578 from one hundred thousand.
4. The product of three numbers is 4301, two of them are 11 and 17, find the third number.
5. Express as decimals: **(i)** $\frac{7}{20} + \frac{3}{8}$ **(ii)** 102.5%
6. **(a)** Write in figures the predecessor of fifty eight lakh seventy nine thousand four hundred.

   (b) What is one seventh of the place value of 7 in the result.
7. Find the H.C.F. of 84, 180, 264.
8. Find the L.C.M. of 48, 56, 112.
9. **Simplify:** $\frac{25}{27} \div \left(\frac{3}{7} + \frac{4}{9}\right) \times 3\frac{1}{7}$
10. What fraction of 1 km is 450 m?
11. What is the volume of a cube whose length is $\frac{1}{4}$ cm?
12. The income of a family is Rs 16,000 per month. Each month 80% of the income is spent. What does the family save each month ?
13. Arrange the following numbers in descending order : $\frac{5}{6}, \frac{3}{8}, \frac{3}{4}, \frac{2}{3}$.
14. Convert $\frac{3}{8}$ into decimal.

15. Find the perimeter of a triangle of sides 9 cm, 14 cm and 16 cm.

16. Subtract Rs 549.12 from the sum of Rs 627.38 and Rs 493.85.

17. The cost of tiling a wall 8 m by 4 m is Rs 7,400. What is the cost per square metre ?

18. Divide 0.1524 by 1·2.

19. Express 0.06 as a fraction in its lowest terms.

20. The perimeter of a square is 24 cm. Find its area.

21. Find the volume of a cuboid whose length, width and height are 10 cm, 8 cm and 6 cm.

22. A boy ran 5 times round a football field of sides 90 m by 72 m. What distance did he run in kilometres ?

23. What must be added to $\frac{3}{8}$ of $\frac{5}{7}$ to make $\frac{9}{14}$ of $1\frac{3}{4}$?

24. Round off 78,27,59,751 to the nearest crore.

25. Show the reflection of the figure B along the line.

26. **Is it possible to have a triangle whose angle measures are:**

(a) 55°, 65°, 60° **(b)** 90°, 50°, 90° **(c)** 120°, 100°, 10° **(d)** 78°, 62°, 40° ?

27. Draw a circle of diameter 9 cm. Find its circumference.

28. The cost price of a sofa is Rs 8,500 and its selling price is Rs 10,200. Find the profit %.

29. What time in 12-hour notation is 5 hours 16 minutes before 3 : 05 p.m.?

30. **(a)** Find the interval in minutes between 6:00 a.m. and 9:30 a.m.

(b) Convert 158°F to °C.

KNOW YOUR NUMBERS

The Number Eight

SPIDER

SCORPION

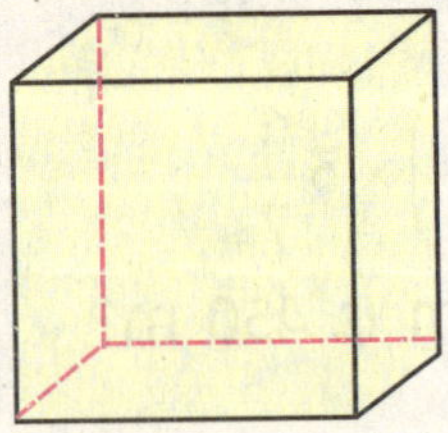

CUBE

8 = 2 X 2 X 2 so
8 is the cube of 2

A cube has eight corners.
Eight is a number that stays the same when written upside down.
There are eight legs on a spider. Scorpions also have eight legs.
There are eight people in a tug-of-war team.

19 PATTERNS

A pattern is formed by repeating a unit, which may be a set of numbers or shapes, according to a rule.

SQUARE NUMBERS:

Patterns of squares using dots can be made in two ways as :

I. Ayushree uses coloured dots to make patterns of squares as :

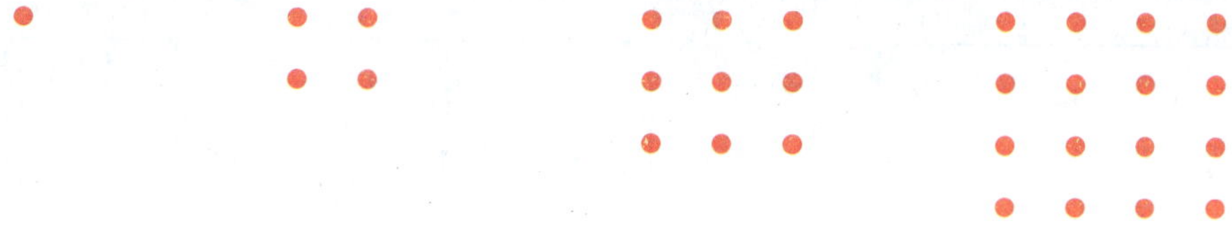

1 4 9 16

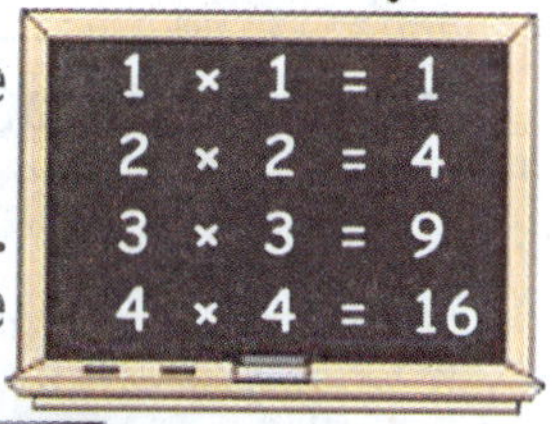

The number of dots in each square names a square number and it are so called because the dots are arranged in a square shape.

A square number is obtained by multiplying the number by itself.

Now can you name the next two square numbers and draw the square patterns ?

II. Abhinav makes the square patterns as :

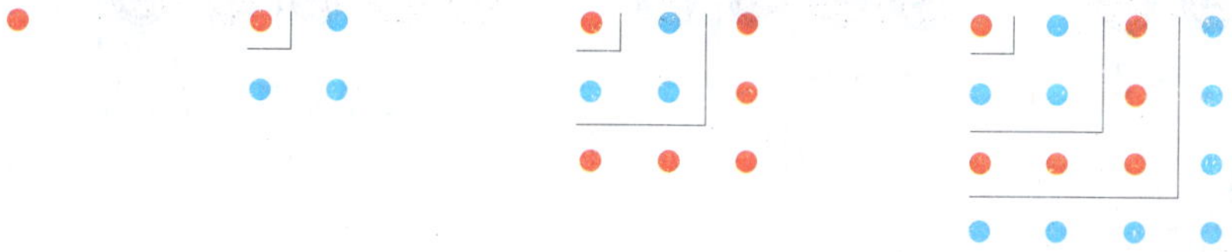

1 4 9 16

1 1 + 3 1 + 3 + 5 1 + 3 + 5 + 7

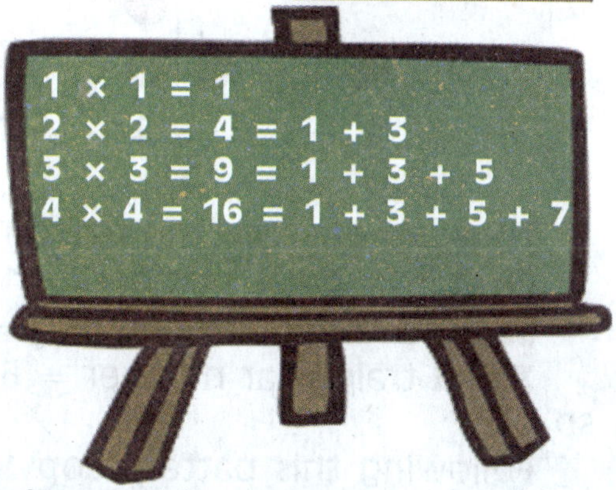

Thus, we observe that

First Square number = 1 = **First** odd number

Second Square number = 4 = 1 + 3 = Sum of **first two** odd numbers

Third Square number = 9 = 1 + 3 + 5 = Sum of **first three** odd numbers

Fourth Square number = 16 = 1 + 3 + 5 + 7 = Sum of **first four** odd numbers

Following this pattern, copy and complete Abhinav's table :

Odd numbers	As a Sum	As a product
1 + 3 + 5 + 7 + 9	25	5 × 5
1 + 3 + 5 + 7 + 9 + 11		
1 + 3 + 5 + 7 + 9 + 11 + 13		
1 + 3 + 5 + 7 + 9 + 11 + 13 + 15		

Without adding can you name the sum of first 10 odd numbers ?

TRIANGULAR NUMBERS

Triangular patterns using dots can be made in two ways as given below. The number of dots in each triangle is known as a triangular number.

I. Rohit uses coloured balls to make a triangular pattern as :

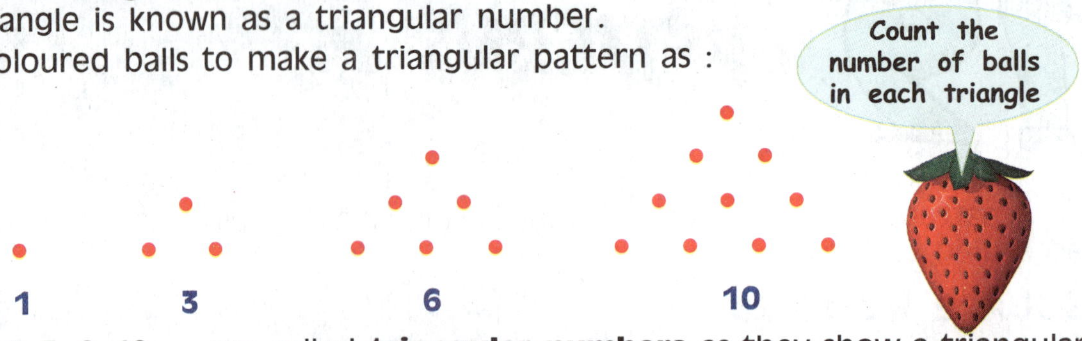

These numbers 1, 3, 6, 10, ... are called **triangular numbers** as they show a triangular shaped picture of dots.

Following the above pattern draw the next two triangular numbers using dots.

II. Rita uses buttons to make triangular patterns as :

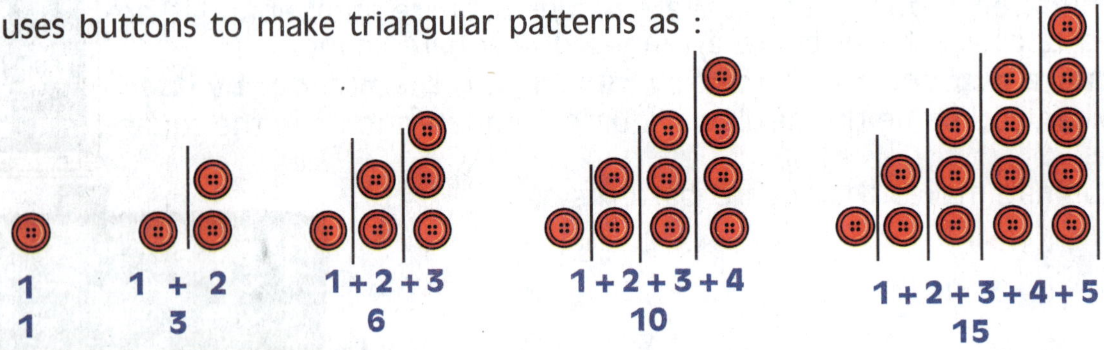

Thus we observe the pattern as :

First triangular number = 1 = **First** natural number.

Second triangular number = 3 = 1 + 2 = Sum of **first two natural** numbers.

Third traingular number = 6 = 1 + 2 + 3 = Sum of **first three natural** numbers and so on.

Following this pattern copy and complete Rita's Table :

Natural numbers	Sum
1 + 2 + 3 + 4	
1 + 2 + 3 + 4 + 5	
1 + 2 + 3 + 4 + 5 + 6	
1 + 2 + 3 + 4 + 5 + 6 + 7	

Now, see what happens if you add two consecutive triangular numbers ?

1 + 3 = 4
3 + 6 = 9
6 + 10 = 16
10 + 15 = 25

} square numbers

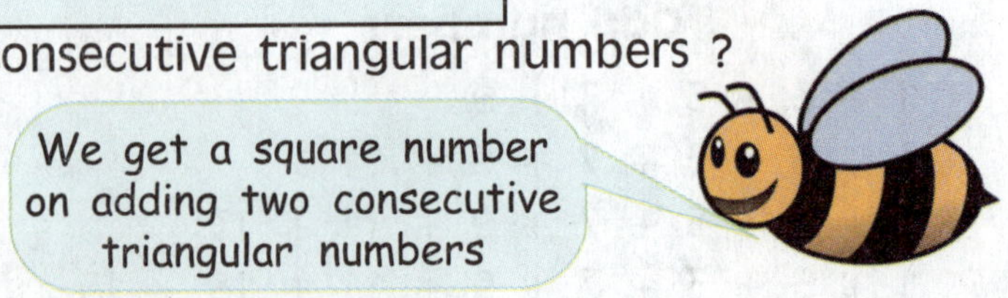

We get a square number on adding two consecutive triangular numbers.

You can also clearly see the square numbers in a multiplication square. They lie on a diagonal line.

1	2	3	4	5	6	7	8	9	10
2	4	6	8	10	12	14	16	18	20
3	6	9	12	15	18	21	24	27	30
4	8	12	16	20	24	28	32	36	40
5	10	15	20	25	30	35	40	45	50
6	12	18	24	30	36	42	48	54	60
7	14	21	28	35	42	49	56	63	70
8	16	24	32	40	48	56	64	72	80
9	18	27	36	45	54	63	72	81	90
10	20	30	40	50	60	70	80	90	100

Skill Practice — **Class Work**

1. Identify the type of number which each of the following sums names :

(i) 1 + 3 + 5 ______ **(ii)** 1 + 3 + 5 + 7 + 9 ______

(iii) 1 + 2 + 3 ______ **(iv)** 1 + 2 + 3 + 4 + 5 ______

(v) 6 + 10 ______ **(vi)** 15 + 21 ______

2. Name two square numbers whose sum names another square number. ______

3. Express each of the following as a sum of consecutive triangular numbers.

(i) $4 \times 4 = 16 =$ ______ + ______ **(ii)** $5 \times 5 = 25 =$ ______ + ______

(iii) $7 \times 7 = 49 =$ ______ + ______ **(iv)** $10 \times 10 = 100 =$ ______ + ______

4. Name a triangular number that is a square number also. ______

BORDER PATTERNS

In carpets, quilts, floors etc. we see border patterns enhancing their original beauty.

These border patterns are formed by repeating a certain shape or unit in one direction following a basic symmetry which may be translation, vertical reflection or horizontal reflection.

For better understanding let us take an example. Consider the letter **F** as the basic unit or shape in the border pattern. Thus we can get the following patterns when F follows the different symmetries as :

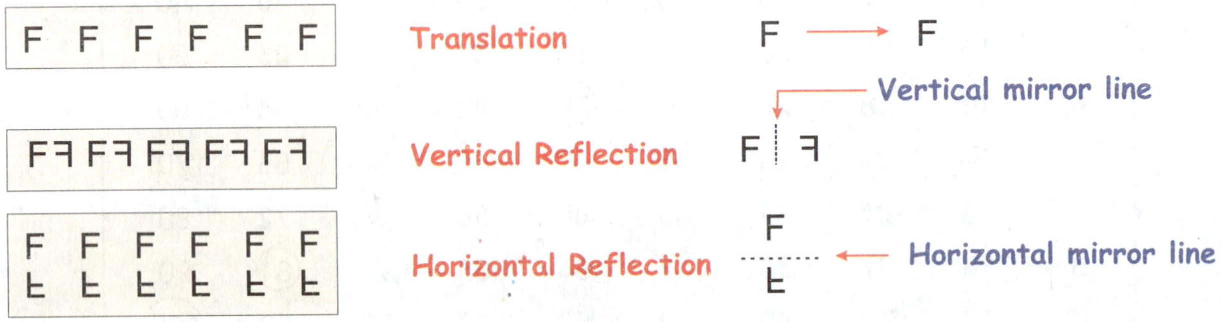

Another example is, taking  as the unit shape. Then the border patterns using the above three symmetries are :

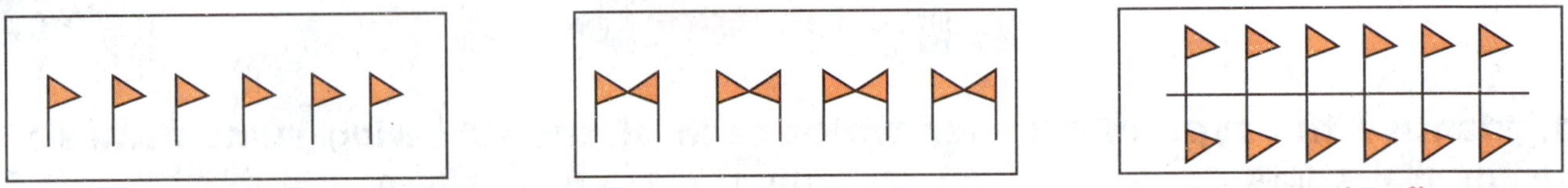

Translation Vertical reflection Horizontal reflection

TILING PATTERNS

A tiling pattern is formed by the repetition of a single unit or shape that, when repeated, fills the plane with **no gaps** and **no overlaps.** The most familiar examples of translations are the patterns by pairing stones or bricks and the cross-section of beehives.

Are there any gaps between the bricks ? Are they overlapping ? Look at some more examples given below :

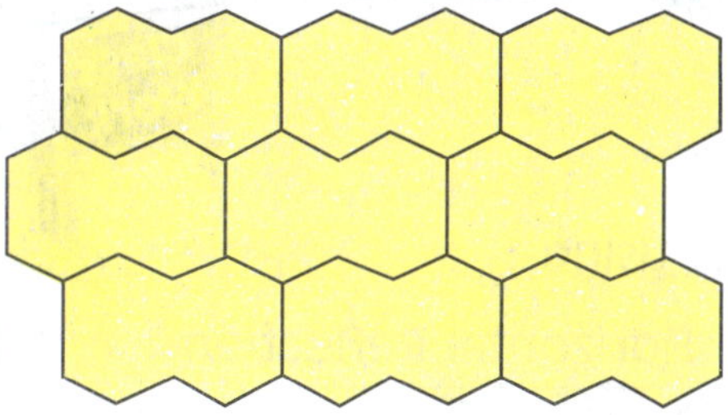

 is the unit shape used to make the tiling pattern in Figure 1.

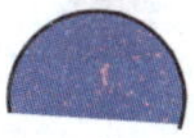 is the unit shape used to make the tiling pattern in Figure 2.

What can you say about the patterns given below ? Are they tiling patterns ? If not, then why not?

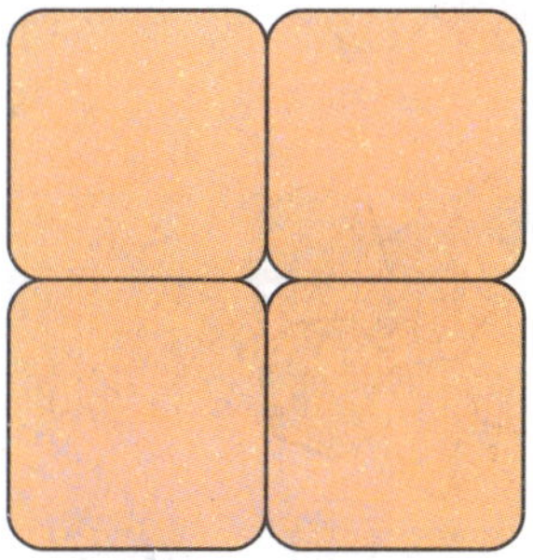

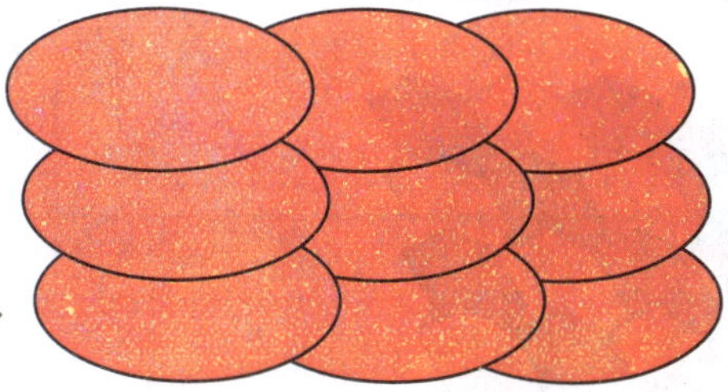

Tiling patterns are found in things around us.

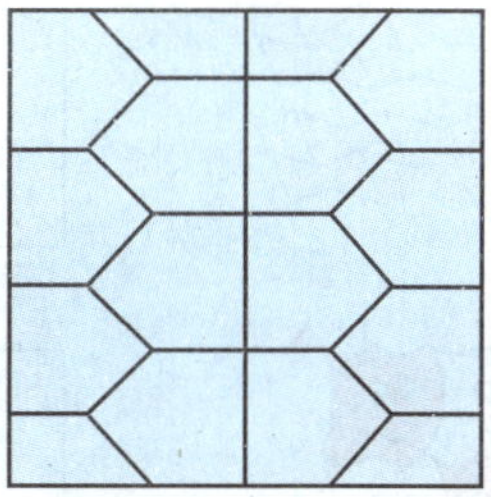

Window Panes

Carpets

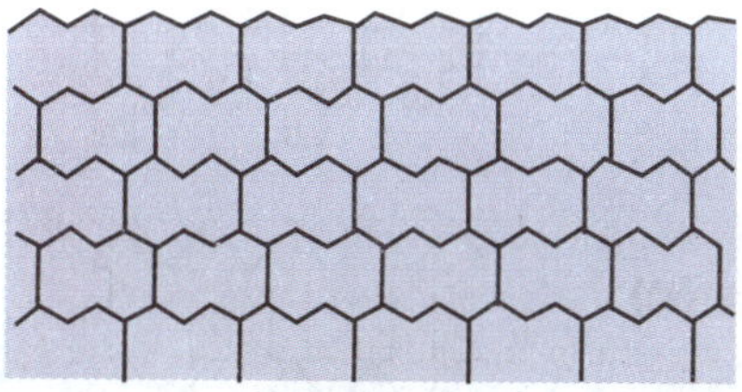

Floor Tiles

Multiple tiling patterns can be made using a single shape say 

1. Taking 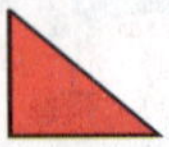 as the basic shape, draw border patterns using :

Translation	Vertical Reflection	Horizontal Reflection
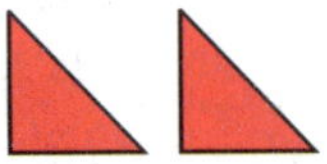		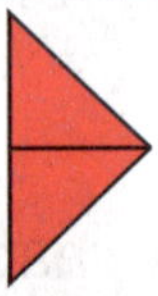

2. Which of the following patterns are tiling patterns ? (Yes/No)

(a) 

(b) 

(c) 

(d) 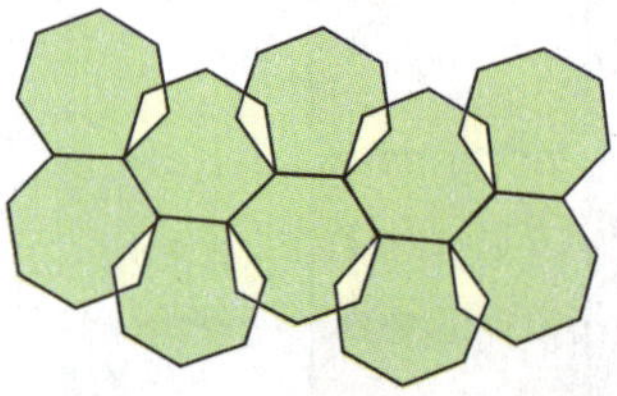

3. Identify the unit shape that is used in each of the tiling patterns given below.

(a) 

(b) 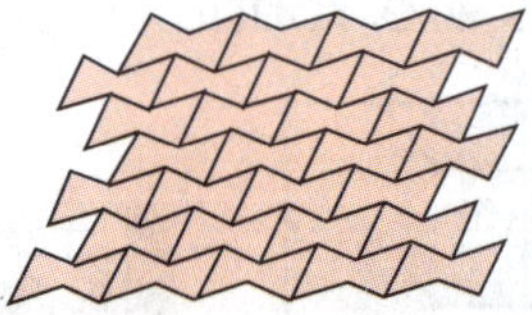

(c) 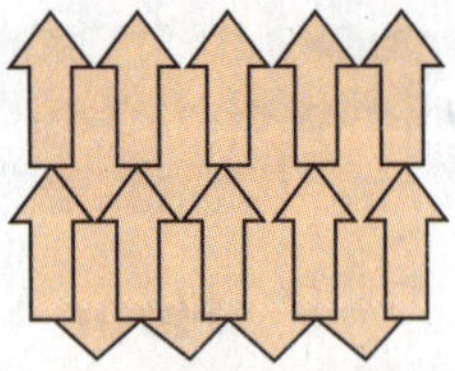

(d) 

4. Tiling patterns can also be made from 2 shapes. Identify the unit shapes used in making the tiling pattern given below:

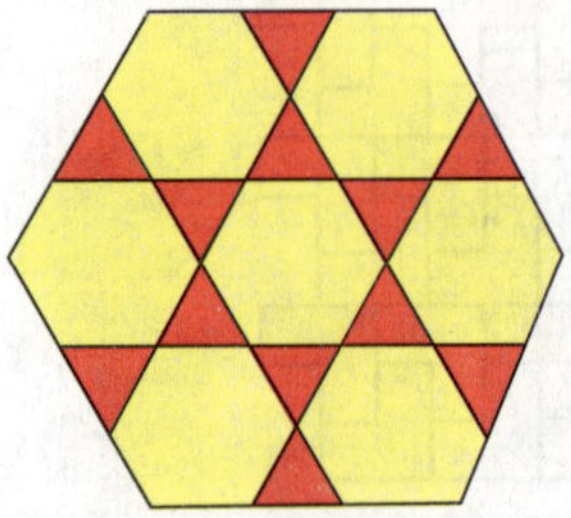

20 DATA HANDLING

DATA

When you ask your friends about their test marks in the recently held Maths Test, or how many hours each of them spends in doing homework each day, or record the daily temperature in your city over a period of say one month, or write down ages of your classmates, you are **collecting information.** All this information is numerical in nature.

> **A collection of information in the form of numerical figures is called data.**

Data can be arranged and displayed in the form of tables, graphs and charts.

The mathematics used for collecting, organizing and studying data is called **statistics.**

Data can be collected by using existing records or by survey. For example, you can collect information about which sport each of your classmates play :

(i) from school records **(ii)** by asking each of them (survey)

After the data has been collected, it needs to be classified and tabulated so that it can be read and interpreted easily. One way of presenting data in an organized manner is by drawing a **tally chart.**

Example: A survey was done on a class of 25 students to find about the favourite sport of each of them and the result was :

Student	Favourite sport	Student	Favourite sport	Student	Favourite sport
Sameer	Cricket	**Deepak**	Basketball	**Vimal**	Cricket
Ramesh	Tennis	**Rajeev**	Basketball	**Vishal**	Football
Harish	Football	**Ramesh**	Cricket	**Ashok**	Football
Sanjay	Basketball	**Rita**	Basketball	**Pushkar**	Basketball
Navin	Cricket	**Hema**	Basketball	**Anil**	Cricket
Amit	Tennis	**Vijay**	Badminton	**Suresh**	Football
Ranjan	Football	**Anurag**	Football	**Gaurav**	Football
Anwar	Cricket	**Sunil**	Cricket	**Shaleen**	Tennis
Suyash	Football				

This data has to be arranged in a meaningful manner. We construct a tally chart as shown below.

Favourite Sport	Tally Marks	Number of Students
Basketball	卌 I	6
Football	卌 III	8
Tennis	III	4
Cricket	卌 II	7
Badminton	I	1

We make a stroke to show the favourite sport of each student and complete a 'gate' to show a group of 5. Thus, from this table we can clearly see that football is the most liked sport followed by cricket.

You have already learnt about Pictographs and bar graphs in class IV. You can draw a bar graph and pictograph of the above given data as shown:

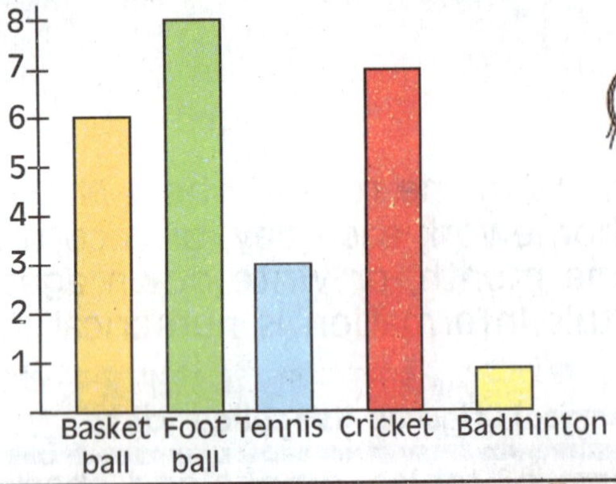

1. A survey was done on a class of 30 students to find about their favourite fruit and the student-wise response was :

Apple, Pear, Orange, Banana, Grapes, Banana, Apple, Pear, Orange, Orange, Chiku, Banana, Orange, Banana, Grapes, Grapes, Apple, Orange, Apple, Orange, Pear, Orange, Chiku, Banana, Orange, Watermelon, Apple, Banana, Pear, Orange

(a) Construct a tally chart for the above given information.

(b) Also draw a pictograph and bar chart to represent this data.

Tally Chart

Fruit	Tally Marks	Number of students
Apple		8
Pear		4
Orange		9
Banana		6
Grapes		3
Chiku		2
Watermelon		1

Answer the following questions :

(i) Which is the most popular fruit ?

(ii) Name the fruit that is liked the least ?

(iii) Name the fruits in the order of liking from least liked to most liked.

BAR GRAPH	PICTOGRAPH

READING A BAR GRAPH

The following bar graph shows the weather in December in New Delhi. Looking at the graph answer the questions given below:

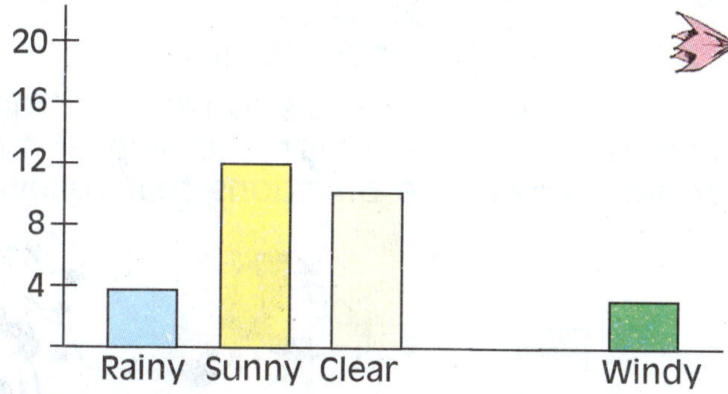

1. How many sunny days were there in December in New Delhi ? **(Ans : 12)**
2. How many windy days were there ? **(Ans : 4)**
3. How many more clear days than rainy days were there ? **(Ans : 6)**

EXERCISE 57

1. Draw a pictograph to show the number of stories you read from your favourite story book each day in a week.

Day	Sunday	Monday	Tuesday	Wednesday	Thursday	Friday	Saturday
Number of stories you read	4	1	2	3	1	2	5

Take  = 1 story.

2. Draw a pictograph to show the number of students who attended a dog show in school from each class of the Junior school.

Class	Class I	Class II	Class III	Class IV	Class V
Number of Students	15	20	10	30	5

Take  = 5 students

3. Draw a bar chart to represent the number of students who visited the school library in a week. Use a graph paper to draw the graph

Day	Monday	Tuesday	Wednesday	Thursday	Friday	Saturday
Number of Students	27	32	45	54	60	15

4. A survey was done on 25 people on their choice of drinks and the individual response was as given below:

Coffee, Tea, Tea, Tea, Milk, Hot chocolate, Coffee, Tea, Milk, Tea, Others, Tea, Coffee, Tea, Milk, Hot Chocolate, Coffee, Tea, Milk, Milk, Tea, Coffee, Tea, Coffee, Others.

Tabulate the given data and draw a bar chart to show the given information.

5. The students of class 5 were given their choice of animals to research for a project in science. The results of their choices were displayed with the help of a bar graph. Look at the bar graph given below and answer the questions that follow :

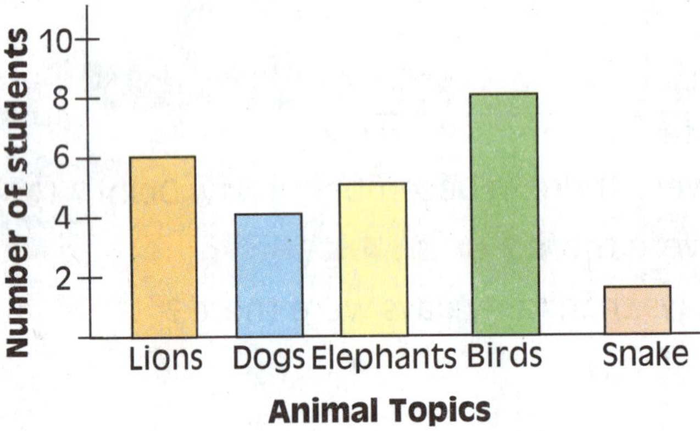

(i) Which is the most popular topic?

(ii) Which is the least popular topic?

(iii) The number of students who chose dogs and snakes is equal to the number of students who chose what other topic?

(iv) What is the total number of students in the class?

(v) Express the number of students who chose snakes as a fraction of the whole class.

6. The following graph shows the future careers of middle school students of a school. Use the bar graph to answer the questions given below.

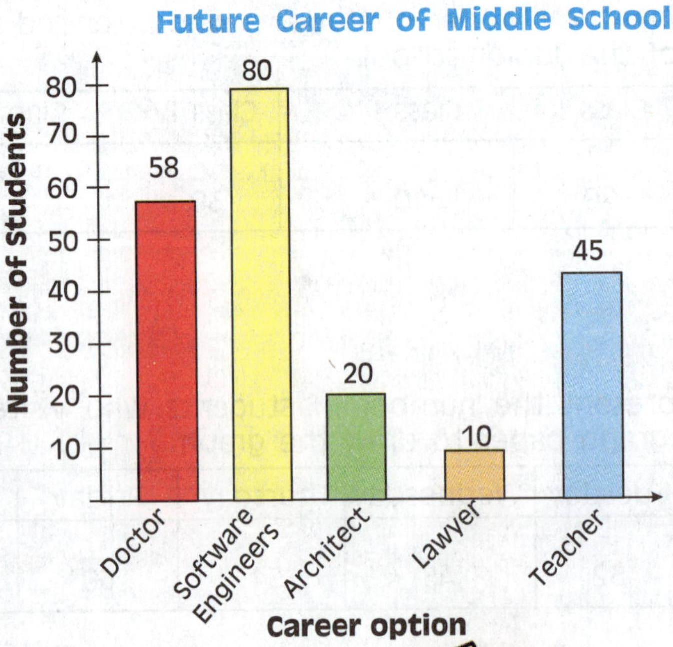

(i) Name the most popular career option chosen.

(ii) How many more students chose doctor as their career option than the number of students who chose Teacher as their career?

(iii) Find the total number of students in the middle school?

(iv) List the careers chosen from most to least popular.

DOUBLE BAR GRAPHS

To compare data we use **double bar graphs.**

Anita conducted a survey in her class to find out which colours the boys and girls of her class liked the most. She recorded the results in a tabular form as :

Colour	Boys	Girls
Blue	7	4
Pink	2	8
Red	5	5
Yellow	3	6
Green	6	2

To draw a double bar graph, we follows these steps :

Step 1 : Take two axes.

Mark the vertical axes as **number of students**

Mark the horizontal axes as **colour.**

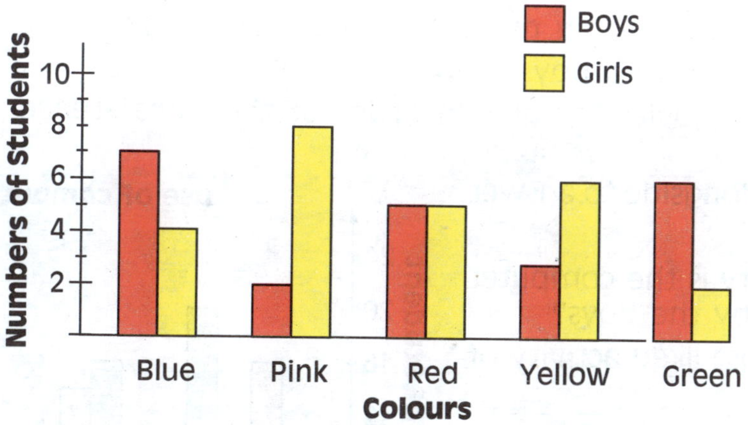

Step 2 : Choose an appropriate scale for the vertical axes. Here you may take 1 cm = 2 students.

Step 3 : For each colour, draw one graph for boys and one for girls. Make a key to show the different colours used for boys and girls bars.

Step 4 : Give a title to the graph.

Home Work

EXERCISE 58

1. Draw a double bar graph to compare the favourite ice-cream flavours of boys and girls in a class whose data is as given below:

Flavours	**Vanilla**	**Chocolate**	**Strawberry**	**Butterscotch**
Boys	9	6	4	8
Girls	5	10	12	7

2. The following graph shows the favourite kinds of movies of boys and girls of class 5 in a school. Look at the graph and answer the questions given below:

Favourite movies of students of class 5

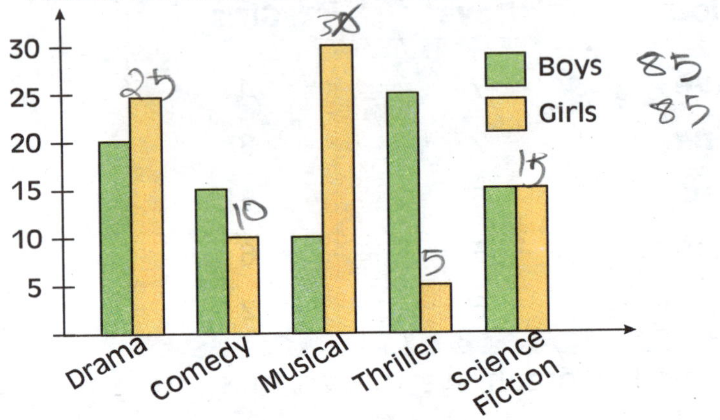

KINDS OF MOVIES

(i) Which kind of movies do boys prefer the most?

(ii) Which kind of movies do girls prefer the most?

(iii) Which kind of movies are liked by both equally?

(iv) How many more girls watch musical movies than boys?

(v) Which kind of movies do boys prefer more than girls do?

(vi) List the kinds of movies for girls in order of their preference from least to most.

3. Use the graph given alongside to answer the given questions :

(i) For which activity is the computer used the most by the boys?

(ii) Which is the most liked activity of the girl?

(iii) Which activity has the least difference in terms of numbers between boys and girls?

(iv) Which activity has the most difference in terms of numbers between boys and girls?

(v) List the computer activities in order of liking for (a) boys (b) girls

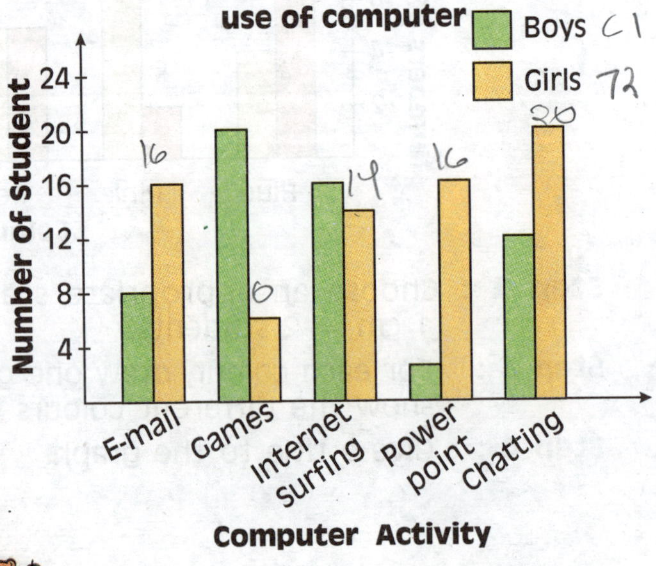

ANSWERS

Page 2 **Exercise 1**

1. 1 **2.** 0

3.

Nature of counting numbers	Numbers from __ to __	Smallest number	Greatest number
one digit numbers	1-9	1	9
2 digit numbers	10-99	10	99
3 digit numbers	100-999	100	999
4 digit numbers	1000-9999	1000	9999
5 digit numbers	10000-99999	10000	99999
6 digit numbers	100000-999999	100000	999999

4. **(a)** 13,586 **(b)** 54,907 **(c)** 80,700 **(d)** 49,200
(e) 60,005 **(f)** 5,43,956 **(g)** 8,90,300

5. **(a)** Fifty nine thousand three hundred eighty **(b)** Eighty three thousand six
(c) Four lakh twenty five thousand six hundred seven
(d) Three lakh five thousand seven hundred eighty four **(e)** One lakh thirty two
(f) Nine lakh eight **(g)** Nine lakh ninety nine thousand nine hundred ninety nine.

6. **(i)** 8723, 8728, 8733, 8738, 8743 **(ii)** 15910, 15915, 15920, 15925, 15930
(iii) 71837, 71842, 71847, 71852, 71857
(iv) 80989, 80994, 80999, 81004, 81009, 81014

7. **(a)** 7000 **(b)** 800 **(c)** 60000 **(d)** 50

8. **(i)** 1538, 1835, 1853, 3581, 3815, 5813, 8135, 8153
(ii) 80008, 80088, 80808, 88008, 88808

9. **(i)** 87465, 86754, 76845, 67854 **(ii)** 427245, 423675, 421032, 420000

10. **(i)** 69,930 **(ii)** 396 **(iii)** 7 92 000

11. **(a)** 97 430 **(b)** 8 76 320 **(c)** 9 85 321 **(d)** 8 74 210

12. **(a)** 10 345 **(b)** 1 25 679 **(c)** 2 03 457

Page 3 **Exercise 2**

1. **(a)** 1 63 394 **(b)** 99 542 **(c)** 2 35 907 **(d)** 2 03 276
(e) 6 54 332 **(f)** 9 20 225

2. **(a)** 54 637 **(b)** 21 739 **(c)** 0 **(d)** 16 307

3. 87080   4. Rs 1,02,440

5. (i) 50487   (ii) 4715   (iii) 83972   (iv) 1,45,865

(v) 8709   (vi) 1,05,108   (vii) 5,12,797   (viii) 1,32,410

6. 8997   7. (a) 0   (b) 30 029   (c) 23, 568

8. 10,238   9. Rs 11, 763   10. Rs 7, 454·72

Page 4 — Exercise 3

1. (a) 3 × 2   (b) 6 × 5   (c) 4 × 0   (d) 4 × 137   (e) 5 × 194

2. (a) 530, 5300, 53000, 530000   (b) 40 800   (c) 1 95 000

3. (a) 18   (b) 0   (c) 825   (d) 0

4. (a) 18   (b) 0   (c) 879   (d) 0   (e) 729   (f) 0

5. (a) 3,72,934   (b) 2,45,904   (c) 81,05,790   (d) 16,71,184

6. 13,81, 890   7. 1, 31, 600   8. 2,10,000 km   9. No need   10. 28,600

Page 5 — Exercise 4

1. (a) 88 r 11   (b) 2250 r 21   (c) 1354 r 8   (d) 679 r 15

2. (a) 8594 r 25   (b) 3939 r 47   (c) 4357 r 25   (d) 5270 r 47

3. (a) 21 r 16   (b) 69 r 25   (c) 626 r 11

4. 220   5. (c) and (d)   6. 12 ; 2 will be left over.

7. 28 complete rows and 18 stitches more   8. 496 km   9. 768   10. 73

Page 6 — Mental Maths–1

1. 25 years   2. 30   3. Rs 8   4. 70   5. 3   6. 67 cm

7. Rs 7200   8. 9673   9. Rs 800

10. Two lakh sixty three thousand nine hundred eighty five.   11. 50

12. 103579   13. 20 days   14. 41 years   15. 600   16. 275   17. 1200

18. 180 minutes   19. 9   20. Rs 100   21. 500 shares

22. <   23. Quotient = 57, Remainder = 8   24. 10   25. 40.

Page 12 — Skill Practice

1. 72, 34, 46, 960; 72 34 46 960

2. (a) Nine crore twenty five lakh eighty six thousand one hundred ninety.

(b) Sixteen crore forty eight lakh nine

3. (a) 12, 07, 02, 094   (b) 50, 00, 00, 505

4. (a) 60,00, 000 or sixty lakh   (b) 6,00,00,000 or 6 crore

5. (a) 6,00,00,000 + 7000000 + 500000 + 80000 + 6000 + 400 + 10 + 2

(b) 200000000 + 30000000 + 4000000 + 800000 + 7

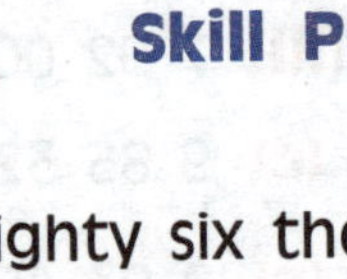

Page 13

Exercise 5

1. **(a)** 65 98 312, Sixty five lakh, ninety-eight thousand, three hundred twelve;

   (b) 13 08 264, Thirteen lakh, eight thousand, two hundred sixty four.

2. **(i)** Four crore, fifty seven lakh, twenty nine thousand, three hundred eighteen.

   (ii) Eight crore, sixty **(iii)** Fifty crore, twenty six thousand, three.

   (iv) Seventy two crore, eighty lakh, sixty six thousand, five hundred fifty.

   (v) Eighty three crore, twenty two thousand, forty nine.

   (vi) Seventy crore, six lakh eighty.

   (vii) Forty eight crore, ninety nine lakh, eighty five thousand, three hundred sixty one.

   (viii) Ninety crore, two lakh, five hundred ninety **(ix)** Ninety crore, nine.

3. **(i)** 58,07,009 **(ii)** 7,02,75,019 **(iii)** 19,84,07,300 **(iv)** 5,08,00,004

   (v) 89,65,59,618 **(vi)** 5,00,00,005 **(vii)** 99,00,00,000 **(viii)** 77,00,00,717.

4. Fourteen crore, eighty eight lakh.

5. **(i)** crores **(ii)** lakhs **(iii)** thousands **(iv)** ten-thousands **(v)** ten-crores **(vi)** crores.

6. **(a)** 4 00 00 000 **(b)** 20 00 000 **(c)** 60 000 **(d)** 300

   (e) 80 00 00 000 **(f)** 50 000.

7.

	Period	Place value		Period	Place value
(a)	thousands	70 000	(d)	10-crores	60 00 00 000
(b)	10 lakhs	20 00 000	(e)	Tens	30
(c)	crores	1 00 00 000	(f)	thousands	2000

8. **(i)** 7 00 000 + 50 000 + 2 000 + 400 + 10 + 8

   (ii) 30 00 000 + 6 00 000 + 10 000 + 2 000 + 900 + 20 + 0

   (iii) 6 00 00 000 + 90 00 000 + 8 00 000 + 40 000 + 5 000 + 100 + 30 + 7

   (iv) 20 00 00 000 + 6 00 00 000 + 10 00 000 + 8 00 000 + 40 000 + 4000 + 500 + 0 + 9

   (v) 70 00 00 000 + 3 00 00 000 + 50 00 000 + 9 00 000 + 60 000 + 5 000 + 600 + 10 + 4

   (vi) 6 00 00 000 + 90 00 000 + 8 00 000 + 0 + 5 000 + 700 + 10 + 4

   (vii) 9 00 00 000 + 0 + 0 + 80 000 + 7 000 + 0 + 20 + 3

   (viii) 20 00 00 000 + 4 00 00 000 + 0 + 0 + 0 + 3 000 + 100 + 0 + 6.

9. **(a)** 58,24,97,631 **(b)** 72,15,83,846 **(c)** 90,54,90,607

   (d) 6,40,08,090 **(e)** 40,00,70,008 **(f)** 7,09,60,085.

10. 99 99 999 **11.** 1 00 00 000 **12.** 1 00 00 002 **13.** 9 99 99 987

14.(i) 59,60,73,876 **(ii)** 6,79,83,970 **(iii)** 5,80,76,276
(iv) 73,00,00,000 **(v)** 6,88,00,000

15. (i) 58,99,99,999 **(ii)** 7,56,87,599 **(iii)** 20,71,57,829
(iv) 70,00,35,999 **(v)** 9,89,99,999 **(vi)** 452699989

16. 50 09 998, 50 10 000, 50 10 002 **17.** 70 09 909, 70 10 009; 70 10 109, 70 10 209

18. 86 59 317, 86 60 317, 86 61 317, 86 62 317, 86 63 317

19. 1 00 00 004, 1 00 00 001, 99 99 998, 99 99 995 **20.** 69 99 300 **21.** 1,60,000

Page 16 — Skill Practice

1. (a) $>$ **(b)** $<$ **2.** 3289762, 4763875, 4893970, 12587689

3. 581234890, 82935871, 82811121, 64812661

Page 16 — Exercise 6

1. (i) $<$ **(ii)** $>$ **(iii)** $<$ **(iv)** $<$ **(v)** $>$ **(vi)** $>$ **(vii)** $<$ **(viii)** $>$

2. (a) 19 95 764, 28 28 335, 29 89 632, 35 37 286
(b) 87 98 422, 3 94 00 316, 5 46 19 717, 9 01 00 510
(c) 8 91 300, 99 56 130, 4 91 82 766, 7 61 80 542

3. (a) 2 00 16 930, 1 98 72 003, 20 03 568, 2 79 868
(b) 3 98 76 542, 2 23 98 761, 50 18 960, 40 00 196
(c) 5 51 09 872, 87 65 014, 56 88 764, 36 10 303

4. (a) Smallest 10, 45, 678, Greatest 87, 65, 410
(b) Smallest 1, 02, 45, 689, Greatest 9, 86, 54, 210
(c) Smallest 10, 23, 45, 689, Greatest 98, 65, 43, 210
(d) Smallest 2, 03, 46, 789, Greatest 9, 87, 64, 320

5. (a) Greatest 41,82,60,329, smallest 5,463,978
(b) Greatest 22,01,00,043, smallest 1,95,01,872
(c) Greatest 9,51,73,694, smallest 80,48,784
(d) Greatest 9,68,80,808, smallest 80,00,008

6. (a) Tokyo, Japan, New York city, U.S **(b)** New York city, Mexico city, Sao Paulo, Mumbai, Tokyo **(c)** Sao Paulo, Mumbai, Tokyo **(d)** Sao Paulo.

7. (a) 4 **(b)** 2 **(c)** 5 **(d)** 5

Page 20 — SKILL PRACTICE

1. (a) 23, 487, 900; Twenty three million four hundred eighty seven thousand nine hundred.

(b) 370, 008, 003; Three hundred seventy million eight thousand three.

2. (a) 65, 205, 746 (b) 200, 070, 032

3. Indian System : 81, 40, 08, 250; Eighty one crore forty lakh eight thousand two hundred fifty.

International System: 814, 008, 250; Eight hundred fourteen million eight thousand two hundred fifty.

Page 20

Exercise 7

1. (a) 183 497 026, One hundred eighty three million four hundred ninety seven thousand twenty six.

   (b) 463 721 985; Four hundred sixty three million, seven hundred twenty one thousand, nine hundred eighty five.

   (c) 5 086 432; Five million, eighty six thousand, four hundred thirty two.

   (d) 46 300 529; Forty six million, three hundred thousand, five hundred twenty nine.

   (e) 700 083 460; Seven hundred million, eighty three thousand, four hundred sixty.

   (f) 810 000 008 Eight hundred ten million eight

2. (a) 85 405 900 (b) 200 817 510 (c) 530 685 939

   (d) 8 455 102 (e) 507 006 002.

3.

(a) **429012795**	42,90,12,795	Forty two crore ninety lakh twelve thousand seven hundred ninety five
	429 012 795	Four hundred twenty nine million twelve thousand, seven hundred ninety five
(b) **501043768**	50,10,43,768	Fifty crore ten lakh forty three thousand seven hundred sixty eight.
	501 043 768	Five hundred one million, forty three thousand, seven hundred sixty eight.
(c) **321543006**	32,15,43,006	Thirty two crore fifteen lakh forty three thousand six.
	321 543 006	Three hundred twenty one million, five hundred forty three thousand, six
(d) **349100860**	34,91,00,860	Thirty four crore ninety one lakh eight hundred sixty.
	349 100 860	Three hundred forty nine million, one hundred thousand, eight hundred sixty.
(e) **92600512**	9,26,00,512	Nine crore twenty six lakh five hundred twelve.
	92 600 512	Ninety two million, six hundred thousand, five hundred twelve.
(f) **99000017**	9,90,00,017	Nine crore ninety lakh seventeen
	99 000 017	Ninety nine million, seventeen

4.

Country	Population		
(a) Afghanistan	**27089593**	2, 70, 89, 593	Two crore seventy lakh eighty nine thousand five hundred ninety three.
		27, 089, 593	Twenty seven million eighty nine thousand five hundred ninety three.
Bangladesh	**137493990**	13,74,93,990	Thirteen crore seventy four lakh ninety three thousand nine hundred ninety
		137, 493, 990	One hundred thirty seven milion four hundred ninety three thousand nine hundred ninety.
Indonesia	**224481720**	22,44,81,720	Twenty two crore forty four lakh eighty one thousand seven hundred twenty.
		224, 481, 720	Two hundred twenty four million, four hundred eighty one thousand, seven hundred twenty.
Japan	**128646345**	12,86,46,345	Twelve crore eighty six lakh forty six thousand three hundred forty five.
		128, 646, 345	One hundred twenty eight million six hundred forty six thousand three hundred forty five.
Malaysia	**28294120**	2,82,94,120	Two crore eighty two lakh ninety four thousand one hundred twenty.
		28, 294, 120	Twenty eight million two hundred ninety four thousand one hundred twenty.
Nepal	**25874519**	2,58,74,519	Two crore fifty eight lakh seventy four thousand five hundred nineteen.
		25,874,519	Twenty five million eight hundred seventy four thousand five hundred nineteen.
Pakistan	**167806831**	16,78,06,831	Sixteen crore seventy eight lakh six thousand eight hundred thirty one.

		167,806,831	One hundred sixty seven million eight hundred six thousand eight hundred thirty one.
Sri Lanka	19796874	1,97,96,874	One crore ninety seven lakh ninety six thousand eight hundred seventy four
		19,796, 874	Nineteen million seven hundred ninety six thousand eight hundred seventy four.
Singapore	3654103	36, 54, 103	Thirty six lakh fifty four thousand one hundred three
		3,654,103	Three million six hundred fifty four thousand one hundred three.
Thailand	67249456	6, 72, 49, 456	Six crore seventy two lakh forty nine thousand four hundred fifty six.
		67, 249, 456	Sixty seven million two hundred forty nine thousand four hundred fifty six.

(b) Indonesia, Pakistan, Bangladesh, Japan, Thailand, Malaysia, Afghanistan, Nepal, Sri Lanka, Singapore.

(c) Indonesia **(d)** Singapore.

5. **(a)** Indian System

Largest Number	98,74,43,210	Ninety eight crore seventy four lakh forty three thousand two hundred ten.
Smallest Number	10,23,44,789	Ten crore twenty three lakh forty four thousand seven hundred eighty nine.

(b) International System

Largest Number	987, 443, 210	Nine hundred eighty seven million four hundred forty three thousand two hundred ten
	102,344,789	One hundred two million three hundred forty four thousand seven hundred eighty nine.

6. 999 995, 999 996, 999 997, 999 998, 999 999, 1 000 000, 1 000 001, 1 000 002, 1 000 003, 1 000 004, 1 000 005, 1 000 006, 1 000 007, 1 000 008, 1 000 009, 1 000 010.

Page 22

Mental Maths–2

2. 8040213 **3.** 8000000 **4.** 5 **5.** 718, 264, 499 **6.** 20

7. Hundred million, ten crore **8.** 70 crore **9.** 20 **10.** 200000

Page 24 — **Skill Practice**

1. 11,79,34,737 2. 42,81,36,301 3. 38,02,929 4. 30,77,88,119

Page 24 — **Exercise 8**

1. 1,60,65,679 2. 1,30,93,421 3. 9,32,46,956 4. 44,75,66,765
5. 85,47,84,990 6. 49,27,41,028 7. 14,94,70,717 8. 17,99,32,204
9. 1,85,14,25,727 10. 97,90,69,016 11. 13,23,65,107 12. 25,27,48,249
13. 12,67,33,819 14. 7,19,36,787 15. 41,06,29,209 16. 48,61,55,852
17. 7,75,35,805 18. 88,73,500 19. 72,16,07,514 20. 1,83,32,267
21. 40,000,001 22. 5,94,22,000 23. 37,640 24. 10,94,65,563
25. 86,45,336
26. West Bengal, by 1,98, 36, 053
(ii) (a) 24,68,15,277 (b) 1,07,65,565 (c) 24,34,35,663
(iii) (a) True (b) True
(iv) Assam, Chhatisgarh, Haryana, Jammu & Kashmir, Jharkhand, Kerala, Orissa, Punjab, 19,83,32,529.
(v) Uttar Pradesh, 166,052,859, one hundred sixty-six million, fifty two hundred thousand, eight hundred fifty nine
27. (b) 4 398 137 (c) 2 515 913 (d) 71,96,438
(e) 5,76,95,104 (f) 285 838 702 (g) 66 175 862
29. Fifteen lakh forty four thousand one hundred seventy six.
30. Cost of flat is Rs 89,55,445.

Page 27 — **Skill Practice**

1. 4,20,000 2. 1,65,00,000 3. 24,000 4. 5,60,000 5. 1,50,00,000
6. 60,00,000 7. 300 8. 50,000

Page 29 — **Skill Practice**

1. 138408747 2. 6807856 3. 35,88,9920

Page 29 — **Exercise 9**

1. (i) 8450 (ii) 957200 (iii) 605000 (iv) 9870000
(v) 54900000 (vi) 237000000 (vii) 20850 (viii) 658800
(ix) 11076000
2. 7278538 3. 8338287 4. 58987350 5. 693492368
6. 471497280 7. 96990300 8. 54368175 9. 337103025
10. 395494873 11. 41281053 12. 150101553 13. 810441039
14. 263368000 15. 247000000 16. 66590250 17. 827099100
18. 281530000 19. 148712060 20. Rs 3,90,00,000 21. Rs 18,07,68,840
22. Rs 81,29,40,800 23. Rs 2,53,28,550 24. Rs 5,11,47,525

Page 31 — **Skill Practice**

1. Q = 8, R = 5930 2. Q = 41, R = 5158 3. Q = 938 R = 95
4. Q = 12361, R = 2 5. Q = 3245, R = 325 6. Q = 14, R = 983
7. Q = 8, R = 93,425

Page 32 **Skill Practice**

1. 2259 R 1 2. 5025 R 92 3. 5009 R 42

Page 33 **Exercise 10**

1. 898 R7 2. 4305 R 20 3. 14810 R 29 4. 79828 R 47 5. 44 R 111
6. 34 R 115 7. 22 R 70 8. 41 R 8 9. 38 R8 10. 57 R 4
11. 656 R 495 12. 239 R 9 13. 770 R 571 14. 27092 R 243 15. 6187 R 5

Page 33 **Exercise 11**

1. 8 2. 973 3. 1572 shelves 4. Rs 15,954 5. 1500 minutes
6. Q = 100100 R = 99 7. 10010 8. 1888
9. 3597, 17 bricks will remain 10. 54018

Page 34 **Mental Maths–3**

1. 5 2. 64000 3. 500 4. 360
5. 72000000, 72 crore 20 lakh, 72 million 6. 1 7. 8
8. 40 9. 300 10. 199910, one lakh ninety nine thousand ten

Page 38 **Skill Practice**

1. 60,000 2. 3,30,000 3. 5,00,000 4. 16,00,000 5. 50,00,000
6. 70,00,000 7. 7,00,00,000 8. 12,00,00,000

Page 38 **Exercise 12**

1. (a) 50000 (b) 80000 (c) 20000 (d) 280000 (e) 540000
(f) 1650000 (g) 2510000 (h) 1470000
2. (a) 400000 (b) 600000 (c) 3500000 (d) 8700000 (e) 2500000
(f) 3800000 (g) 7500000 (h) 2400000
3. (a) 2000000 (b) 3000000 (c) 1000000 (d) 21000000 (e) 75000000
(f) 62000000 (g) 32000000 (h) 92000000
4. (a) 70000000 (b) 30000000 (c) 20000000 (d) 300000000 (e) 110000000
(f) 50000000 (g) 320000000 (h) 300000000
5. (a) 593702160 (b) 593702200 (c) 593702000 (d) 593700000 (e) 59 3700 000
(f) 594000000 (g) 590000000
6. (a) 17 crore (b) 8 crore (c) 8 crore (d) 8 crore (e) 4 crore
(f) 6 crore
7. 103,00,00,000 crore.

Page 40 **SKILL PRACTICE**

1. 1,2,3,6,9,18,27, 54 2. (a) Prime number are 7, 11, 23, 31, 37, 43
(b) Sample answer : 4 and 7 and 9, 9 and 11, 9 and 23, 9 and 25

Page 41 **SKILL PRACTICE**

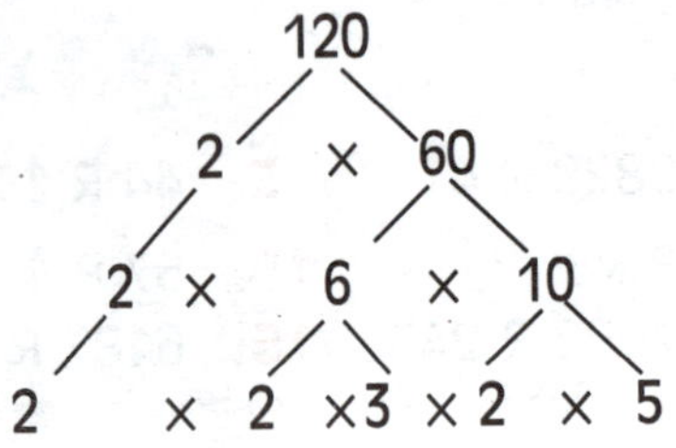

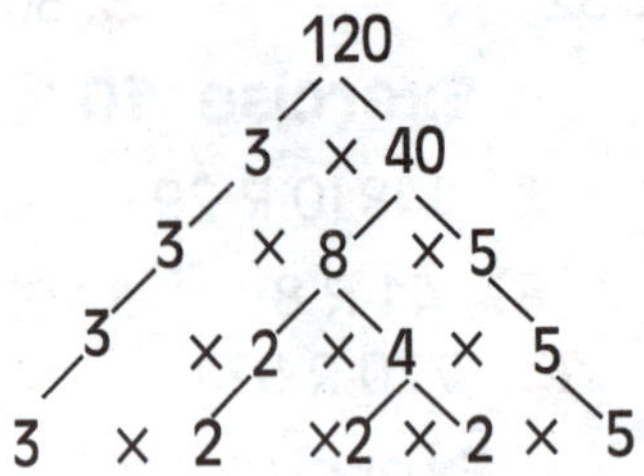

2	120
2	60
2	30
3	15
	5

$120 = 2 \times 2 \times 2 \times 3 \times 5$

Page 42 **Skill Practice**

1. 4 2. 6 3. 9 4. 6

Page 43 **Skill Practice**

1. 7 2. 32 3. 75

Page 44 **Skill practice**

1. 36 2. 240 3. 1008 4. 150

Page 45 **Exercise 13**

1. (a) 1,11; Prime (b) 1, 2, 4, 8, 16, 32; Composite
   (c) 1, 2, 4, 5, 8, 10, 20, 40; Composite (d) 1, 19; Prime
   (e) 1, 2, 3, 6, 9, 18; Composite (f) 1, 41; Prime
2. Sum = 93, No
3. (a)

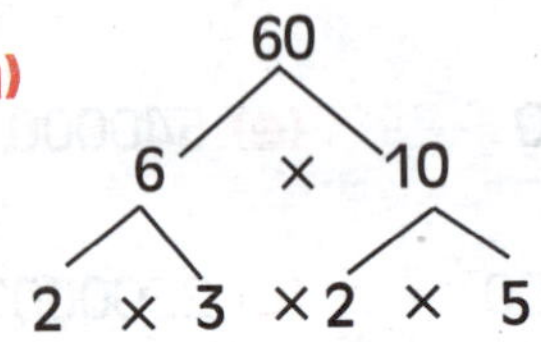

(b)

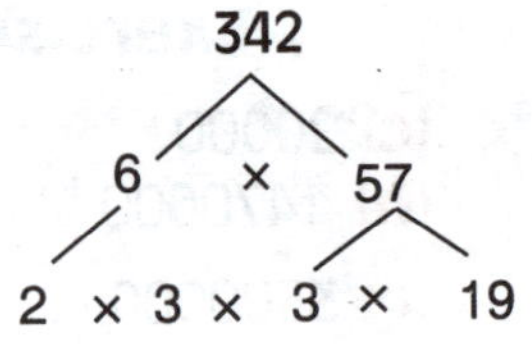

(c)

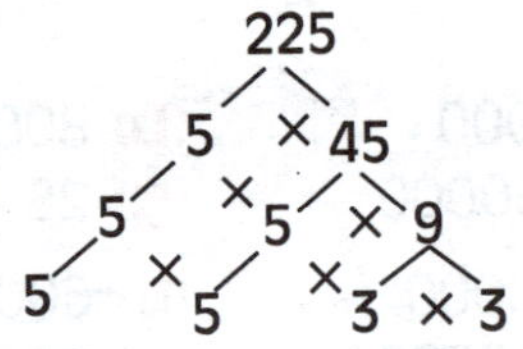

4. (a) $216 = 2 \times 2 \times 2 \times 3 \times 3 \times 3$ (b) $150 = 2 \times 3 \times 5 \times 5$
   (c) $2 \times 2 \times 3 \times 7$ (d) $1080 = 2 \times 2 \times 2 \times 3 \times 3 \times 3 \times 5$
5. (a) 5 (b) 1 (c) 6 (d) 2
6. (a) 12 (b) 1 (c) 6 (d) 5 (e) 4 (f) 9 (g) 14 (h) 38
7. (a) 64 (b) 5 (c) 3 (d) 45 (e) 12 (f) 34 (g) 65 (h) 5 (i) 2
8. (a) 35 (b) 30 (c) 48 (d) 50 (e) 42 (f) 90 (g) 56 (h) 45
9. (a) 24 (b) 30 (c) 36 (d) 96 (e) 220 (f) 260 (g) 96 (h) 200
10. (a) 60 (b) 180 (c) 1232 (d) 3510 (e) 1200 (f) 13650 (g) 1260 (h) 4536

Page 47 **Skill Practice**

1. 2, 4, 5, 10 2. factors 3. 3 and 11 4. No.

Page 47 **Exercise 14**

1. (a) Yes (b) Yes (c) No (d) Yes (e) Yes (f) No (g) Yes (h) Yes
   (i) Yes (j) No (k) Yes (l) No

Page 47 **Mental Maths–4**

1. 15987 2. 38790 3. 5938000 4. 31,37 5. 2 6. 4th 7. 21, 28, 35
8. 2, 5 and 7 9. 6 10. 75

Page 51 — Skill Practice

Mixed No.	Improper fraction
1. $3\frac{3}{4}$	$\frac{15}{4}$
2. $1\frac{5}{6}$	$\frac{11}{6}$
3. $2\frac{3}{10}$	$\frac{23}{10}$

Page 52 — Skill Practice

1. (a) $\frac{1}{4}=\frac{2}{8}$ (b) $\frac{3}{7}=\frac{6}{14}=\frac{9}{21}$

2. (a) $\frac{2}{5}=\frac{4}{10}=\frac{6}{15}=\frac{8}{20}$ (b) $\frac{5}{8}=\frac{10}{16}=\frac{15}{24}=\frac{20}{32}$

Page 53 — Skill Practice

1. $\frac{8}{9}$ 2. $\frac{4}{7}$ 3. $\frac{6}{7}$ 4. $\frac{11}{14}$

Page 56 — Skill Practice

1. $<$ 2. $>$ 3. $<$ 4. $>$ 5. $\frac{3}{4}, \frac{1}{2}, \frac{2}{5}$

Page 56 — Skill Practice

1. $\frac{5}{7}$ 2. $\frac{1}{3}$ 3. 1 4. $\frac{1}{3}$ 5. $\frac{1}{5}$ 6. $\frac{2}{5}$

Page 57 — Exercise 15

1. $\frac{8}{11}$ 2. 1 3. $\frac{2}{3}$ 4. $\frac{3}{4}$ 5. $\frac{3}{13}$ 6. $\frac{1}{3}$

7. $\frac{1}{2}$ 8. $\frac{5}{21}$ 9. $\frac{13}{17}$ 10. $1\frac{3}{14}$ 11. $\frac{9}{22}$ 12. $\frac{1}{5}$

13. $\frac{4}{7}$ 14. $\frac{7}{15}$

Page 59 — Skill Practice

1. $\frac{2}{3}$ 2. $\frac{17}{30}$ 3. $\frac{1}{30}$ 4. $\frac{5}{36}$

Page 59 — Exercise 16

1. $\frac{1}{2}$ 2. $\frac{5}{6}$ 3. $\frac{8}{9}$ 4. $\frac{11}{12}$ 5. $\frac{13}{21}$ 6. $\frac{14}{15}$

7. $\frac{29}{35}$ 8. $1\frac{5}{9}$ 9. $1\frac{17}{24}$ 10. $1\frac{31}{40}$ 11. $1\frac{17}{28}$ 12. $1\frac{11}{30}$

13. $\frac{101}{144}$ 14. $\frac{11}{12}$ 15. $1\frac{4}{5}$ 16. $\frac{29}{30}$ 17. $2\frac{1}{6}$ 18. $1\frac{13}{15}$

19. $1\frac{23}{28}$ 20. $\frac{1}{12}$ 21. $\frac{5}{24}$ 22. $\frac{11}{24}$ 23. $\frac{1}{10}$ 24. $\frac{7}{26}$

25. $\frac{7}{20}$ 26. $\frac{2}{27}$ 27. $\frac{13}{36}$ 28. $\frac{11}{48}$ 29. $\frac{1}{42}$ 30. $\frac{1}{12}$

31. $\frac{13}{72}$ 32. $\frac{1}{5}$ 33. $\frac{1}{18}$ 34. 0

Page 61 **Skill Practice**

1. $4\frac{2}{3}$ 2. $10\frac{3}{5}$ 3. $6\frac{1}{2}$ 4. $4\frac{1}{4}$

Page 62 **Skill Practice**

1. $3\frac{5}{6}$ 2. $9\frac{7}{12}$ 3. $3\frac{11}{12}$ 4. $7\frac{1}{10}$

Page 62 **Exercise 17**

1. 8 2. 7 3. $5\frac{4}{5}$ 4. $23\frac{4}{7}$ 5. 9 6. 20

7. $325\frac{3}{5}$ 8. $23\frac{13}{19}$ 9. 30 10. $37\frac{1}{3}$ 11. $6\frac{1}{2}$ 12. $6\frac{1}{10}$

13. $7\frac{3}{14}$ 14. $6\frac{17}{24}$ 15. $3\frac{37}{100}$ 16. $9\frac{37}{40}$ 17. $7\frac{13}{15}$ 18. $3\frac{11}{18}$

Page 64 **Skill Practice**

1. $2\frac{2}{7}$ 2. $3\frac{3}{13}$ 3. $1\frac{1}{4}$ 4. $3\frac{1}{12}$

Page 65 **Skill Practice**

1. $\frac{3}{5}$ 2. $\frac{13}{21}$ 3. $2\frac{3}{7}$ 4. $6\frac{5}{8}$ 5. $2\frac{2}{7}$

Page 65 **Exercise 18**

1. $\frac{1}{4}$ 2. $\frac{5}{12}$ 3. $\frac{22}{31}$ 4. $1\frac{4}{7}$ 5. $2\frac{7}{12}$ 6. $4\frac{1}{8}$

7. $5\frac{2}{3}$ 8. $1\frac{3}{10}$ 9. $2\frac{1}{5}$ 10. $3\frac{7}{13}$ 11. $1\frac{1}{3}$ 12. $5\frac{2}{9}$

13. $5\frac{1}{13}$ 14. $\frac{3}{8}$ 15. $33\frac{1}{3}$

Page 66 — Skill Practice

1. $\frac{2}{7}$ 2. $\frac{3}{22}$ 3. $3\frac{1}{8}$ 4. $11\frac{1}{2}$ 5. 0 6. $\frac{1}{3}$ 7. $\frac{1}{6}$

Page 68 — Exercise 19

1. $1\frac{5}{8}$ kg 2. one-sixth 3. $3\frac{1}{4}$ kg 4. No 5. $\frac{3}{4}$ 6. $\frac{3}{10}$ kilometre

7. $2\frac{5}{6}$ m 8. $1\frac{5}{6}$ litres 9. $\frac{1}{6}$ km 10. $\frac{7}{24}$ of an hour 11. $72\frac{3}{10}$ kg 12. $\frac{1}{5}$

Page 70 — Skill Practice

	$\frac{2}{7}\times\frac{3}{5}$	$\frac{1}{2}\times\frac{5}{6}$	$\frac{7}{9}\times\frac{2}{3}$	$\frac{5}{11}\times\frac{5}{11}$	$\frac{3}{4}\times\frac{3}{4}$
Product of numerators	6	5	14	25	9
Product of denominator	35	12	27	121	16
fractions Product of	$\frac{6}{35}$	$\frac{5}{12}$	$\frac{14}{27}$	$\frac{25}{121}$	$\frac{9}{16}$

Page 71 — Exercise 20

1. (a) $\frac{1}{2}$ (b) $\frac{2}{3}$ (c) $2\frac{1}{3}$ (d) $\frac{3}{20}$ (e) $\frac{2}{5}$ (f) $\frac{12}{25}$

(g) $\frac{3}{5}$ (h) $\frac{34}{315}$ (i) $\frac{5}{18}$ (J) $\frac{1}{3}$

2. (a) $\frac{5}{6}$ (b) $\frac{3}{22}$ (c) $\frac{4}{5}$ 3. (a) 6 (b) $91\frac{2}{3}$ (c) 25

(d) $7\frac{1}{2}$ (e) $6\frac{2}{9}$ (f) $7\frac{1}{2}$ (g) $5\frac{1}{3}$ (h) 40

4. (a) $84\frac{3}{20}$ (b) $4\frac{2}{3}$ (c) 5

Page 72 — Exercise 21

1. $6\frac{2}{3}$ 2. 1 3. $4\frac{1}{5}$ 4. $11\frac{1}{4}$ 5. $11\frac{1}{3}$ 6. $3\frac{1}{3}$

7. $8\frac{2}{3}$ 8. $14\frac{1}{4}$ 9. $5\frac{1}{4}$ 10. $2\frac{2}{3}$ 11. 12 12. 35

page 73 — Exercise 22

1. $\frac{1}{2}$
2. She drank $\frac{1}{16}l$; $\frac{9}{16}l$ of juice is left.
3. $\frac{6}{11}$ of an hour.
4. Rs. $134\frac{3}{8}$
5. Rs. 459 [(Easy and Qukck way : Total contributing = Rs (15×30) + Rs $15\times\frac{3}{5}$]
6. $1\frac{1}{2}$m
7. $\frac{4}{7}$kg
8. 33 m

Page 74 — Skill Practice

Number	Reciprocal	Number	Reciprocal	Number	Reciprocal
$\frac{2}{3}$	$\frac{3}{2}$	6	$\frac{1}{6}$	$\frac{11}{9}$	$\frac{9}{11}$
$\frac{5}{6}$	$\frac{6}{5}$	$\frac{1}{7}$	7	$\frac{1}{9}$	9
4	$\frac{1}{4}$	$\frac{22}{7}$	$\frac{7}{22}$		

Page 76 — Skill Practice

1. 32
2. $\frac{1}{12}$
3. $2\frac{2}{3}$
4. $1\frac{19}{25}$

Page 76 — Exercise 23

1. 12
2. 10
3. 16
4. 6
5. $\frac{1}{9}$
6. $\frac{2}{5}$
7. $\frac{3}{32}$
8. $1\frac{1}{35}$
9. 1
10. $\frac{2}{9}$
11. $2\frac{1}{4}$
12. $\frac{27}{392}$
13. 3
14. $\frac{5}{24}$
15. $\frac{3}{4}$
16. $\frac{7}{31}$
17. 45
18. $\frac{2}{15}$
19. 0
20. 3
21. $1\frac{3}{5}$
22. $1\frac{1}{8}$
23. $6\frac{1}{3}$
24. $7\frac{1}{2}$
25. $1\frac{7}{8}$
26. $1\frac{5}{9}$
27. $1\frac{1}{2}$
28. $2\frac{1}{2}$

Page 77 — Exercise 24

1. $\frac{1}{9}$
2. $\frac{1}{14}l$
3. 6
4. $\frac{4}{5}$ of an hour
5. 5 days
6. 7 days
7. 6 km
8. $1\frac{3}{10}$cm
9. 18 sheets
10. 26 minutes

Page 79 — Skill Practice

1. 4
2. 24
3. 12 kg
4. 10 days

Page 79

Exercise 25

1. (a) 8 (b) 9 (c) 14 (d) 12 (e) 36 (f) 9 (g) 9 (h) 150
2. (a) 16 hours (b) 10 minutes (c) 8 km (d) 2 days
   (e) 35 seconds (f) 55 cm
3. 20 apples
4. 40 stamps in his album; Gane 8 stamps to his friend.
5. Ayushree had 40 story books, 20 other books.
6. 210 km.

Page 80

Mental Maths–5

1. $\frac{1}{8}, \frac{1}{6}, \frac{1}{5}, \frac{1}{4}, \frac{1}{3}, \frac{1}{2}$
2. $\frac{2}{7} = \frac{6}{\boxed{21}} = \frac{\boxed{10}}{35} = \frac{12}{\boxed{42}}$
3. 21
4. $\frac{1}{20}$
5. $\frac{3}{4}$
6. 75
7. 5
8. $13\frac{1}{5}$
9. False
10. $\frac{2}{3}$
11. 44
12. $\frac{1}{4}$
13. $\frac{2}{5}$
14. 5
15. $\frac{4}{3}$
16. $\frac{2}{15}$
17. 15 girls
18. $5\frac{3}{5}$

Page 85

Skill Practice

2. $\frac{65}{100} = 0.65$ zero point six five
3. $\frac{435}{1000} = 0.435$ zero point four three five
4. $\frac{237}{100} = 2.37$ Two point three seven
5. $\frac{153}{10} = 15.3$ Fifteen point three
6. $145\frac{695}{1000} = 145.695$ one hundred forty five point six nine five

Page 86

Skill Practice

I 2. 6 tenths 3. 5 thousandths 4. 1 tens 5. 5 ones
6. 8 hundreths

II 1. 425.347 2. 8607.321

III 1. 100 + 70 + 8 + 0.3 + 0.04 + 0.005

2. $20 + 6 + \frac{8}{10} + \frac{9}{1000}$

Page 87

Skill Practice

1. 0.09 2. 4.1 3. 0.027 4. 2.73 5. 15.3 6. 1.704
7. 0.75 8. 0.80 9. 0.35 10. 0.375 11. 0.08 12. 0.38

13. $\frac{4}{5}$ 14. $\frac{9}{20}$ 15. $1\frac{1}{5}$ 16. $3\frac{1}{2}$ 17. $2\frac{7}{10}$ 18. $3\frac{2}{5}$

Page 88 **Skill Practice**

1. 6.800, 7.830, 12.040 2. 0.900, 12.825, 1.500, 6.830 3. 6.500, 6.080, 60.930, 61.405

Page 89 **Skill Practice**

1. (a) $>$ (b) $=$ (c) $<$ (d) $>$ (e) $<$ (f) $>$ (g) $<$ (h) $<$ (i) $<$

Page 89 **Exercise 26**

1. (a) 5.13, 5.3, 5.31 (b) 0.72, 7, 7.2 (c) 0.103, 1.03, 10.002
   (d) 60.000, 60.060, 60.60 (e) 5.330, 5.331, 6.033, 6.303
   (f) 0.009, 0.109, 0.901, 0.910, (g) 0.334, 0.343, 0.434, 0.443
2. 16.02, 15.99, 15.87, 15.78, 15.7, 15.69, 15.49, 15.1
3. A $\rightarrow$ 1.21 m, B $\rightarrow$ 1.20 m, C $\rightarrow$ 1.18 m, D $\rightarrow$ 1.02 m, E $\rightarrow$ 0.87 m.
4. Asia, Africa, Europe, North America, South America, Antarctica
5. Justin Gatlin, Francis Obikwelm, Shaw Garford, Asafa Powell; Justin Gatlin
6. (a) T (b) F (c) T (d) F (e) F (f) F

Page 92 **Skill Practice**

1. 46.398 2. 133.835 3. 5.876 4. 54.702

Page 92 **Exercise 27**

1. 17.01 2. 45.28 3. 43.812 4. 524.54 5. 185.526
6. 50.418 7. 33.431 8. 724.03 9. 65.485 10. 102.714
11. 807.088 12. 1.9094 13. Rs. 221.03 14. 128.03 cm 15. Rs. 223.87
16. 33.19 cm

Page 92 **Exercise 28**

1. 4.575 2. 14.115 3. 7.957 4. 252.82 5. 120.716 6. 4.005
7. 41.87 m 8. 104.04 ml 9. Rs. 119.92 10. 3.767 11. Rs. 10.63
12. 60.48 m 13. 4.246 14. 659.829

Page 93 **Mental Maths–6**

1. (a) 12.7 (b) 370.52
2. $5\frac{3}{5}$ 3. 30 4. 8.4 5. (a) 4.7 (b) 9.35 (c) 84.01
6. $\frac{1}{3}$ 7. 100 8. 0.007 9. 49.99 , Forty nine point nine-nine

Page 95 **Skill Practice**

1.

Decimal	× 10	× 100	× 1000
2.7	27	270	2700
3.06	30.6	306	3060
0.519	5.19	51.9	519
0.23	2.3	23	230
18.25	182.5	1825	18250
3.007	30.07	300.7	3007
16.73	167.3	1673	16730
148.954	1489.54	14895.4	148954
24.06	240.6	2406	24060
100.035	1000.35	10003.5	100035

Page 95 **Exercise 29**

I. **1.** 10 **2.** 100 **3.** 100 **4.** 1000 **5.** 1000 **6.** 1000
7. 100 **8.** 10

III. **1.** × 10 **2.** × 10 **3.** ×10 **4.** × 100 **5.** × 100 **6.** ×1000
7. ×1000 **8.** ×1000

Page 96 **Skill Practice**

1. 12.46 **2.** 64.265 **3.** 49.92 **4.** 274.275

Page 97 **Exercise 30**

1. 1.06 **2.** 4.25 **3.** 0.24 **4.** 3.6 **5.** 14.01 **6.** 37.1
7. 47.44 **8.** 168.42 **9.** 595.84 **10.** 132 **11.** 495.75 **12.** 283.8
13. 47993.4 **14.** 375.525 **15.** 2609.502 **16.** 64350.56
17. **(a)** 35.202 **(b)** 352.02 **(c)** 3520.2 **(d)** 3.5202
18. Rs 1133.50 **19.** 26.28 kg **20.** Rs 27510.10

Page 98 **Skill Practice**

1. 0.14 **2.** 11.312 **3.** 630 **4.** 2.4

Page 98 **Exercise 31**

1. 0.24 **2.** 0.72 **3.** 0.0042 **4.** 0.791 **5.** 16.84 **6.** 0.0576
7. 0.0052 **8.** 19.125 **9.** 493.83 **10.** 609.62 **11.** 0.02826
12. 20.23 **13.** 0.312 **14.** 16.8705 **15.** 14.801 **16.** 1134.886
17. 422.51 **18.** 0.008 **19.** 0.0024 **20.** 0.000027 **21.** 0.00000006
22. 0.000000024 **23.** 15920 **24.** 455840
25. **(a)** 51.33 **(b)** 513.3 **(c)** 51.33 **(d)** 5.133 **(e)** 0.5133
(f) 5.133 **(g)** 0.0005133

26. 66.141 (a) 661.41 (b) 6614.1 (c) 66141 (d) 6.6141 (e) 0.66141
(f) 0.066141

27. 139.4 km 28. 76.305 kg

Page 100 **Skill Practice**

Number	÷10	÷100	÷1000
5	0.5	0.05	0.005
18	1.8	0.18	0.018
389	38.9	3.89	0.389
0.8	0.08	0.008	0.0008
0.92	0.092	0.0092	0.00092
4.532	0.4532	0.04532	0.004532
87.1	8.71	0.871	0.0871
93.85	9.385	0.9385	0.09385
105.293	10.5293	1.05293	0.105293
60.07	6.007	0.6007	0.06007

Page 100 **Exercise 32**

I. 1. 10 2. 10 3. 100 4. 10 5. 100 6. 1000
7. 1000 8. 100

II. 2.4 → 0.024, 16 → 0.16, 421.5 → 4.215, 645 → 6.45, 0.08 → 0.0008

III. 1. ÷ 100 2. ÷ 10 3. ÷10 4. ÷ 1000 5. ÷ 100 6. ÷ 1000
7. ÷ 100 8. ÷ 100

Page 103 **Skill Practice**

1. 3.37 2. 5.32 3. 0.89 4. 1.565 5. 2.37

Page 103 **Exercise 33**

1. 4.63 2. 5.9 3. 5.7 4. 25.58 5. 4.3 6. 0.57
7. 0.032 8. 0.0043 9. 0.00043 10. 2.115 11. 1.91 12. 6.6
13. 0.203 14. 1.3 15. 0.562 16. 4.52 17. 2.175 18. 15.325
19. 5.874 20. 7.51 21. 0.9 22. 0.09 23. 0.0205 24. 0.3
25. 0.21 26. 0.032 26. 0.007 28. 0.02 29. 0.049 30. 9.0545
31. 1.3524 32. 0.15612 33. 0.25 *l* 34. Rs. 9.54

Page 105 **Skill Practice**

1. 3. 2. 27 3. 120 4. 50

Page 106 **Exercise 34**

1. 7 2. 4 3. 8 4. 54 5. 540 6. 1.6 7. 0.63
8. 600 9. 4 10. 125 11. 890 12. 630 13. 14507 14. 0.87

15. 14.6 16. 15100 17. 50 18. 40 19. 40 20. 16000
21. 0.027 cm 22. Rs. 2.43 per hour 23. 38 24. 17 25. 34

Page 107 **Skill Practice**

1. 0.80 2. 0.375 3. 1.75 4. 0.55 5. 5.3125

Page 108 **Skill Practice**

1. 0.45 m 2. 0.06 m 3. 2.96 m 4. 0.275 km 5. 0.043 km 6. 0.009 km
7. 4.05 m 8. 19.46 m 9. 2.525 km 10. 18.075 km 11. 600.005 km
12. 16.08 m 13. 0.089 kg 14. 0.925 kg 15. 3.006 kg 16. 21.054 kg 17. 14.425 kg
18. 1.700 kg 19. 0.425 *l* 20. 0.065 *l* 21. 1.125 *l* 22. 5.005 *l* 23. 10.025 *l*
24. 16.720 *l* 25. 3.05 *l* 26. 4.19 m 27. 2.455 kg 28. 4 km

Page 111 **Skill Practice**

1. 10 2. 44 3. 3 4. 13 5. 14 6. 101
7. 1000 8. 110 9. 63 10. 3.3 11. 4.9 12. 9.5
13. 12.5 14. 111.2 15. 17.2 16. 2 17. 29.4 18. 100
19. 0.32 20. 4.79 21. 2.85 22. 16.56 23. 0.01 24. 99.99
25. 100.00 26. 6.75 27. 6.00

Page 112 **Exercise 35**

I. 1. 0.468 2. 0.781 3. 0.003 4. 0.582 5. 0.795 6. 18.497
7. 21.001 8. 99 9. 41.563 10. 927.488
II. 1. 17.274 2. 6.83 3. 4.638 4. (a) 130 (b) 130.4 5. (a) 1.8 (b) 2.1
III 1. Rs. 5 2. 9 kg 3. 2 kg 4. Rs. 12 5. 4 m 6. 9 kg
7. 7 cm 8. 5*l* 9. 16 *l* 10. 6 cm

Page 115 **Skill Practice**

A. 1. Point P 2. Ray AB 3. Line segment MN

B. 1. 

2. 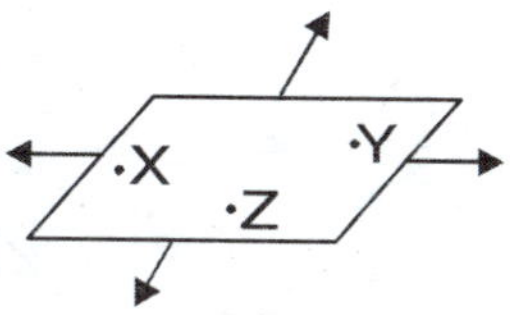

3. 

Page 116 **Exercise 36**

1. (c)
2. (a) Line, $\overleftrightarrow{XY}$ (b) Line segment, $\overline{AB}$ (c) Ray, $\overrightarrow{EF}$ (d) Line segment, $\overline{PQ}$ (e) Ray, $\overrightarrow{CD}$
(f) Line, $\overleftrightarrow{LM}$
3. (a) F (b) T (c) T (d) F (e) F
4. (a) 8; AB, BC, CD, DE, EF, FG, GH, HA
(b) 12; AB, BC, CD, DE, EF, FG, GH, HI, IJ, JK, KL, LA
(c) 6; AF, FE, ED, DC, CB, BA

5. Rays — Line Segments

	Rays	Line Segments
(a)	$\overrightarrow{BA}$, $\overrightarrow{DE}$	$\overline{BC}$, $\overline{CD}$
(b)	$\overrightarrow{PS}$ $\overrightarrow{PT}$, $\overrightarrow{QS}$, $\overrightarrow{RT}$	PR, PQ, QR
(c)	$\overrightarrow{BA}$, $\overrightarrow{AB}$, $\overrightarrow{CD}$, $\overrightarrow{DC}$	EF

Page 117 — Skill Practice

1. **(a)** Point P **(b)** ∠QPR **2.** **(a)** ∠MLN *or* ∠NLM **(b)** Point L **(c)** $\overrightarrow{LM}$ and $\overrightarrow{LN}$

3. **(a)** ∠BDC, ∠CDB, ∠EDC, ∠CDE **(b)** Vertex-Point B /Arms – AB and BC **(c)** ∠BDA

4. **(a)** ∠YVZ **(b)** G, S, H

Page 119 — Skill Practice

1. 30° **2.** 3600° **3.** $\frac{1}{2}, \frac{1}{4}, \frac{1}{6}, \frac{1}{8}, \frac{1}{360}$ **4.** 90° **5.** **(a)** 72° **(b)** 360° **(c)** 144° **(d)** 216°

Page 123 — Skill Practice

1. 140° **2.** 50° **3.** 80° **4.** 90° **5.** 50° **6.** 140° **7.** 10° **8.** 90°

9. 70° **10.** 110 **11.** 180° **12.** ∠KBC and ∠ABI, ∠KBF and ∠ABF

Page 125 — Skill Practice

	Type of angle	Measure of angle	Name of angle (1)	Name of angle (2)
1.	Right angle	90°	∠AOB	∠BOA
2.	Obtuse angle	120°	∠RST	∠TSR
3.	Acute angle	45°	∠PQR	∠RQP
4.	Acute angle	25°	∠LNM	∠MNL
5.	Obtuse angle	130°	∠ZYX	∠XYZ
6.	Right angle	90°	∠DEF	∠FED

Page 126 — Exercise 38

1. **(i)** Acute angle **(ii)** Right angle **(iii)** Obtuse angle **(iv)** Straight angle **(v)** Reflex angle

2. **(a)** **(i)** Acute **(ii)** Right **(iii)** obtuse **(iv)** Reflex **(v)** acute

(b) **(ii)** and **(v)** are fixed, **(i)**, **(iii)** and **(iv)** can change size

3. **(a)** 90° **(b)** 180° **(c)** 360° **4.** 2 turns

5. **(a)** Acute **(b)** Obtuse **(c)** Right **(d)** Straight **(e)** Obtuse **(f)** Acute

6. **(a)** 1 **(b)** 2 **(c)** 3 **(d)** 4 **(e)** 1 **(f)** 2 **(g)** 4 **(h)** 1 **(i)** 1

8. No, because the sum of two acute angles is less than 180°.

10. **(a)** Acute **(b)** Acute **(c)** Obtuse **(d)** Acute **(e)** Obtuse **(f)** Straight.

11. **(a)** A, D, E, F, G, H, I, J, K, L, M, N, P, R, T, V, W, X, Y, Z
(b) **(i)** A, K, M, N, R, V, W, X, Y, Z **(ii)** A, K, Y **(iii)** E, F, H, L, T

Page 129 — Skill Practice

1. AB || DC, AD || BC, AB ⊥ AD, AB ⊥ BC, DC ⊥ AD, DC ⊥ BC
2. PQ || SR, SP ⊥ PQ, PS ⊥ SR **3.** RS ⊥ RT **4.** LM || ON, MN || LO
5. PQ || SR, QR || PS, PQ ⊥ PS, PQ ⊥ QR, PS ⊥ SR, QR ⊥ RS
6. AB || ED, BC || FE, CD || AF
7. **(a)** F **(b)** T **(c)** T **(d)** F **(e)** T **(f)** T

Page 132 — Skill Practice

I. (2) and (5) **II.** **(a)** LM, MN, NL **(b)** ∠LMN, ∠MNL, ∠NLM or simply, ∠L, ∠M, ∠N.
(c) L, M, N **(d)** Δ LMN

Page 133 — Skill Practice

3. **(a)** Vertices – A, B, C; Sides – AB, BC, AC; Angles – ∠ABC, ∠BAC, ∠BCA or ∠A, ∠B *and* ∠C
(b) Vertices – P, Q, R; Sides – PQ, QR, RP; Angles – ∠P, ∠Q and ∠R
(c) Vertices – X, Y, Z; Sides – XY, YZ, XZ; Angles – ∠X, ∠Y and ∠Z
4. **(a)** Scalene **(b)** Isosceles **(c)** Equilateral **(d)** Isosceles **(e)** Scalene **(f)** Equilateral **(g)** Isosceles
5. **(a)** Acute Triangle **(b)** Obtuse Triangle **(c)** Right Triangle
6. **(a)** ΔABC **(b)** ΔADE **(c)** ΔABD, ΔABM, ΔABE, ΔAEC, ΔAMC, ΔADC, ΔADM, ΔAEM
(d) ΔABE, ΔADE, ΔADC **(e)** ΔABD, ΔAEC **(f)** ΔAMD, ΔAMB, ΔAME, ΔAMC

Page 136 — Skill Practice

1. p = 90° **2.** f = 56°, h = 56°, g = 34° **3.** a = 65°, b = 40°

Page 136 — Exercise 39

1. **(a)** 10° **(b)** 70° **(c)** 42° **(d)** 30° **(e)** 40° **(f)** 64°
2. **(a)** 90° **(b)** 80° **(c)** 90° **(d)** 100° **(e)** 90° **(f)** 70°
Triangles (a), (c) and (e) are right triangles.
3. 110°, **4.** 60° **5.** 23° **6.** 62° **7.** No **8.** No
9. No **10.** Yes

Page 140 — Skill Practice

I. **1.** Reflected **2.** Rotated **3.** Rotated **4.** Reflected
5. Rotated **6.** Reflected

II. **1.** 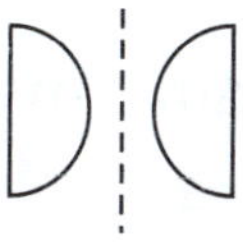 **2.** 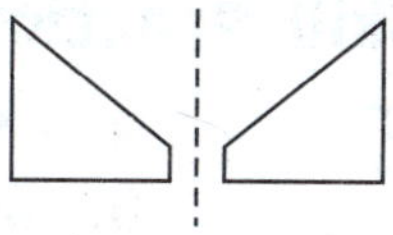 **3.** 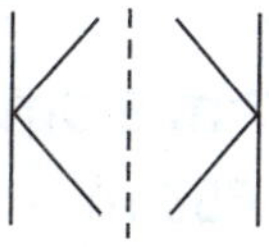

III **1.** 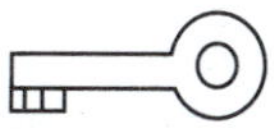 **2.**  **3.** 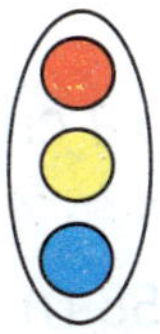 **4.** 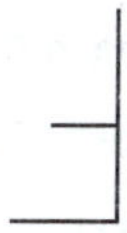

Page 142 — Skill Practice

I. 1. No 2. Yes 3. No 4. No 5. Yes 6. Yes

II. 1. Yes 2. Yes 3. Yes 4. No 5. Yes 6. No 7. Yes
8. Yes

Page 144 — Skill Practice

I. 1. – E 2. –D 3. –B 4. –A 5. –C

Page 145 — Skill Practice

I. 1. diameter 2. Chord 3. Radius 4. Chord 5. Radius
6. Radius 7. Centre 8. Radius

II. 1. 16 cm 2. 8.6 cm 3. 9 cm 4. 8.2 cm 5. 5 m 6. 3.5 m
7. 22.50 cm 8. 6.25 m

Page 147 — Skill Practice

I. 1. 12.56 cm 2. 31.4 cm 3. 62.8 cm 4. 6.28 cm 5. 28.26 cm 6. 75.36 cm

II. 1. 7cm 2. 3 cm 3. 40 cm

III. 1. 5 cm 2. 4 cm 3. 10 cm

Page 148 — Mental Maths–7

1. (a) Acute (b) Obtuse (c) Right (d) Straight
2. 720° 3. (a) ∠ABC, ∠GHI (b) ∠DEF, ∠LKJ
4. Triangle 5. isosceles 6. True 7. 5 cm
8. (a) $\overline{AB}, \overline{CD}$; AC,BD (b) AB ⊥ AC, BD ⊥ CD (answers may vary)
9. (d) 10. diameter

Page 153 — Skill Practice

1. 8000 m 2. 126 hm 3. 165 m 4. 3200 cm 5. 450 cm 6. 9000 mm
7. 18400 m 8. 7m 9. 1.8 m 10. 0.2 m 11. 34 hm 12. 5 km
13. 2.5 dam 14. 6.750 km

Page 155 — Skill Practice

1. 10 2. 100 3. 1000 4. 100 5. $\frac{1}{1000}$ 6. $\frac{1}{100}$ 7. $\frac{1}{1000}$
8. $\frac{1}{10}$ 9. 1180 10. 15000 11. 6250 12. 18 13. 0.153 14. 9.25
15. 2500 16. 25 17. 1.4 18. 18.5 19. 1.9 20. 0.419

Page 156 — Skill Practice

1. 412 m 92 cm 2. 27 m 8 dm 9 cm 3 mm 3. 6 km 8 hm 9 dam 3 m
4. 12.45 m 5. 2.653 m 6. 160 .895 km

Page 156 — Exercise 40

1. (a) 382 m 14 cm (b) 15 m 8 cm (c) 895 m 40 cm (d) 500 m (e) 69 cm
(f) 3 cm
2. (a) 5.13 m (b) 721.58 m (c) 29.17 m (d) 173.02 m

(e) 2.01 m (f) 0.38 m (g) 0.04 m (h) 0.5 m (i) 9.38 m

3. (a) 549 m 3 dm 2 cm 1 mm (b) 7 dm 3 mm (c) 605 m 2 dm 3 mm
(d) 9 cm 2 mm (e) 7 dm 5 cm 9 mm (f) 354 m 1 cm 8 mm (g) 8 mm
(h) 935 m 1 mm (i) 300 m

4. (a) 327.568 m (b) 0.004 m (c) 723.005 m (d) 0.407 m (e) 402.079 m
(f) 0.314 m (g) 0.503 m

5. (a) 430 km 7 hm 5 dam 6 m (b) 8 hm 7 m (c) 902 km 4 hm 5 m
(d) 7 dam 8 m (e) 6 hm 5 dam 2 m (f) 230 km 4 dam 8 m (g) 8 m
(h) 603 km 1 m (i) 600 km

6. (a) 540.736 km (b) 0.005 km (c) 857.002 km (d) 0.603 km
(e) 703.085 km (f) 839.401 km (g) 0.659 km (h) 0.076 km

Page 159 — Skill Practice

1. 1000 2. 1000 3. 10 4. 1000 5. 100 6. 10 7. $\frac{1}{1000}$

8. $\frac{1}{10}$ 9. $\frac{1}{10}$ 10. $\frac{1}{100}$

11.

kg	hg	dag	g	dg	cg	mg
13.7	137	—	13700	137000	—	13700000
—	45	450	4500	—	450000	4500000
0.02178	—	2.178	21.78	217.8	2178	—
0.7002	7.002	70.02	700.2	—	70020	—
0.08	—	8	—	800	8000	80,000
0.0014572	0.014572	—	1.4572	—	145.72	1457.2
1.5	15	150	—	15000	—	1500000

12. 96000 grams 13. 3.25 kg 14. 0.750 grams

15. (a) 451 kg 3 hg 7 dag 8 g (b) 0 kg 6 hg 0 dag 2 g (c) 0 kg 0 hg 5 dag 6 g
(d) 0 kg 0 hg 0 dag 2 g

16. (a) 418.259 kg (b) 0.487 kg (c) 23.008 g (d) 0.042 kg

Page 162 — Skill Practice

1. F 2. T 3. F 4. T 5. T 6. T 7. F 8. T
9. 8000 10. 1.5 11. 1600 12. 1.875 13. 4.85 14. 12 15. 6254 16. 15000
17. 70 18. 24.75 19. 25 20. 1.94 21. 30,000 22. 12 *l* 23. 0.5 *l*

24. (a) 529 *l* 1 dl 8 cl 3 *ml* (b) 3dl 0 cl 8 *ml* (c) 6 cl 2 *ml*

25. (a) 12.452 *l* (b) 40.069 *l* (c) 0.19 *l* (d) 215.005 *l*

Page 164 — Exercise 41

1. 17 m 7 dm 2. 32 cm 1 mm 3. 33 dm 3 cm 4. 138 km 262 m 5. 10 cm 9 mm
6. 11 m 85 cm 7. 76 km 750 m 8. 6 dm 4 cm 9. 101 kg 120 g 10. 12 g 273 mg
11. 137 kg 857 g 12. 3 g 730 mg 13. 175 *l* 5 *ml* 14. 28 kl 15 *l* 15. 22 *kl* 879 *ml*
16. 109 *l* 900 *ml* 17. 0.875 m 18. 34.23 m 19. 78 mm

20. 1 km 120 m; 380 m **21.** 90 kg 350 g **22.** 1 kg 185 g **23.** 9 kg 652 g
24. **(a)** 150 kg **(b)** 150000 grams; 375 g **25.** 3 *l* 550 *ml*
26. 40 glasses **27.** **(a)** 12 weeks **(b)** 7 weeks

Page 167 — Skill Practice

I. **1.** 15 cm **2.** 1 m **3.** 5 kg **4.** 70 g **5.** 5 *ml* **6.** 40 km
7. 30 mm **8.** 500 mg **9.** 400 g **10.** 40 *l* **11.** 119 m
II. **1.** mm **2.** km **3.** g **4.** *ml* **5.** *l* **6.** mg **7.** t **8.** *l* **9.** kg **10.** m **11.** mm
12. kl

Page 169 — Exercise 42

1. 250 mg **2.** 60 *cl* **3.** 800 g **4.** 750 m **5.** 20 dal **6.** 500 mg **7.** 4 *hl*
8. 80 cm **9.** 75 *ml* **10.** 50 m **11.** 50 g **12.** 1 dl **13.** 25 cm
14. 600 *ml* **15.** 494 g **16.** 348 *ml* **17.** **(a)** 1450 m **(b)** 1.45 km **18.** 250 g
19. 160 g **20.** 525 mm

Page 170 — Mental Maths–8

1. 1700 cm **2.** right, 100 **3.** 0.658 m **4.** 100 milligrams **5.** True
6. 90 hecto litres = 9 kilolitres **7.** 2 cm **8.** cm
9. **(b)** **10.** **(a)** hm **(b)** dg **(c)** dm **(d)** *cl*

Page 174 — Skill Practice

1. 300 s **2.** 960 min **3.** 6 min **4.** 11 h **5.** 282 s **6.** 165 min

Page 174 — Exercise 43

1. **(a)** 360 s **(b)** 900 s **(c)** 36 s **(d)** 7200 s **(e)** 436 s **(f)** 635 s
(g) 44 s **(h)** 34 s
2. **(a)** 480 min **(b)** 30 min **(c)** 45 min **(d)** 325 min **(e)** 552 min
(f) 570 min **(g)** 445 min **(h)** 532 min
3. **(a)** 8 min **(b)** 12 min **(c)** 2 min **(d)** 2 min 25 s **(e)** 5 min 30 s
(f) 15 min **(g)** 19 min **(h)** 31 min 40 s
4. **(a)** 3 h **(b)** 7h **(c)** 1h 30 min **(d)** 3h 25 min **(e)** 2h 15 min
(f) 5 h **(g)** 17 h 30 min

Page 175 — Skill Practice

1. 1 h 40 min **2.** 1 min 40 s **3.** 23 min 37 s **4.** 18 h 25 min
5. 4h 6 min 31 s **6.** 8 h 1 min

Page 176 — Skill Practice

1. 50 min **2.** 3 min 40 s **3.** 2 min 40 s **4.** 2h 30 min **5.** 1 h 56 min 37 s
6. 3h 23 min 43 s

Page 177 — Exercise 44

1. 29 min 7 s **2.** 12 h **3.** 19 h 9 min **4.** 6h 14 min 17 s **5.** 7 h 24 min
6. 1 min 58 s **7.** 55 min **8.** 1 h 22 min **9.** 21 h 58 min 59 s **10.** 2 h 49 min 55 s
11. 1 h 15 min **12.** 3h 3 min **13.** 17 h 47 min 15 s **14.** 38 min 38 s

Page 179 — Exercise 45

1. (a) 11 h 30 min (b) 6 h 40 min (c) 10 h 57 min (d) 12 h 45 min
2. (a) 12:07 p.m. (b) 10:37 p.m. (c) 10:40 a.m. (d) 2205 hours (e) 0020 hours (f) 0300 hours
3. 4 hours 20 min
4. 1715 hours or 5:15 p.m.
5. 2h 59 min

Page 181 — Exercise 46

1. 17 years 5 months
2. 23 years 3 months
3. 4 weeks 5 days
4. 10 years 10 months
5. 3 weeks 6 days
6. 6 years 11 months
7. 1 week 3 days
8. 2 weeks
9. 6 years 8 months
10. 10 years 3 months
11. 31 years 11 months
12. 21 years 9 months
13. 49 years 8 months
14. 8 years 5 months

Page 181 — Mental Maths–9

1. hours
2. 00:00 hours
3. 126 months
4. 6 hours 25 minutes
5. 16 March
6. 150 minutes
7. 9 a.m.
8. 12 midnight
9. 10 years
10. 3600 s
11. 1812 hours
12. 2:25 p.m .
13. 23:25 hours.

Puzzle : Jha

Page 183 — Skill Practice

1. 60° C
2. 10°
3. 90°
4. 60°
5. 30°
6. 30°
7. 42°

Page 184 — Skill Practice

1. (a) 32°C (b) 40°C (c) 20°C (d) 9°C (e) 100°C (f) 0°C
2. (i) (a) (ii) (b) (iii) (c) (iv) a
3. (a) 6°C (b) 40°C (c) 15°C (d) 35°C
4. (a) Agra (b) Gangtok (c) 18°C
   (d) Ahmedabad, Coimbatore, Delhi; Dehra Dun, Bangalore; Chennai, Hyderabad, Jaipur

Page 185 — Exercise 47

1. (a) 38.5°C (b) Indicates normal body temperature
   (c) This is the range of body temperature when alive.
2. (a) 38°C (b) 37.7°C (c) 37.4°C (d) 38.2°C
3. (a) 86°F (b) 50°F (c) 104°F (d) 212°F
4. (a) 100°C (b) 140°C (c) 35°C (d) 235°C
5. (a) 20°C (b) 36.7°C

Page 187 — Mental Maths–10

1. 212°F
2. 0°C
3. 100°C
4. 98.6°F or 37°C
5. $(T°F-32) \times \frac{5}{9}$
6. Normally in Farenheit. Clinical thermometers with Celsius Scale and also used.
7. Yes
8. Cylinder
9. 51 months
10. 15 minutes
11. Saturday
12. 29

Page 188 **Try This**

1. 21°C 2. 27°C 3. 32°C 4. 35°C 5. 36°C

Page 190 **Skill Practice**

2. 40% 3. 11% 4. 25% 5. 43% 6. 58% 7. 65% 8. 100%

Page 190 **Skill Practice**

1. $\frac{7}{100}$ 2. $\frac{19}{100}$ 3. $\frac{63}{100}$ 4. $\frac{1}{50}$ 5. $\frac{1}{10}$ 6. $\frac{1}{4}$ 7. $\frac{8}{25}$ 8. $\frac{27}{50}$ 9. $\frac{19}{20}$

Page 191 **Skill Practice**

2. 0.05 3. 0.01 4. 0.75 5. 0.9 6. 1.25 7. 0.062 8. 0.283

Page 192 **Skill Practice**

1. (b) 72% (c) 6% (d) 13.8% (e) 180% (f) 100%
2. (c) 3. (c) 4. (b) 5. (d)
6.

Fraction	Decimal	Per cent
(b) $\frac{1}{4}$	0.25	25%
(c) $\frac{1}{2}$	0.5	50%
(d) $\frac{3}{4}$	0.75	75%

7. (a) hundred (b) 48 per hundred (c) 69% (d) per cent

Page 194 **Skill Practice**

2. 50% 3. 2% 4. 96% 5. 20% 6. 50%

Page 194 **Skill Practice**

2. 240 3. 200 4. 300 g

Page 195 **Exercise 48**

1. (a) 60% (b) 20% (c) 20% (d) 20% (e) 25% (f) 50% (g) 25%
2. 80% 3. 75%
4. (a) 36 (b) 11 (c) 300 g (d) 15 s (e) 360 ml (f) 7.2 h (g) 20 cm
   (h) Rs 2000 (i) 984 ml
5. 240 children come by bus.

Page 195 **Mental Maths–11**

1. 600 2. 45% 3. $\frac{1}{4}$ 4. 100% 5. 0.325 6. Re 1 7. 70% 8. $\frac{1}{2}$
9. 20% 10. 40%

Page 197 **Skill Practice**

1. Rs 30 2. Rs 24

Page 198 **Skill Practice**

1. 20 stickers for Rs 60 2. 10 bananas for Rs 15

Page 198 **Exercise 49**

1. Rs 42 2. Rs 6500 3. Rs 451.50 4. Rs 15 5. Rs 10.25 6. Rs 129
7. Rs 102 8. Rs 240 9. 3 kg for Rs 105 10. 5 / pack for Rs 310

Page 200 **Skill Practice**

2. Profit = Rs 38 3. Loss = Rs 13 4. Profit = Rs 2.65

Page 201 **Skill Practice**

1. Rs 425 2. Rs 164.85 3. Rs 720 4. Rs 1092.10

Page 201 **Exercise 50**

1. (a) Profit = Rs 125 (b) Loss = Rs 13.25 (c) Profit = Rs 115.50 (d) Loss = Rs 10.60
2. Profit = Rs 1510 3. Loss = Rs 80 4. Profit = Rs 6640 5. Profit = Rs 7
6. Profit = Rs 1.50 per pen
7.

	Cost price	Selling price	Cartage Octroi, etc.	Profit or Loss
(a)	Rs 170	Rs 200	Rs 10	Rs 20 (Profit)
(b)	Rs 508	Rs 615	Rs 12	Rs 95 (Profit)
(c)	Rs 860	Rs 710	Rs 20	Rs 170 (Loss)
(d)	Rs 325	Rs 310	Rs 15	Rs 30 (Loss)
(e)	Rs 57.23	Rs 56.10	Rs 3.15	Rs 4.28 (Loss)
(f)	Rs 1600	Rs 1780	Rs 18	Rs 162 (Profit)
(g)	Rs 188.80	Rs 217.68	Rs 7.13	Rs 21.75 (Profit)

8. C.P. = Rs 1987 9. S.P. = Rs 370 10. S.P. = Rs 85.15 11. C.P = Rs 2051.

Page 203 **Skill Practice**

1. profit 10% 2. loss 20% 3. profit 20%

Page 204 **Exercise 51**

1. 10% 2. 60% 3. $16\frac{2}{3}\%$ 4. 8% 5. 5% 6. Profit $11\frac{1}{9}\%$

Page 204 **Mental maths–12**

1. Rs 5 2. Rs 5; Rs 75 3. Rs 100 4. Rs 600 5. Re 1; 25%
6. Rs 80 7. Rs 100; 20 8. Rs 300 9. Rs 210 10. 40 cm

Page 205 **Skill Practice**

1. 19.1 cm 2. 24 cm 3. 20 cm

Page 207 **Skill Practice**

2. 18 cm

Page 207 **Exercise 52**

1. (a) 56 cm (b) 24 mm (c) 18.8 m (d) 8.68 km (e) 35.6 cm (f) 0.76 m
2. (a) 26 cm (b) 60 m (c) 19 mm (d) 20 cm (e) 4.4 m (f) 30.6 km
3. 43 m 4. Rs 52 5. 2.60 m; Rs. 20.8 6. 6 m

7. (a) Rectangle = 50 cm; square 24 cm (b) Rectangle by 26 cm
8. 2 rounds
9. (a) Breadth 15 m (b) Length = 9 cm (c) Breadth = 5 mm (d) Length = 8 cm
10. (a) 3 cm (b) 2.6 cm (c) 9 m (d) 1.8 m

Page 211 **Skill Practice**

1. 28 m^2 2. 4.08 cm^2 3. 4.2 cm^2 4. 64 cm^2 5. 2.56 m^2 6. 16 mm^2

Page 211 **Exercise 53**

1. (a) 160 sq cm (b) 140 sq km (c) 39 sq m (d) 125 sq m (e) 1500 mm^2
2. (a) 54 cm^2 (b) 115.5 cm^2 (c) 1.8 sq m (d) 81.801 sq m
3. (a) 12 sq cm (b) 48 sq cm (c) 300 sq cm (d) 588 sq cm (e) 1.4 sq m.
4. (i) 9 sq cm (ii) 25 sq cm (iii) 169 sq cm (iv) 1.44 sq cm
   (v) 6.25 sq cm (vi) 900 sq m (vii) 179.56 sq km (viii) 10000 sq m
   (ix) 147.8656 sq m
5. (a) sq cm (b) sq mm (c) sq m (d) sq km (e) sq m (f) sq mm (g) sq km
   (h) sq m
6. (i) a (ii) b (iii) b 7. 150 8. 575 sq mm. 9. Square, by 137 sq cm
10. Square, by 4 sq cm.
11. (i) yes (ii) Rectangular field, 200 metres more.

Page 213 **Skill Practice**

1. Length = 9 cm 2. Breadth = 12 cm 3. Breadth = 1.7 cm
4. Perimeter = 52 m, Area = 160 sq m. 5. 81 sq. cm
6. Breadth = 4 cm, Perimeter = 18 cm

Page 214 **Exercise 54**

1. (a) 20 cm (b) 45 cm (c) 18 cm (d) 6 cm
2. (a) 11 cm (b) 15 mm (c) 22 m (d) 100 m
3. (a) 130 cm (b) 18 cm (c) 280 m (d) $7\frac{5}{18}$ mm (e) 15.4 cm (f) 1960.166 m
   (g) 1 Km

Page 216 **Skill Practice**

1. (a) 3 cm^3 (b) 4 cm^3 (c) 5 cm^3 (d) 6 cm^3 (e) 10 cm^3 (f) 8 cm^3
   (g) 14 cm^3 (h) 15 cm^3
2. (a) cm, cm^2, cm^3, (b) mm, mm^2, mm^3 (c) m, m^2, m^3.

Page 219 **Skill Practice**

1. 100 cm^3 2. 64m^3 3. 512 cm^3 4. 80 mm^3 5. $\frac{1}{3}m^3$

Page 219

Exercise 55

1. (a) 1440 cu cm (b) 12.6 cu cm (c) 38 cm^3 (d) 1.728 cm^3
2. (a) 24 cu cm (b) 0.6 cu cm (c) 52 cu cm (d) 4.8 cu cm (e) 9 cu m
3. (a) 27 cu cm (b) 125 cu cm (c) 1.331 cu cm (d) 343 cu cm (e) 512000 cu mm (f) 79.507 m^3
4. 1000000 cu cm
5. second box
6. 135 cu dm
7. (i) 135 cu cm (ii) 52 cu m
8. 459 cu m
9. (b), (f)
10.

Cuboid	I	II	III	IV	V	VI
Length	9	9	6	12	4	18
Breadth	4	2	3	3	3	2
Height	1	2	2	1	3	1

11. cube
12. 1,25,000 cu cm
13. (a) 50 m (b) 150 sq m (c) 1200 cum
14. One centimetre denotes length, one square centimetre denotes area and one cubic centimetre denotes volume.
15. 120 m^3
16. (a) H = 3 cm (b) H = 2 cm (c) L = 6 mm (d) 0.6 cm (e) 5.2 mm
17. (a) Volume (b) equal (c) 14 cu cm (d) 1 cu cm (e) $l \times b \times h$
18. (a) m^3 (b) cm^3 (c) cm^3 (d) m^3 (e) mm^3 (f) cm^3 (g) m^3 (h) cm^3

Page 222

Exercise 56

1. (a) 72 m^3, 72000 l (b) 9m^3, 9000 l (c) 8000 cm^3, 8 l (d) 12000 cm^3, 12 l
2. (a) 24000 cm^3 (b) 24 litres (c) Rs. 379.20
3. (a) 4500 cm^3 (b) 4500 ml; 4.5 l
4. (a) (i) 60,000 cm^3 (ii) 60 litres (b) Rs. 1770
5. (a) 1000000 cm^3 (b) 1000000 ml (c) 1000 litres

Page 222

Mental Maths–13

1. 10 cm
2. 8 cu units
3. 10 cm
4. 8m^2
5. 30 cm^3
6. 64 cm^3
7. 1 litre
8. 4000 cm^3
9. E
10. Rs. 30
11. Profit = Rs. 11
12. $\frac{35}{100} = 0.35 = 35\%$

Page 223

Mixed Review 3

1. 1,16,30,535
2. 2,19,74,568; 1,81,608
3. 80,422
4. 23
5. 0.725; 1.025
6. (a) 58,79,399; (b) 10000
7. 12
8. 336
9. $3\frac{1}{3}$
10. $\frac{9}{20}$
11. $\frac{1}{64}$ cu cm.
12. Rs. 3200
13. $\frac{5}{6}, \frac{3}{4}, \frac{2}{3}, \frac{3}{8}$
14. 0.375
15. 39 cm
16. Rs. 572.11
17. Rs. 231.25
18. 0.127
19. $\frac{3}{50}$
20. 36 sq cm
21. 480 cu cm
22. 1.620 km
23. $\frac{6}{7}$
24. 78 crore
25. ᗺ
26. (a) Yes (b) No (c) No (d) Yes

27. 28.26 cm 28. 20% 29. 9:49 a.m. 30. (a) 210 minutes (b) 70°C

Page 227

Skill Practice

1. (i) square (ii) square (iii) triangular (iv) triangular (v) square (vi) square
2. 9 and 16
3. (i) 6 + 10 (ii) 10 + 15 (iii) 21 + 28 (iv) 45 + 55
4. 36

Page 230

Skill Practice

2. (a) Yes (b) No (c) Yes (d) No
3. (a)  (b)  (c)  (d) 
4. 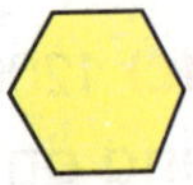 

Page 232

Skill Practice

Fruit	Tally Marks	Number of students
Apple	𝍸	5
Pear	IIII	4
Orange	𝍸 IIII	9
Banana	𝍸 I	6
Grapes	III	3
Chiku	II	2
Watermelon	I	1

(i) Orange
(ii) Watermelon
(iii) Watermelon < Chiku < Grapes < Pear < Apple < Banana < Orange

Page 233

Exercise 57

5. (i) Birds (ii) Snake (iii) Lions (iv) 25 (v) $\frac{2}{25}$
6. (i) Software Engineers (ii) 13 students (iii) 213 students
   (iv) Software Engineers > Doctor > Teacher > Architect > Lawyer.

Page 236

Exercise 58

2. (i) Thriller (ii) Musical (iii) Science Fiction (iv) 20
   (v) Comedy and Thriller (vi) Thriller < Comedy < Science fiction < Drama < Musical.
3. (i) Games (ii) Chatting (iii) Internet Surfing (iv) Games
   (v) (a) Games > Internet Surfing > Chatting > E-mail > Power point
   (b) Chatting > Power point > E-mail > Internet Surfing > Games.